SOIL PHYSICS

L. D. BAVER

Department of Agronomy
Ohio State University

WALTER H. GARDNER

Department of Agronomy
Washington State College

WILFORD R. GARDNER

Department of Soil Science
University of Wisconsin

SOIL PHYSICS

Fourth Edition

JOHN WILEY & SONS, Inc., New York

London • Sydney • Toronto

Library of Congress Cataloging in Publication Data:

Baver, Leonard David, 1901–

Soil physics.

Includes bibliographies.
1. Soil physics. I. Gardner, Walter H., joint author. II. Gardner, Wilford R., joint author.

S592.3.B38 1972   631.4'3   72-5318
ISBN 0-471-05974-9

Printed in the United States of America

10 9 8 7 6 5 4 3 2 1

This edition of SOIL PHYSICS is dedicated to the memory of

WILLARD GARDNER

in appreciation of his many pioneering contributions to the study of soil water and his training of soil physicists.

Preface

To the Fourth Edition

The fourth edition of *Soil Physics* represents an expanded approach to the problems associated with the physical properties of soils. It combines the concepts and experiences of three soil physicists who have devoted considerable study to their special interests in the different phases of soil physics. The text has been reorganized to give better coordination of the subject matter among the various chapters. Increased emphasis is placed upon soil water and the theoretical aspects involved. The fundamental principles of the retention and movement of soil water are presented and interpreted in terms of the field moisture regime. Applications to the practical problems of irrigation, drainage, and water erosion are then made.

The importance of clay mineralogy to the colloidal and physical properties of soils is accentuated in Chapter 1. Chapter 2 discusses the viscosity and swelling of colloidal clays as indexes of particle hydration. The dynamic properties of soils are analyzed in Chapter 3, where the basic concepts in soil physics that are related to soil mechanics, soil engineering, and soil tillage are discussed. The important subject of soil structure has been divided into two chapters to ensure adequate coverage of the agricultural significance of structure in the areas of soil tilth, root development, crust formation, and plant growth. The original soil temperature chapter has been revised and expanded to cover the thermal regime of the soil; this not only involves soil temperature but also the heat balance of soils. The contents of the former chapter on soil tillage have been coordinated with the appropriate subject matter in Chapters 3 and 5.

Emphasis in this book is on the simple and clear presentations of ideas rather than upon the chronology of discovery. Where citations appear, their intent is to lead the reader into some of the literature rather than to give a chronology of development in the field of soil physics. We believe that although some ideas may be traced directly to the genius of an individual, more often a development or discovery arises from a compounding

of the efforts of many scholars culminating finally in its first articulate
expression by one or several of those scholars.

L. D. Baver

Walter H. Gardner

Wilford R. Gardner

Columbus, Ohio
Pullman, Washington
Madison, Wisconsin
May 1972

Preface

To the Third Edition

The third edition of *Soil Physics* attempts to incorporate the many excellent ideas that have been communicated to the author from soil scientists all over the world. One such suggestion, for example, has resulted in a reorganization of the original Chapters 2, 3, and 4 into three better-coordinated chapters with different headings. It is hoped that this new organization presents the various properties of clays in a more closely related sequence. Another has changed the references to the date of publication rather than designating them by number. The chapter on Soil Aeration has been moved ahead of Soil Water to follow Soil Structure. Two new chapters have been added, Principles of Soil Irrigation, Chapter 8, and Principles of Soil Drainage, Chapter 9. Major changes in the other chapters include the elimination of some of the older work on mechanical analyses of soils and the wide usage of pF in the discussions on soil water; additions include discussions of soil puddlability in Chapter 4, of the effect of chemical soil conditioners on soil structure in Chapter 5, of more recent contributions of the diffusion process in soil aeration in Chapter 6, of hydraulic conductivity and of soil moisture stress and plant growth in Chapter 7, of the importance of compaction in soil tillage in Chapter 11, and of wind-erosion processes in Chapter 12. An enlarged table of contents is used to show more clearly the contents of the various chapters.

The author is indebted to many individuals for their suggestions to improve *Soil Physics*. Particularly he desires to mention the following: Dr. Don Kirkham and Mr. Jan van Schilfgaarde of Iowa State College, Dr. C. E. Marshall of the University of Missouri, Dr. M. B. Russell of the University of Illinois, Dr. L. A. Richards of the United States Regional Salinity Laboratory, Dr. C. H. M. van Bavel of North Carolina State College, and Dr. Cecil Wadleigh of the United States Department of Agriculture.

L. D. BAVER

Honolulu, Hawaii
January, 1956

Preface

To the Second Edition

This revision of *Soil Physics* follows the same general pattern as the original edition. The author has attempted to incorporate in this new edition the major contributions to the field of soil physics that have taken place during the past seven years. Major additions to the book are centered around the results of the electron microscope on the shape of soil particles, the many advances in our knowledge of soil-moisture relationships, certain new developments in the field of soil structure and soil aeration, additional information on plowing, and the effect of raindrops on soil erosion.

The author is indebted to Dr. Byron T. Shaw, Agricultural Research Administration, United States Department of Agriculture; Dr. L. A. Richards, United States Regional Salinity Laboratory, United States Department of Agriculture; and Mr. G. W. Musgrave, Soil Conservation Service, United States Department of Agriculture; for their constructive suggestions in the revision of the book; to Dr. Byron T. Shaw, for the electron microscope photographs showing the shape of clay minerals; to Mr. W. D. Ellison, Soil Conservation Service, United States Department of Agriculture, for the photograph on the effect of raindrops on soil erosion; and to Mrs. Nettie Haywood Metts, for the typing and assembling of the revised manuscript.

L. D. BAVER

Honolulu, Hawaii
May 1, 1948

Preface

To the First Edition

Soil physics is a phase of soil science that has been receiving increasing interest and attention within the last twenty years. Numerous technical contributions have appeared in English, French, German, and Russian scientific journals. In almost every instance, the individual has been interested in only one particular aspect of the physical properties of the soil. The field is so large that it limits the scope of activity of any one person.

Although much research in the field of soil physics has been accomplished, teachers of the subject have been handicapped by a lack of suitable instructional material. Unless the teacher happens to be closely associated with soil-physics research and has access to the numerous foreign publications, the preparation of a comprehensive course in soil physics is a difficult task.

The author has taught a course in this subject for the past nine years and has been fortunate in having had to review most of the foreign work in conjunction with his research projects. Practical experience in various aspects of soil-physics research, in addition to a rather wide coverage of French, German, and Russian literature, has resulted in the preparation of a complete set of notes which have been used to build up a course in soil physics that has been rather favorably received by the students, especially those of graduate standing. Incidentally, the student reactions to the philosophy and content of the course have contributed much to the final shaping of the material into a form that can be easily and clearly presented.

An attempt has been made to discuss the various phases of soil physics from the point of view of the teacher explaining them to his class. Simple analogies and often extremely detailed discussions have been used to illustrate a significant point. The author has done this purposely, because it is his firm conviction that a subject cannot be taught successfully by assuming that the student is fully aware of the implications of the subject that seem so obvious to the highly specialized teacher. The book is designed primarily for the advanced undergraduate students of fairly good caliber and for graduates. There are several sections that undoubtedly will prove too involved for strictly undergraduate classes. However, they may be omitted without materially affecting the continuity of thought in the book. Detailed mathematical discussions were purposely avoided for the sake of clarity. They may be added by the teacher who desires to use them.

The book does not propose to give complete citation of all published literature on a particular subject. The references at the end of each chapter have been selected because the author felt that they represented the more important points of view. Considerable reference has been made to some of the classical soil-physics research that was carried out during the last twenty-five years of the past century. This has been deemed essential, since many of these publications are not universally accessible.

In many instances, experimental evidence does not permit the formulation of an exact statement of fact concerning a particular property. The author has attempted to present the different viewpoints as he sees them and has stated his own personal opinions regarding these viewpoints. His opinions, of course, are subject to change as soon as further research data warrant.

It is realized that other investigators may view a given phenomenon differently from the way it is discussed in this book. Nevertheless, the material is presented to represent the author's concept of the subject. Differences in interpretation are usually only differences in points of view. The author has tried to develop a concept of the fundamental aspects of soil physics and their practical application. Special attention has been given to the practical interpretation of each of the different phases of the subject. It is believed that this philosophy will make the book not only more interesting but also more usable.

The author is deeply indebted to the following individuals who have helped to make this book possible: Professors G. W. Conrey and R. M. Salter, Department of Agronomy, Ohio State University, for their many suggestions relating to the subject matter discussed and for reading certain chapters of the manuscript; Professors J. A. Slipher, Department of Agronomy (Extension), and C. O. Reed, Department of Agricultural Engineering, Ohio State University, for their contributions to the chapter on soil tillage; Byron T. Shaw, Department of Agronomy, Ohio State University, for his critical reading of the manuscript and his aid in developing the mathematical phases of the book; Professor H. J. Harper, Oklahoma Agricultural and Mechanical College, for soil-structure photographs; Professor C. E. Marshall, University of Missouri, for the use of the sketches and pictures of the crystal-lattice makeup of clays; I. F.

Reed, Alabama Agricultural Experiment Station, for the photograph of the primary shear planes of the plow; Mrs. Bertha Fischer, for her cooperation in assembling and typewriting the major portion of the manuscript; and Robert G. Parks for helping with the different illustrations.

L. D. BAVER

Raleigh, North Carolina
October 1, 1940

Contents

SOIL PHYSICS

CHAPTER 1

The Soil as a Disperse System

The soil is a very complex system. A given volume of soil is made up of solid, liquid, and gaseous material. The solid phase may be mineral or organic. The mineral portion consists of particles of varying sizes, shapes, and chemical compositions. The organic fraction includes residues in different stages of decomposition as well as live active organisms. The liquid phase is the *soil water* which fills part or all of the open spaces between the solid particles and which varies in its chemical composition and the freedom with which it moves. The gaseous or vapor phase occupies that part of the pore space between the soil particles that is not filled with water; its composition may change within short intervals of time. The chemical and physical relationships among the solid, liquid, and gaseous phases are affected not only by the properties of each but also by temperature, pressure, and light.

The dispersed or solid phase predominates, and the dispersion medium, soil water, provides the aqueous films around the individual particles and helps to fill the pores between the solid separates. Moreover, the solid phase consists of particles of varying degrees of subdivision, which range from the lower limits of the colloidal state to the coarsest fractions of sand and gravel. These different particles, especially those of colloidal dimensions, may be found in a state of almost complete dispersion or in a condition of nearly perfect aggregation, or granulation. In most soils, however, there is only a partial aggregation of the various particles.

The completely dispersed, individual or primary particles are usually referred to as *textural separates*. The aggregates or secondary particles, which are formed by a grouping together of mechanical separates, are generally considered the *structural units*.

1

CHARACTERISTICS OF THE DISPERSED PHASE

The solid phase is constituted of the products of weathering of parent rocks or material and the minerals they contain. In analyzing the characteristics of the dispersed phase, one is interested in the size and shape of the individual particles resulting from this weathering and their chemical and mineralogical composition as they affect the nature and properties of the surfaces involved.

Size of Particles

The different textural separates in soils are classified into various sized groups on the basis of their equivalent diameters. It is recognized that soil particles are not spherical but anisometric. Consequently, equivalent diameter for larger particles that are separated by mechanical sieving refers to the diameter of a sphere that will pass through a given sized opening. For small particles that are separated by sedimentation techniques, equivalent diameter refers to the diameter of a sphere that has the same density and velocity of settling in a liquid medium. At present, there are several systems of classification of soil separates that have evolved over a period of years. Two of the five classifications shown in Figure 1-1 are used by agriculturists. One is that of the United States

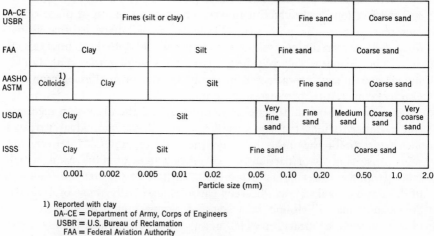

1) Reported with clay
DA–CE = Department of Army, Corps of Engineers
USBR = U.S. Bureau of Reclamation
FAA = Federal Aviation Authority
AASHO = American Association of State Highway Officials
ASTM = American Society for Testing and Materials
USDA = U.S. Department of Agriculture
ISSS = International Society of Soil Science

Fɪɢ. 1-1. Classification of soil separates <2.0 mm on the basis of particle size. (Adapted from USDA, 1951, and Portland Cement Association, 1962.)

Department of Agriculture and the other that of the International Society of Soil Science. It is to be noted that only the upper size limit of clay, 0.002 mm, is common to both classifications. The International classification was suggested by Atterberg (1912), who established the upper limit of clay on the basis that particles smaller than 0.002 mm ($2\,\mu$) exhibited Brownian movement in aqueous suspension. Capillary movement of water was very slow for particles of this size and the properties of stiff clays were strongly manifested. This definition of clay was given further justification by mineralogical studies which showed (as will be discussed later) that relatively few unweathered primary minerals existed in fractions smaller than $2\,\mu$. Prior to 1938, the United States Department of Agriculture classification had used 0.005 mm ($5\,\mu$) as the upper limit of clay; this limit had been established arbitrarily in 1896 on the basis of the limitations of the microscopic equipment that was being used to observe the size of the particles. In 1938, the $2\,\mu$ limit of clay was accepted from the Atterberg classification without changing the size limits of the other separates. The two systems are not too dissimilar except that the United States Department of Agriculture makes more separations in the sand fractions.

The key particle size in these two classifications has been the upper limit of the clay fraction, based primarily on the expression of the physical and chemical properties associated with surface activity. Clay is a surface-active fraction with a high degree of chemical and physical activity for reasons that will be discussed in subsequent paragraphs. Sand and silt separates, on the other hand, do not exhibit marked physical activity and may be considered the skeleton of the soil. Therefore, from a textural or particle size point of view, sandy soils usually tend to be droughty and subject to excessive leaching. Clay soils have a tendency to be plastic and sticky when wet, to swell on wetting, and to shrink on drying; they are likely to have a slow permeability to water and to be poorly aerated.

The Chemical and Mineralogical Nature of Particles

It is not the purpose of this book to discuss the chemical and mineralogical nature of soil particles in detail. There are adequate sources for such details (Brown, 1961; Grim, 1953, 1962, 1968; Jackson, 1964; Marshall, 1964; van Olphen, 1963). Nevertheless, it is necessary to have a general concept of the constitution of the active clay fraction in order to understand its physical behavior. It will suffice to indicate the mineralogical variation in the coarser separates.

The Sand and Silt Fractions

It is recognized that the sand and silt separates contain many primary minerals that have considerable importance from the standpoint of soil weathering and development as well as soil chemistry. Some have direct impacts upon the mineralogical nature of the clay minerals formed in the weathering process. However, due to the small amount of surface exposed per unit weight, their influence on the physical properties of soil, which are usually associated with surface phenomena, is limited. Most of the minerals in the coarser fractions consist of quartz and aluminosilicates, primarily feldspars.

Mineralogical analyses have shown that the feldspars are not found in abundance in fractions smaller than 2μ, although there are particles of quartz present along with intermediate products of weathering that fall between the feldspars and the clay minerals. Mineralogical analyses of different New Zealand soils (Fieldes, 1962) indicated the presence of 15 to 21 species of minerals in the coarser fractions and 8 to 12 different types in the clay. Although quartz predominated in the sand fraction of nonvolcanic soils, there were many more minerals present; most of them were aluminosilicates and oxides. The acid feldspars ranked high as a constituent of sand. Soils derived from schists that have a high content of micas in the sand fraction contain a considerable amount of illite in the clay fraction. This is in confirmation of the weathering sequence of mica → illite → vermiculite → montmorillonite (Fieldes, 1962; Jackson, 1964). Volcanic basalt weathers to allophane and gibbsite in the clay fraction. These examples should suffice to illustrate the fact that the minerals of the coarser soil separates can have an indirect effect upon the properties of the soil as they may relate to the formation of the type of clay minerals present.

The Clay Fraction

A chemical analysis of the clay fraction shows that it is composed primarily of Si_2O, Al_2O_3, Fe_2O_3, and H_2O along with varying amounts of TiO_2, CaO, MgO, MnO, K_2O, Na_2O, and P_2O_5. The question naturally arises as to the types of chemical compounds or substances in the soil that may contain these various elements and exhibit the colloidal properties of clays. The older concept of clay visualized it as a mixture of hydrated oxides of silicon, aluminum, and iron, with the bases present in the adsorbed state.

It took the application of X-ray and petrographic techniques to mineralogical studies of clays to provide a clearer concept of the nature of clay (Ross and Shannon, 1926; Pauling, 1930; Kelley, Dore, and Brown, 1931; Hofmann, Endell, and Wilm, 1933; Marshall, 1935; Hendricks, 1942).

These investigations proved that clays are constituted primarily of distinctly crystalline minerals, although there may be certain amounts of noncrystalline material present. The major building stones are silicon, aluminum, ferrous and ferric iron, magnesium, and oxygen atoms plus hydroxyl groups. Two basic units are responsible for the patterns of construction of the different clay minerals. The first is the silicon tetrahedron in which oxygen atoms are arranged in such a way that each forms a corner of a tetrahedron held together by a silicon atom in the center (see Figure 1-2a). It is possible for the aluminum atom to substitute for silicon in the tetrahedron. The significance of this substitution will be discussed later.

In an idealized structure, the tetrahedra thus formed are linked together in a sheet known as the silica or tetrahedral sheet. In this sheet, the three oxygen atoms that form the base of the tetrahedron are jointly shared by three adjacent tetrahedra. This is clearly illustrated in Figure 1-2b, which shows a basal view of the silica sheet with the oxygen atom attached to the apex of the tetrahedron pointed downward below the oxygens in the plane of the paper. It should be noted that the holes in this sheet are ringed by six oxygen atoms in a symmetrical hexagonal arrangement. The unit cell of this sheet in layer-lattice clay consists of four silica [Figure 1-2b, tetrahedra 3 (oxygens shared with tetrahedra, 1, 2, and 4), 4 (oxygens shared with tetrahedra 3, 5, and 6), and ½ (5, 6, 7, 8)] and six oxygen atoms (Figure 1-2b, 3 from tetrahedron 4 and ½ each of 2 atoms from tetrahedra 3, 7, 8). Thus it is seen that the base of the silica sheet is an oxygen surface.

The second unit is the aluminum octahedron in which six hydroxyl groups (OH) or oxygen atoms are so arranged that each forms a corner of an octahedron held together by an aluminum atom in the center (see Figure 1-2c). It is possible for magnesium, ferrous and/or ferric iron atoms to substitute for aluminum, the significance of which will be discussed later. The octahedra thus formed are linked together in a sheet known as the alumina or octahedral sheet. In this sheet, the six OH groups that form the octahedron are jointly shared by three adjacent octahedra. This is illustrated in Figure 1-2d, which shows a top view of the octahedral sheet in the gibbsite linkage. It should be noted that six OH groups form a symmetrical hexagonal pattern to give a closely packed arrangement with another OH in the center. The unit cell consists of four aluminum atoms and six OH groups. Thus the top and bottom of the alumina sheet are hydroxyl surfaces.

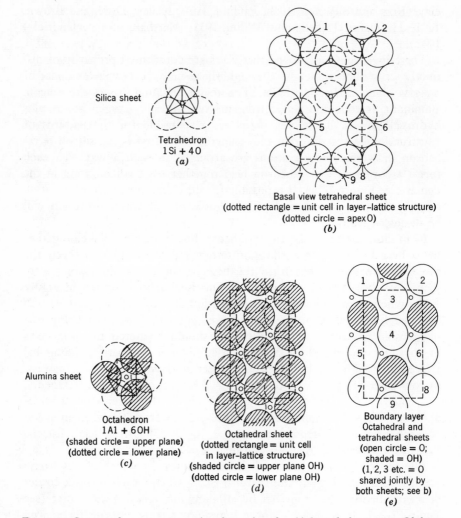

Silica sheet

Tetrahedron
$1 Si + 4 O$
(a)

Basal view tetrahedral sheet
(dotted rectangle = unit cell in layer-lattice structure)
(dotted circle = apex O)
(b)

Alumina sheet

Octahedron
$1 Al + 6 OH$
(shaded circle = upper plane)
(dotted circle = lower plane)
(c)

Octahedral sheet
(dotted rectangle = unit cell
in layer-lattice structure)
(shaded circle = upper plane OH)
(dotted circle = lower plane OH)
(d)

Boundary layer
Octahedral and
tetrahedral sheets
(open circle = O;
shaded = OH)
(1, 2, 3 etc. = O
shared jointly by
both sheets; see b)
(e)

FIG. 1-2. Structural arrangements in clay minerals. (Adapted from van Olphen, 1963.) (Used with permission of John Wiley and Sons.)

Now let us put the silica and alumina sheets together to form a clay mineral. In order to accomplish this, two steps must be taken. First, the OH groups in the upper plane of Figure 1-2d, corresponding to positions 1 to 9 in Figure 1-2e, are removed. Second, the silica sheet shown in Figure 1-2b is placed on top of the octahedral sheet so that the oxygen atoms on the apex positions 1 to 9 occupy the corresponding positions in Figure 1-2e. This means that the two sheets are held together by jointly sharing oxygen atoms that are on the apexes of the silica tetrahedra. Figure 1-2e shows only the oxygen-hydroxyl boundary layer between the alumina and silica sheets, along with the alumina atoms of the accompanying octahedra. It is to be noted that the aluminum in each octahedron is now surrounded by four OH groups and two oxygen atoms. The OH groups in the boundary layer are found directly opposite the holes in the silica sheet. When the top silica-oxygen and the bottom hydroxyl layers are added, the resultant structure is a 1:1 clay lattice mineral.

The schematic drawing of a 1:1 basic structural unit is shown in Figure 1-3. The oxygen surface of the tetrahedral arrangements has six atoms,

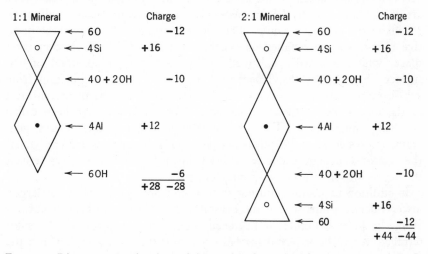

FIG. 1-3. Diagrammatic sketches of layers in clay minerals. (Courtesy of L. P. Wilding.)

the hydroxyl surface of the octahedral pattern has six OH groups, and the boundary layer has four oxygen atoms and two OH groups. This sketch shows that the unit cell has 28 positive and 28 negative charges to make it electrostatically neutral.

A second silica sheet can be added to the bottom of the octahedral sheet in the manner just described to give 2:1 lattice clay minerals. A schematic drawing of such a mineral is shown in Figure 1-3. There are now two boundary layers and two oxygen surfaces. The unit cell has 44 positive and 44 negative charges to make it electrostatically neutral.

The structures of these clay minerals are ideal arrangements showing smooth, clear-cut surfaces and electrostatically neutral units, although this is not exactly true in nature. First, as has been previously mentioned, there are isomorphic substitutions in the crystal that can change the properties of the clay mineral (Marshall, 1935b). The trivalent aluminum atom with a radius of 0.57 Å (angstrom units) can substitute for the tetravalent silicon atom with a radius of 0.39 Å in the silica sheet. The larger diameter of the aluminum atom will force the four oxygen atoms of the tetrahedron farther apart, thus introducing a strain in the crystal. This substitution also decreases the positive charge on that particular tetrahedron by one, which increases the negative charge by a like amount. This negative charge must be balanced either by a positive charge nearby or by an extra cation.

Similarly, the divalent magnesium and ferrous iron atoms and trivalent ferric iron can substitute for aluminum in the alumina sheet. Their radii are 0.78, 0.83, and 0.67 Å, respectively. These larger atoms will cause strains in the octahedral units. The divalent atoms will increase the negative charge of the octahedron where the isomorphic substitution takes place, which is usually balanced by an extra cation adsorbed on the surface. The cations that balance the negative charges originating from substitutions within the crystal can be replaced from the planar surfaces of the clay minerals by other cations. Therefore they have been called "exchangeable cations." The total number of exchangeable cations, generally expressed as milliequivalents per 100 g of clay, is referred to as the "cation-exchange capacity" (CEC) and in a number of clays reflects the degree of substitution within the crystal.

In addition to strains resulting from isomorphic substitutions of larger, lower-valence atoms within the crystal, there can be physical departures from the ideal structure. It has been suggested that a few of the silica tetrahedra may be inverted instead of having the apex pointing into the octahedral sheet (Edelman and Favejee, 1940). Such a variation in arrangement would change the nature of the crystal surface. The strains in the crystal set up by such inversions along with the introduction of larger atoms in the tetrahedra and octahedra would tend to produce a rough instead of a smooth surface. Moreover, the edges of the crystals may not be straight and uniform since electron microscope observations have indicated both beveled and frayed situations (Jackson, 1964).

Radoslovich (1960, 1962) has proposed that the silica tetrahedra can rotate freely to distort the ideal hexagonal arrangement so as to give a ditrigonal surface symmetry. Such rotations are caused by the misfit of the larger tetrahedral layer onto a smaller octahedral one. The strains thus set up are partially relieved by a contraction of the tetrahedral layer through rotation of the basal tetrahedra. Such rotations can be as much as 30°. The interlayer cation can play a major role in determining the unit-cell dimensions. The silica tetrahedra can rotate until the six nearest oxygen atoms are in contact with the interlayer cation. The K ion can have a great impact upon mica-type structures, since the hole left within the ditrigonal arrangement is too small to accommodate it completely. It therefore holds the layers slightly apart.

The basic principles of the interlinking of a specific number of sheets of silica tetrahedra and alumina octahedra into various layer-lattice structures along with the origin and nature of isomorphous substitution of proxy atoms in the crystal account for the occurrence of many types of clay mineral with varying physical and chemical properties. Crystalline clay minerals may be divided into four major groups (Marshall, 1949, 1964):

1. Kaolin group with a 1:1 lattice type.
2. Hydrous mica group with a 2:1 lattice type.
3. Montmorillonite group with a 2:1 expanding-lattice type.
4. Palygorskite or fibrous clay group with a 2:1 modified lattice type.

Two other types, chlorites (2:2 lattice type) and regular or random interstratified clay minerals of the foregoing components have also been identified (Brown, 1961).

The kaolin group is characterized by having one silica and one alumina sheet. Members are kaolinite, dickite, nacrite, and halloysite (Brindley, 1951). There is little, if any, isomorphous substitution in the crystal lattice of these clay minerals. The two unit layers are held together through hydrogen bonding between the hydrogen atoms in the hydroxyl surface of one unit with oxygen atoms in the oxygen surface of the adjacent unit. The unit layers are bonded together so tightly that ions or water molecules cannot permeate the interlayer positions between adjacent kaolinite layers. This means that their colloidal properties are determined by external surfaces only. There is little swelling or shrinkage and low plasticity. The unsatisfied valences resulting from broken bonds on the edges of the clay mineral are responsible for ionic reactions. The CEC is less than 10 me per 100 g of clay. A diagrammatic edge view of two unit layers of kaolinite is shown in Figure 1-4.

The hydrous mica group consists of two silica sheets to one of alumina. The two major subgroup members are illite (Grim, Bray, and Bradley, 1937) and vermiculite (Gruner, 1934). According to present concepts, illite may be an interstratified mineral. There is isomorphous substitution of Al for Si in the silica sheets and Mg and Fe for Al in the alumina sheet. The major substitutions in illite occur in the silica sheets. The negative charges thus produced are balanced by potassium ions. Since the radius of the potassium ion is 1.33 Å, it has been assumed that the

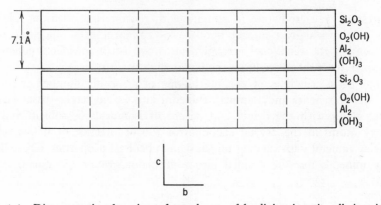

Fɪɢ. 1-4. Diagrammatic edge view of two layers of kaolinite six unit cells in width. (Courtesy of C. E. Marshall, 1964.) (Used with permission of John Wiley and Sons.)

K ion just fits into the holes of the hexagonal oxygen rings in the silica sheets. The unit layers therefore would be held together by these nonexchangeable potassium ions which are surrounded by 12 oxygen atoms.

Apparently, the K ion does not fit completely in the basal network because of the rotated tetrahedra (Radoslovich and Norrish, 1962). It is in a sixfold rather than a twelvefold coordination with the oxygen atoms. Nevertheless, the unit layers are held together by the potassiums ions, even though their size does not permit as close a spacing as would be expected if the ions fitted completely in the holes. The CEC varies between 20 and 40 me per 100 g. Since the unit layers are held so tightly together by the potassium ions, this exchange capacity resides on the external surfaces or on frayed or scrolled crystal edges, except where there may be an interstratification with a nonpotassium, hydrated layer that would be responsible for some internal surface. This is one mineral in which the CEC does not reflect the degree of isomorphous substitution in the crystal. The chemical and physical properties of illite will be discussed in later paragraphs.

Vermiculites differ from the illites primarily because magnesium, rather than potassium, is the interlayer exchangeable cation that helps to balance the negative charges on the lattice arising from a sizable substitution of Al for Sl in the silica sheet. The magnesium ions are highly hydrated. Consequently, there is both an interlayer of exchangeable cations and of water that hold the unit layers together. The lattice has only limited expansion, depending primarily upon the size of the exchangeable ions present in the interlayer. The CEC varies between 100 and 150 me per 100 g, which includes both external and internal surfaces.

The montmorillonite group (Hofmann, Endell, and Wilm, 1933) is characterized by having two silica and one alumina sheets in which the crystal lattices expand and contract depending on the amount of water and interlayer cations present. The most important members of this group are montmorillonite, beidellite, and nontronite. In montmorillonite there is some substitution of Al for Si in the silica sheet and Fe and Mg for Al in the alumina sheet. Negative charges originate in both the tetrahedral and octahedral layers. These negative charges are balanced by exchangeable cations that are found between the unit layers. The hydration of these cations along with the adsorption of water molecules on the oxygen surfaces of the silica sheets through hydrogen bonding causes interlayer swelling and an expansion of the crystal lattice. The extent of this expansion or contraction of the lattice varies with the nature of the exchangeable cation and the degree of hydration of the internal surfaces. A diagrammatic edge view of montmorillonite with the interlayer of water and cations between the basic units of the crystal is shown in Figure 1-5. The CEC of montmorillonite varies between 80 and 150 me per

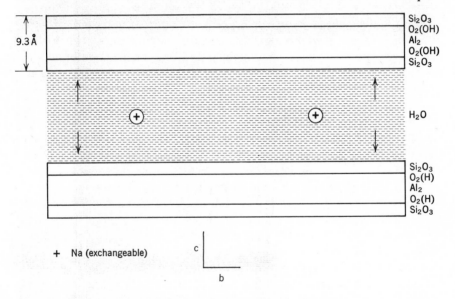

Fig. 1-5. Diagrammatic edge view of montmorillonite. (Courtesy of C. E. Marshall, 1964.) (Used with permission of John Wiley and Sons.)

100 g, including both internal and external surfaces. As will be discussed later, this high exchange capacity and expanding nature of the crystal lattice have tremendous impacts on viscosity, swelling, plasticity, and other physical properties.

Beidellite is one of the end members of the montmorillonite group in which most of the negative charge arises in the silica sheets due to the substitution of Al for Si. There is some substitution of Fe and Mg for Al in the octahedral sheet. The CEC is less than that of montmorillonite, ranging from about 65 to 90 me per 100 g. In nontronite, Fe replaces Al in the alumina sheet.

The palygorskite or fibrous clay group has a 2:1 lattice structure in which the silica tetrahedra occur in alternate strips on both sides of the basal oxygen plane resulting from the apexes of tetrahedra being systematically but alternately inverted. This produces a box or chainlike structure (Bradley, 1940). There are two members of this group, attapulgite and sepiolite.

The chlorite mineral has a structure much the same as vermiculite except that the exchangeable magnesium and water interlayer between the basic unit layers is replaced by an interlayer crystalline sheet in which Mg instead of Al is in octahedral coordination with OH groups. There is substitution of Al for Si in the silica sheets to give them a negative charge. There is substitution of Al and Fe for Mg in the interlayer crystalline sheet to give it a positive charge. Consequently, there is an electrostatic binding together of the structural units. Essentially, then, there are two silica, one alumina, and one magnesia sheet in the chlorite structure. Due to the electrostatic balance within the crystal, the CEC ranges between 10 and 40 me per 100 g. One might expect its physical properties to be similar to those of illite.

The clay minerals just discussed, with the exception of members of the fibrous clay group, all have had layered structures of either tetrahedral or octahedral arrangements of oxygen atoms or OH groups with atoms of Al, Si, Mg, or Fe. This has been true of the kaolinite, illite, vermiculite, montmorillonite, and chlorite minerals. One should expect, therefore, to find clay minerals in which there is a mixed stacking of unit layers in the crystal instead of a uniform stack of the same units. When this occurs, the minerals are called interstratified or mixed-layer clays. Such interstratification can change the surface properties usually associated with the original clay mineral, particularly with respect to ionic exchange and hydration.

The reader is referred to previous references for greater details of the structure and chemistry of the crystalline materials that constitute the clay fraction of soils. It suffices for the purposes of this text to realize that differences in the crystal-lattice makeup of various clays will have an impact upon their physical and chemical properties. Such effects will be discussed in subsequent paragraphs and chapters.

In addition to the crystalline silicate minerals in clay, there are noncrystalline materials present in varying amounts, depending upon the nature of the soil and the conditions under which it has formed. For example, the volcanic ash soils of Hawaii, Japan, and New Zealand contain the amorphous mineral allophane. It is not known whether allophane is a mixture of hydrated silica and alumina gels or if there is enough order in their arrangement in the soil to consider them as hydrous aluminosilicates. There has been the suggestion that random arrangements of tetrahedral silicon and octahedral aluminum make up the allophane mineral (Grim, 1953). Since it borders on the amorphous state, it possesses a very large specific surface, has high cation and anion exchange capacities, and contains tremendous quantities of water.

The free oxides of aluminum, iron, titanium, and silica are widely distributed in most soils. This is especially true in the tropics where they occur in great abundance under the conditions of lateritic weathering. They may be present in either the crystalline or hydrated gel forms and play an important part in the chemical and physical properties of many soils.

The Shape of Clay Particles

Although many attempts have been made to develop concepts of the behavior of soils by assuming spherical particles, most of the experimental evidence indicates that the smaller fractions are distinctly nonspherical. This evidence is found in ultramicroscopic observations of clay sols, the double refraction of clay particles, the layering of particles during deposition, the nature of the clay crystals, and from electron microscopy.

When a beam of light is passed through a colloidal system, part is transmitted and part diffracted. The diffracted light is visible to the observer, and the beam can be seen in the suspension; this is the so-called Tyndall effect. Under the ultramicroscope, a clay sol exhibits scintillating or twinkling types of particles in that the diffracted light comes and goes and is not distinctly visible at all times. This effect is associated with nonspherical particles. The scintillating or twinkling effects are produced by the sudden orientation of the particles in a visible position and their equally sudden disappearance from view as they assume a different position.

Exceedingly good evidence of the disk-shaped* character of clay particles is associated with the double refraction that is exhibited by clay suspensions. If a Tyndall beam is passed through a suspension of colloidal clay at rest and viewed through a microscope with crossed Nicols, the field will be dark. When the suspension is rotated, however, the field becomes bright. In other words, as the particles are oriented with their

---

* Realizing that the particles are not necessarily circular, we use disk-shaped and plate-shaped interchangeably in this discussion.

long axes at right angles to the direction of light propagation, they exhibit double refraction. This phenomenon is known as *streaming double refraction*. Particles may be oriented electrically also to give *electrical double refraction*. Rotation has no effect upon singly refracting particles such as cubes or spheres.

It is often observed that thin layers of deposited clay curl up into minute sheets when dried. These sheets are distinctly platelike in arrangement and are very coherent. Such a phenomenon could take place only with disk-shaped (perhaps rods) particles that settle out with their broad surfaces parallel to each other.

The nature of the crystal lattice of clays suggests a disk-shaped type of particle. It has been shown that the clay minerals are constituted of alternating silica and alumina sheets. This sheetlike structure of the crystal, which is similar to that of mica, would be expected to favor a platelike shape of the entire particle, rather than a cubical, spherical, or rodlike form.

The advent of the electron microscope has made it possible to determine rather accurately the shape of clay particles (Shaw and Humbert, 1941; Humbert and Shaw, 1941; Marshall et al., 1942). Photomicrographs of different clay particles are shown in Figure 1-6. The plate-shaped character of montmorillonite and beidellite is shown by the photomicrographs of the 100–50 mμ fraction of Putnam clay (beidellite) in *a*. The particles are very well-defined and are obviously platelike. The thickness of the disk is about 16 mμ. Montmorillonite, *b*, shows structures ranging from an amorphous appearing material to extremely thin plates.

The hexagonal character of the plates of kaolinite is shown in *c*. It will be observed that the edges of the crystals are sharp and well defined. Halloysite, a 1:1 lattice-type mineral like kaolinite, has definite rod-shaped particles, which appear in *d* to be composed of twin sections.

Thus it is now fairly well established that the majority of the clay minerals found in soils are platelike in character. This fact helps one to visualize the causes of the plasticity, crusting, and compacting of soils. The relation of shape to these properties will be discussed in subsequent chapters.

The Surface of Clay Particles

Relation of Particle Size

The large amount of surface per unit mass is a characteristic property of all disperse systems. Colloidal chemistry is primarily surface chemistry or, more correctly, the chemistry of interfaces. The extent of surface of a disperse system is usually expressed in terms of *specific surface*,

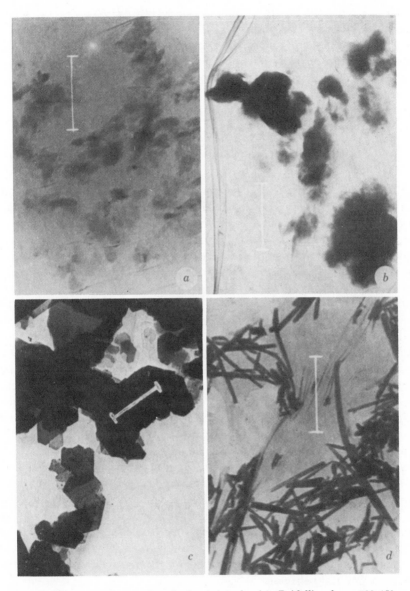

Fig. 1-6. Electron micrographs of clay minerals. (*a*) Beidellite from 100–150 mμ Putnam clay; (*b*) montmorillonite; (*c*) kaolinite; (*d*) halloysite. (Unit of measurement is 1 μ.)

which signifies the square centimeters of surface per gram or per cubic centimeter of dispersed phase.

It is interesting to follow the increase in surface as a single cube is subdivided into smaller ones. For example, visualize a cube of any material whose sides are 1 cm long. The volume of such a cube is 1 cm³; the total surface is 6 cm²; and the specific surface is 6 cm²/cm³. Let us cut this into smaller cubes whose sides are $\frac{1}{10}$ as long, namely, $\frac{1}{10}$ cm or 1 mm. Such particles will be about the size of coarse sand. Each cube will have a volume of $\frac{1}{1000}$ cm³ and a surface of $\frac{6}{100}$ cm². But there will be 1000 cubes, which will make the total surface equal to 60 cm² (1000 $\times$ $\frac{6}{100}$). Let us suppose that the original cube is subdivided into cubes whose sides are $\frac{1}{10,000}$ as long as originally, that is, $\frac{1}{10,000}$ cm, or $\frac{1}{1000}$ mm, or 1 μ (micron). These particles will be about the size of fine clay. Such a subdivision will produce 10^{12} cubes with a total surface of 60,000 cm². In order to carry the subdivision of this hypothetical cube into the colloidal state, let us divide it into cubes whose sides are $\frac{1}{100,000}$ as long as they were originally, namely, $\frac{1}{100,000}$ cm, or $\frac{1}{10,000}$ mm, or $\frac{1}{10}$ μ or 100 mμ (millimicrons). We shall now have 10^{15} cubes with a total surface of 600,000 cm². Thus it may be seen that the specific surface increases from 6 to 600,000 cm²/cm³ as the size of the cubes decreases from 1 cm (gravel) to 100 mμ (colloidal state) on a side.

Inasmuch as the different textural separates in soils are classified into various groups on the basis of their effective diameters, it is of special interest to know the specific surface of each of these groups. The calculations in Table 1-1 point out that a very small amount of material (about 0.52 cm³) may exhibit an extraordinarily large surface area if a sufficiently high degree of subdivision is attained. Several interesting comparisons may be made from these results. For example, a given weight, or volume, of 2-μ clay has 50 times more surface than the same quantity of very fine sand; it has 10 times more surface area than the same weight of silt. Colloidal clay (100 mμ) possesses 20 times the surface area of 2-μ clay and 1000 times the surface of very fine sand. These large differences between the amount of surface per unit mass in clay and sandy soils should permit a clearer understanding of the recognized dissimilarities in their physical behavior.

Relation to Particle Shape

The amount of surface per unit mass, or volume, varies with the shape of the particles. Moreover, the amount of contact per unit surface also changes with shape. It is well known that a sphere has the smallest surface

TABLE 1-1
The Relation of Surface to Particle Size

Diameter of sphere	Textural name	Volume per particle $(\frac{1}{6}\pi D^3)$	Number of particles in $\frac{\pi}{6}$ cc	Total surface $\pi D^2 \times$ number of particles
1 cm	Gravel	$\frac{1}{6}\pi(1)^3$	1	3.14 cm² = 0.49 in.²
0.1 cm (1 mm)	Coarse sand	$\frac{1}{6}\pi\left(\frac{1}{10}\right)^3$	1×10^3	31.42 cm² = 4.87 in.²
0.05 cm (0.5 mm or 500 μ)	Medium sand	$\frac{1}{6}\pi\left(\frac{5}{100}\right)^3$	8×10^3	62.83 cm² = 9.74 in.²
0.01 cm (0.1 mm or 100 μ)	Very fine sand	$\frac{1}{6}\pi\left(\frac{1}{100}\right)^3$	1×10^6	314.16 cm² = 48.67 in.²
0.005 cm (0.05 mm or 50 μ)	Coarse silt	$\frac{1}{6}\pi\left(\frac{5}{1000}\right)^3$	8×10^6	628.32 cm² = 97.34 in.²
0.002 cm (0.02 mm or 20 μ)	Silt	$\frac{1}{6}\pi\left(\frac{2}{1000}\right)^3$	125×10^6	1,570.8 cm² = 1.69 ft²
0.0005 cm (0.005 mm or 5 μ)	Fine silt	$\frac{1}{6}\pi\left(\frac{5}{10,000}\right)^3$	8×10^9	6,283.2 cm² = 6.76 ft²
0.0002 cm (0.002 mm or 2 μ)	Clay	$\frac{1}{6}\pi\left(\frac{2}{10,000}\right)^3$	125×10^9	15,708 cm² = 16.9 ft²
0.0001 cm (0.001 mm or 1 μ)	Clay	$\frac{1}{6}\pi\left(\frac{1}{10,000}\right)^3$	1×10^{12}	31,416 cm² = 33.8 ft²
0.00005 cm (0.0005 mm or 500 mμ)	Clay	$\frac{1}{6}\pi\left(\frac{5}{100,000}\right)^3$	8×10^{12}	62,832 cm² = 67.6 ft²
0.00002 cm (0.0002 mm or 200 mμ)	Colloidal clay	$\frac{1}{6}\pi\left(\frac{2}{100,000}\right)^3$	125×10^{12}	157,080 cm² = 169 ft²
0.00001 cm (0.0001 mm or 100 mμ)	Colloidal clay	$\frac{1}{6}\pi\left(\frac{1}{100,000}\right)^3$	1×10^{15}	314,160 cm² = 338 ft²
0.000005 cm (0.00005 mm or 50 mμ)	Colloidal clay	$\frac{1}{6}\pi\left(\frac{5}{1,000,000}\right)^3$	8×10^{15}	628,320 cm² = 676 ft²

per unit volume. If such a sphere is deformed into a rod or into a disk or plate, the surface increases, with the platelike particles exhibiting the greatest surface area. The calculations shown in Table 1-2 emphasize

TABLE 1-2
Surface Area in Relation to Shape of Particle

Shape	Radius (cm)	Volume (cm³)	Surface (cm²)	Increase in surface (percent)
Sphere	1×10^{-4}	4.2×10^{-12}	1.26×10^{-7}	
Disk				
$h = 1 \times 10^{-4}$ cm	1.155×10^{-4}	4.2×10^{-12}	1.56×10^{-7}	23.8
$h = 5 \times 10^{-5}$ cm	1.67×10^{-4}	4.2×10^{-12}	1.84×10^{-7}	45.8
$h = 2 \times 10^{-5}$ cm	2.58×10^{-4}	4.2×10^{-12}	4.51×10^{-7}	257.8
$h = 1 \times 10^{-5}$ cm	3.65×10^{-4}	4.2×10^{-12}	8.59×10^{-7}	538.9

the importance of shape in determining the specific surface of clay systems. These have been made by using a 2-μ spherical particle as the reference surface. The surface of disks of the same volume but with varying thicknesses have been computed. When the thickness of the disk is the same as the radius of the sphere, there is a 23.8 percent increase in surface over that of the sphere. When the thickness of the disk is $\frac{1}{10}$ that of the sphere, there is a 538.9 percent augmentation in surface area. These calculations show external surface only and do not take into consideration expanding lattices of clay minerals, as will be discussed shortly. It must be kept in mind that, for the purposes of illustration, these computations have been made on the assumption that the particles were uniformly shaped. Electron microscope observations have shown that the leaflike particles are quite irregular in shape. However, these irregularities are unimportant because clay minerals tend to be plate-shaped particles with a definite thickness which is much smaller than the lateral dimensions. This can be illustrated from a calculation of the specific surface of a plate-shaped particle of thickness d and length and width both equal to w The surface area of a planar side is w^2 and of an edge wd. Adding the area of the two sides and the four edges gives a total area of $2w^2 + 4wd$. The volume is w^2d so the ratio of the area to the volume or specific surface is

$$S = 2w^2 + \frac{4wd}{w^2d} = 2w + \frac{4d}{wd} \tag{1-1}$$

In clay plates the thickness d may be of the order of 10 Å (10^{-7} cm) and w of the order of $\frac{1}{10}$–1 μ (10^3–10^4 Å); therefore $2w$ is very much larger than $4d$. Thus

$$S \sim \frac{2}{d} \, \text{cm}^{-1} \qquad (1\text{-}2)$$

In other words, the area of the edges is negligible compared to that on the planar sides. The specific surface is more conveniently expressed as area per unit mass. This is obtained by dividing the area per unit volume by the density of the clay particle. If p is the particle density, the specific surface of a clay is then approximately $2/pd$. For a fully dispersed montmorillonite with $d = 10$ Å and $p = 2.68$ g/cm^3 this gives about 750 m^2/g, which is very close to the specific surface as actually measured.

In considering the specific surface of a soil, it must be kept in mind that the organic fraction often has a highly reactive surface which may exhibit an effective or apparent value as high as 1000 m^2/g.

In addition to the increased surface per unit volume, it should be obvious that disk-shaped particles can be arranged in more intimate contact than spherical ones. Four spheres will make five point contacts when placed in close-packed arrangement. On the other hand, four disks of the same size will make intimate contact with about three-fourths of the total flat surface area, when closely packed together. Spherical particles therefore do not provide favorable conditions of surface contact for great cohesion. Disk-shaped particles, however, cause high cohesion when orientation is at a maximum. In addition, laminar particles can slide over each other under an applied force, whereas spheres will permit a deformation only by a breaking down or rolling apart of the spherical arrangement.

Relation to Clay Mineralogy

The evidence on the structure of clay minerals indicates that all groups have both planar and edge external surfaces. Members of the montmorillonite group and the vermiculites also have planar internal surfaces as a result of the expanding lattices. Data in Tables 1-3 and Table 1-4 show the variations that occur in total surface both between and within the clay mineral groups. The kaolinite arrangement is usually quite thick as a result of stacking of nonexpanding layers and thus has a low specific surface because it lacks interlayer surfaces. The nonexpanding illites have a surface area about ten times that of the kaolin group. The specific surface of the vermiculites is about midway between the illites and the

TABLE 1-3

The Specific Surface, Cation-Exchange Capacity, and Density of Charge of Various Clay Minerals

Clay mineral	Specific surface (m²/g)			Cation-exchange capacity (me/g)	Density of charge (me/m² × 10³)	
Kaolinites	5– 20*			0.03–0.15†	6–7.5‡	
Illites	100–200*			0.10–0.40†	1.0–2.0‡	
Vermiculites	300–500*			1.00–1.50†	3.0–3.3‡	
Montmorillonites	700–800*			0.80–1.50†	1.1–1.9‡	
	N₂	CPB	E.G.		N₂	CPB
Illite-1 §	93	96	91	0.26	2.8	2.7
Illite-2 §	132	138	144	0.41	3.1	2.9
Kaolinite-1 §	17	21	21	0.043	2.5	2.0
Kaolinite-2 §	36	36	—	0.050	1.4	1.4
Kaolinite-3 §	40	9	13	0.030	0.75	3.3
Montmorillonite-1 §	47	800	—	0.98	21.0	1.22
Montmorillonite-2 §	49	600	—	0.98	21.0	1.6
Montmorillonite-3 §	101	800	—	0.99	9.8	1.22

* Fripiat (1964).
† Grim (1962). (Used with permission of McGraw-Hill Book Co.)
‡ Calculated from the data in second and third columns.
§ Greenland and Quirk (1964).
CPB = cetyl pyridinium bromide.
E.G. = ethylene glycol.

TABLE 1-4

Heat of Wetting, Specific Surface, Cation-Exchange Capacity, and Density of Charge of Various Clay Minerals (after Greene-Kelly, 1962)

Clay mineral	Heat of wetting (cal/g)	Specific surface (m²/g)	CEC (me/g)	Density of charge (me/m² × 10³)
Kaolinite-1	1.6	14.8	0.035	2.4
2	1.4	12.0	0.023	1.9
3	1.4	11.0	0.019	1.9
4	2.1	25.0	0.043	1.7
Hydrous mica-1	7.6	150	0.25	1.7
2	4.8	110	0.17	1.5
3	7.9	160	0.30	1.9
4	16.5	250	0.43	1.7
Montmorillonite-1	16.5	690	0.92	1.3
2	17.4	640	0.83	1.3
3	22.2	700	1.13	1.6

expanding-lattice montmorillonites. Although there may be as much as 100 percent or more variation within the same group, the overall magnitude of the surface area is characteristic for the particular type of clay mineral.

The methods for the measurement of specific surface usually make use of the adsorption properties of the mineral surfaces. A brief description of some of these methods follows the discussion of the surface behavior of clay particles.

The Surface Behavior of Clay Particles

Density of Charge

Previous discussions on the structure of clay minerals have shown that the substitution of trivalent cations for tetravalent silicon in the tetrahedra of the silica sheet or divalent cations for trivalent aluminum in the octahedra of the alumina sheet results in an increase in the negative charge of the crystal. These negative charges are balanced by counter cations which are adsorbed on the planar surfaces and are exchangeable by other cations. In addition, there are broken bonds on the edges of the crystals that create negative charges which are also balanced by exchangeable cations. Since there is little, if any, isomorphous substitution in the kaolinite crystal, broken bonds are the major source of the negative charge. These broken bonds are also important with the illite minerals, even though isomorphous substitutions in the silica sheet account for negative charges on the external planar surfaces and the frayed edges of the crystal that are balanced by exchangeable cations. Isomorphous substitutions are primarily responsible for the negative charges in the vermiculite and montmorillonite groups. It has been estimated that 80 percent of the charges come from this source (Grim, 1962).

The forces of attraction resulting from these isomorphous substitutions depend considerably upon the sheet in which the replacements take place. Most of the substitutions in montmorillonite occur in the octahedral sheet. Those in vermiculite are found in the tetrahedral sheet. This means that even though both clay minerals could have the same number of negative charges, the intensity of the charge at the surface would be greater with the vermiculite. It is perhaps for this reason that montmorillonite is characterized by unit layers that are not held tightly together, whereas vermiculite layers are more firmly held together.

The density of charge is determined by measuring both the specific surface and the cation-exchange capacity of the clay and then calculating the milliequivalents per unit surface. It is interesting to note from data in Tables 1-3 and 1-4 that the densities of charge of illites, kaolinites, and montmorillonites do not vary over wide ranges. Even though kaolinite has a relatively low exchange capacity and specific surface, its density of charge is higher than that of montmorillonite or illite. The vermiculites appear to have the largest density of charge of the 2:1 mineral types cited. This can probably be attributed to the sizable substitution of Al for Si in the silica sheet. Illite would have a higher density of charge were it not for the fact that the potassium counter cations that balance the negative charges on the inner silica sheets are nonexchangeable and therefore do not contribute to the total exchange capacity.

It should be recognized that these charges represent the net negative charges of the clay mineral surfaces. In addition to the permanent negative charges on the planar surfaces, there are the broken bond sites. These include both negative and positive exchange sites. In the 1:1 lattice-type mineral such as kaolinite, there should be as many positive as negative sites on the edges of the crystal.

Experimental evidence has shown that some clay minerals exhibit increases in negative charge and CEC as the pH is raised (Schofield, 1949; Coleman, Weed, and McCracken, 1959; Fieldes and Schofield, 1960; Jackson, 1963; de Villiers and Jackson, 1967). This is known as pH-dependent charges of the cation-exchange capacity. The cause of this augmentation in CEC has been attributed to the dissociation of a proton from a hydronium group on the crystal edges (Jackson, 1963).

The pH-dependent negative charges can be affected in several ways, as illustrated in Figure 1-7. Hydroxy aluminum, also known as aluminohexahydronium, may be deposited within the interlayers of expanding-lattice-type clay minerals or on the tetrahedral broken bonds of the silica sheet. Two H ions and an Al are bonded to oxygen in the substituted hydronium. The excess positive charges on this hydroxy aluminum will block the negative charge sites arising either from isomorphous substitution or broken bonds. As the pH increases, the protons are neutralized and the blocking of the negative charges to cation exchange is removed with a concurrent raising of the CEC. This action takes place at acid pH values corresponding to a pK value of the acid of 5. This augmentation of CEC represents the activating of negative charges that had been existing before the deposition of the hydroxy aluminum.

The positive broken-edge hydroxyls on the edges of the crystal represent oxygen atoms that are bonded to both Si and Al with an attached H that has a residual positive charge. As the pH is raised further, there is a dissociation of H ions from these broken-bond hydroxyls to produce an increase in net negative charge and CEC. These reactions occur in the pH range of 5.5 to 8.5, corresponding to a pK value of 7.0 for the

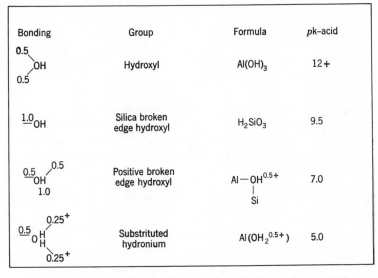

Bonding	Group	Formula	pk–acid
0.5 >OH 0.5	Hydroxyl	$Al(OH)_3$	12+
$\underline{1.0}$OH	Silica broken edge hydroxyl	H_2SiO_3	9.5

0.5 ⟋0.5
OH
  1.0
	Positive broken
	edge hydroxyl
		$Al-OH^{0.5+}$
		    |
		    Si
			7.0

0.25$^+$
$\underline{0.5}$ H
  O H
   0.25$^+$
	Substrituted
	hydronium
		$Al(OH_2^{0.5+})$	5.0

FIG. 1-7. O-H bonding in silicate surfaces. (After Jackson, 1963.)

acid involved. The negative charge sites thus produced are new to the mineral surface.

At higher pH values corresponding to a pK value of 9.5, there can be dissociation of the H ions from the OH groups bound to Si through the oxygen atom in the broken bonds of the silica sheet. Finally, at extremely high pH values, there is the possibility of the dissociation of H ions from OH groups that are bonded to Al only. All of these dissociations contribute to the pH-dependent CEC and result in a higher density of charge.

The amorphous mineral allophane has a large pH-dependent net negative charge and CEC. It arises from Al in the tetrahedral form (Fieldes and Schofield, 1960; de Villiers and Jackson, 1967) as it substitutes for Si in the tetrahedra. Hydroxy aluminum can cover a portion of the surface and block the negative charges from cation exchange. Neutralization of the protons by increasing the pH will release these negative charges to effect a sizable pH-dependent CEC, as has been discussed previously. The pH-dependent negative charges and CEC apparently require a high specific surface of positive edges of hydroxy aluminum on a mineral with a net negative permanent charge.

The relationships of net negative charges of different clay mineral types to pH are shown in Figure 1-8. The negative charges of the 2:1 lattice-type clays, montmorillonite, Rothamsted subsoil (illite), and Taita soil clay (illite) vary little with pH, although there are significant in-

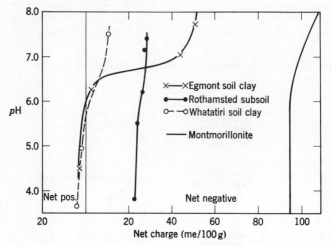

Fig. 1-8. Net electric charges of clays in relation to pH. (After Fieldes and Schofield, 1960.)

creases at the higher pH values. In these clays, particularly montmorillon-ite, the permanent negative charges overshadow the pH-dependent ones. There is a fairly sizable increase in net negative charge above pH 5.5 of the Whatatiri soil clay (metahalloysite, gibbsite). Egmont soil clay, which contains allophane, shows the largest increase in net negative charge (pH 6.0 to 7.6). The curves also point out that the Whatatiri and Egmont clays have isoelectric points below which they have a net positive charge which is important in determining the amount of anion exchange.

Ionic Adsorption

It is not the intent of this discussion to elaborate on the details of the physical chemistry of ion exchange. Excellent treatment of this sub-ject may be found in a number of publications (Mattson, 1929; Jenny, 1932; Kelley, 1948; Bolt, 1955; Marshall, 1964; Wiklander, 1964). How-ever, since many of the physical properties of soils are affected by the amount, nature, and activity of the adsorbed ions, we are interested in the amount of exchangeable ions per unit mass and the energy with which the different ions are held on the surface.

ADSORPTION OF CATIONS. It has been shown that cation adsorption is the result of balancing negative charges arising from isomorphous substi-tutions in the crystal lattices of clays and from the broken bonds occur-ring on the edges of the crystals, especially where these bonds are associ-ated with the tetrahedral sheets. Attention has also been called to the fact that broken bonds can be the source of pH-dependent cation exchange sites. The cation exchange capacities of the different clay mineral groups vary with the nature of the lattice constitution and the degree of isomorphous substitutions. This order of CEC is vermiculites > montmorillonites > illites > kaolinites (Table 1-3).

The cation exchange capacity is related to the amount of internal and external surface and therefore should increase with decreased particle size. This should be especially true of those clay minerals where broken bonds are responsible for many of the exchange sites. One would not normally expect the expanding-lattice-type clays to exhibit increased CEC with decreasing particle size, since 80 percent of the adsorption sites are on the planar surfaces (Grim, 1962). The curves in Figure 1-9 point

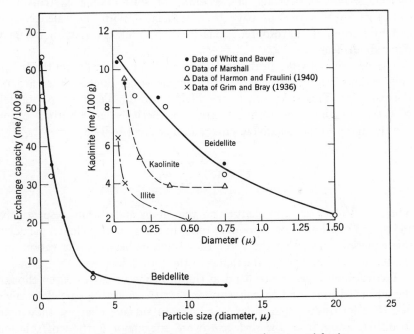

Fɪɢ. 1-9. Relation of cation exchange capacity to particle size.

out these particle size-exchange capacity relationships for Putnam clay (beidellite), illite, and kaolinite. The exchange capacity of beidellite increases rapidly when the particle size becomes smaller than 2 μ. The percentage increases in CEC for beidellite, illite, and kaolinite when the particle size decreases from 500 to 50 mμ were about 55, 100, and 150 percent, respectively. The largest increases occurred with kaolinite and illite where broken bonds play an important role in determining the exchange capacity. It cannot be ascertained if the larger CEC for the colloidal fractions of beidellite was due to more broken bonds or other factors. These data emphasize the importance of particle size and accompanying specific surface in determining the amount of exchangeable ions in a given system.

The amount of exchangeable cations on the surface of clay particles has been shown to be primarily a function of the nature of the clay mineral and the extent of isomorphous substitutions in the lattice (Tables 1-3 and 1-4). It is important to know the factors that affect the energy of adsorption and release of the specific cations involved. It is a well-established fact that various cations have different effects on the zeta potential (migration velocity in an electric field), dispersity, adsorption of water molecules, heat of wetting, and swelling of soil colloidal systems. The causes of these variations have been associated with the valence, size, and hydration of the different ions in their interactions with the clay mineral surfaces.

The negative charges in the surface of the clay minerals and the balancing cations attracted thereto create an electrical double layer around the particles. This is known as the Helmholtz double layer. According to Coulomb's law, the force of attraction between an adsorbed cation and the anion in the negatively charged inner layer will be

$$F = K \frac{e_a e_c}{(r_a + r_c)^2} \tag{1-3}$$

where e_a and e_c are the charges on the anion and cation, respectively, and r_a and r_c are the corresponding radii. In the case of colloidal particles, this equation states that the force of attraction between the cation and the surface will increase as the radius of the cation decreases.

The cationic layer is not fixed in the presence of water. Even though the adsorbed ions tend to concentrate near the negative particle surface due to electrostatic attractions arising from surface charges, they tend to diffuse away from the surface toward the major portion of solution where their concentration is lower. This means that the concentration of cations decreases with distance from the surface. In other words, there is a diffuse double layer surrounding the particle. This layer consists of not only an excess of cations that are attracted to the negative charges but also a deficiency of anions that are repelled by the negative surface. Whereas the cations are positively adsorbed, the anions are negatively adsorbed. This diffuse double layer is known as the Gouy double layer (see Bolt, 1955; van Olphen, 1963). The distribution of positive and negative ions can be calculated by the use of the Poisson-Boltzmann equa-

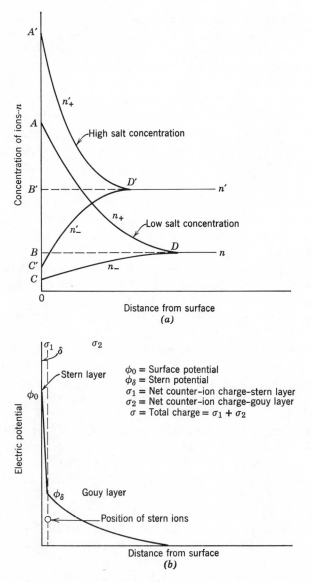

Fig. 1-10. Diffuse double layer relationships. (a) Concentration of ions versus distance from surface; (b) electric potential of Stern and Gouy layers. (Adapted from van Olphen, 1963.) (Used with permission of John Wiley and Sons.)

tion to give curves similar to those in Figure 1-10*a*. The Boltzmann distribution function specifies that the concentration c of a given ion in the double layer will be determined by the potential energy at any given point according to the relation

$$c = c_0 \exp - \frac{ze\psi}{kT} \tag{1-4}$$

where c_0 is the concentration in the equilibrium solution, z is the valence of the ion, e is the electrostatic unit of charge, ψ is the electrical potential (the product $ze\,\psi$ represents the potential energy of the ion in the electrostatic field), k is the Boltzmann constant, and T is the absolute temperature. The concentration of the equilibrium solution is that of the soil water out of the range of influence of the particle surface. Figure 1-10*a* illustrates the distribution of cations and anions for two different equilibrium concentrations.

One observes that the concentrations of cations and anions in the diffuse double layer decrease with distance BD ($B'D'$) from the surface until both are equal at the point D (D'). The rate of change in concentration of cations in the immediate vicinity of the surface is much greater than that of the anions. The excess cations are represented by the area ABD under the curve for low concentration of electrolyte; the deficit of anions is shown by the area CBD. The electric potential curve would be similar to the concentration curve AD.

Addition of an electrolyte to the system will result in a compression of the double layer and a decrease in the electric potential. This is especially true at high concentrations. Compare areas ABD and $A'B'D'$ and diffuse layer thicknesses BD and $B'D'$ in Figure 1-10*a*. Note the compressive effect of the electrolyte which is determined by the valence and concentration of the cations (ions of opposite sign to that on the surface of the particle). The higher the valence and concentration, the greater is the compression.

The total surface charge σ_t is equal to the sum of the excess charges due to the positive adsorption of cations σ_c plus the deficit of charge arising from the negative adsorption of anions σ_a:

$$\sigma_t = \sigma_c + \sigma_a \tag{1-5}$$

This is equivalent to the surface areas ACD and $A'C'D'$ in Figure 1-10*a*.

The thickness of the liquid layer around the soil particle is assumed to be infinite. In reality this is never true. If the "effective thickness" of the double layer is greater than the actual thickness of the liquid layer, a different distribution is obtained. The double layer can never extend beyond the surface of the liquid layer since the cations necessary to neutralize the surface charge must remain within the liquid layer. The double layer is then truncated and the concentration of the soil solution varies from point to point; nowhere does it have the concentration of an equilibrium solution. If water is allowed into the system, the double layer tends to extend itself spontaneously; this results in a swelling of the system. If this swelling is prevented by the application of an external force, a measurable swelling pressure is observed. There are binding forces between the particles in natural soils which prevent the double layer from extending, as would be predicted from theory. A Na-montmorillonite has few binding forces and swells almost as predicted. On the other hand, a Ca-montmorillonite is strongly bound and tends to swell only to a very limited extent. No adequate description of swelling in a system in which there is appreciable binding, such as mixed Na-Ca systems, has yet been worked out. However, some success (McNeal, Norvell, and Coleman, 1966) has been achieved by assuming that a mixed system is constituted of domains in which Na-Ca clays are considered to be made of discrete Na and Ca regions, with only the Na regions swelling at low salt concentrations. The number of each type of domain is a function of the relative abundance of Na and Ca.

The Gouy theory assumes that the charge on the surface is continuous and uniform since the ions are considered as point charges. The adsorbed cations are not related to any negative sites on the surface but are treated on the basis that their proportion in the adsorbed state is equal to that in solution. In other words, there is no consideration of specific surface interactions with the charge-balancing cations. The theory does predict that if there are two ions of different valence present, there will be a larger fraction of the ion with the higher valence near the particle surface.

Important modifications to the Gouy concept were made by Stern (van Olphen, 1962), who considered the impact of the adsorbed ion on the properties of the double layer. First he stated that the size of the cation would limit its approach to the surface. Second he postulated that there could be a specific adsorption of certain cations that would affect their participation in the diffuse phase of the double layer. Thus the liquid charge is divided into two parts. The layer of ions adsorbed on the wall causes the surface charge to be concentrated in a plane at a small distance δ from the charge on the surface of the wall. In this part of the double layer, the electric potential decreases linearly with the distance (like a molecular condenser). Beginning at this distance δ from the surface, the diffuse or Gouy double layer accounts for the liquid charge and the electric potential decreases asymptotically with the distance. These effects are illustrated in Figure 1-10b. The total surface charge is given by

$$\sigma_t = \sigma_c + \sigma_a$$

As the energy of adsorption of the cation increases, the ratio of the Stern layer charge to the diffuse double layer charge increases and the Stern potential $\phi\delta$ decreases. This says that the addition of an electrolyte not only compresses the double layer but also causes the cations to enter the Stern layer from the diffuse layer and thereby decrease the Stern potential. The Stern concept therefore links the energy of adsorption with the characteristics of the adsorbed ions.

Ionic size, hydration, and valence affect the release of exchangeable cations from surfaces of beidellite (Putnam clay). The symmetry concept employed in these data call attention to the variations that exist in the energy of adsorption of different cations. The release of adsorbed ions depends upon the valence as the ease of replacement follows the series monovalent > divalent > trivalent > tetravalent. In the case of the monovalent cations this order is $Li > Na > K = NH_4 > Rb > Cs$. This is an apparent reversal of Coulomb's law since the small Li ion is held with much less energy than the large Cs ion. However, the smaller ions become hydrated in solution and have a larger effective radius than in the nonhydrated form. When the thickness of the water hull is included as part of the radius, the ease of replacement diminishes with decreasing hydrated ion size.

The energy of adsorption may be evaluated by calculating the bonding energies of various cations in homionic systems from data on the activity of the adsorbed cation as determined by clay membrane electrodes (Marshall, 1950). Adsorbed cations are not considered to be completely dissociated from the adsorbing surface. It is only the active fraction that can be measured with the membrane electrodes. This active fraction varies with the interrelated factors of percentage saturation of the clay with the ion in question, its degree of ionization, and its mean free bonding energy. The mean free bonding energy represents the amount of work that must be done to transfer an ion in the adsorbed state on the clay surface to a state of complete dissociation. This can be computed from the formula

$$\Delta F = RT \ln \left(\frac{c}{a}\right) \quad \text{or} \quad RT \ln \frac{1}{f_1} \tag{1-6}$$

where ΔF is the mean free bonding energy of the cation to the surface, c is the total concentration of the exchangeable cation, a is the mean measured activity of the cation, and f_1 is the fraction active. Since a divalent cation would probably react with two negative surface sites, the bonding energy should be twice that of a monovalent cation.

This relationship is affected by the nature of any other cations that may be present in the system. The ion with the higher bonding energy will occupy the exchange sites in greater abundance than one with a lower bonding energy. This is illustrated in Figure 1-11. The bonding

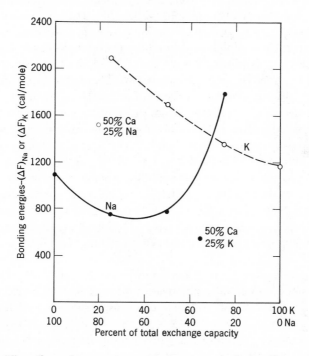

FIG. 1-11. The effect of percentage saturation on cationic bonding energies of Na and K. (After McLean, 1951.)

energy of potassium is higher than that of sodium, except where more potassium than sodium occupies the exchange complex (i.e., 75 percent potassium and 25 percent sodium). Increasing saturation of sodium with concomitant decreases in potassium progressively increases the bonding energy of potassium. On the other hand, potassium decreases the $(\Delta F)_{Na}$ at 25 and 50 per cent saturation but increases it markedly at 75 percent saturation. When the exchange complex contains 75 percent of complementary ions divided into 50 percent Ca and 25 percent Na or K, the $(\Delta F)_{Na}$ and $(\Delta F)_K$ are both decreased below the levels obtained when the other monovalent cation constitutes 75 percent of the base saturation.

This indicates that calcium has a higher bonding energy than sodium or potassium. The divalent cations have bonding energies about twice those of monovalent, as would be expected.

The nature of the clay minerals also affects the bonding energy of the adsorbed cation. In the saturation of the acid form of these minerals with basic cations, the shapes of the bonding energy curves vary with the type of clay mineral (Chatterjee and Marshall, 1950). This is particularly true for the divalent cations. The 2:1 lattice-type clay minerals indicate rising $(\Delta F)_{Ca}$ values with increasing base saturation up to a plateau effect from which the bonding energies decrease gradually as equivalence is approached. On the other hand, kaolinite shows a sharp maximum at about 90 percent saturation. There is a rapid decrease in bonding energy at about 85 percent of equivalence, which is explained by the presence of OH groups on the inner layer as the pH of the system rises (Marshall, 1964). The original increase in bonding energy may be due to exchangeable aluminum effects or to the presence of traces of impurities. The maximum bonding energies for the calcium systems follow the series beidellite $>$ illite $>$ kaolinite $>$ montmorillonite. The $(\Delta F)_{Ca}$ values for the 2:1 lattice-type clays show that montmorillonite is the lowest, a mineral in which the major negative charges arise from isomorphous substitutions in the alumina sheet. One would expect higher bonding energies where the negative charges originate in the silica sheet.

Corrections to the Gouy-Chapman theory to take into account the impact of ionic interactions, polarization of ions, and dielectric saturation are small in the case of colloidal clay suspensions where the density of charge does not exceed 2 to 3×10^{-7} me/cm^2 (Bolt, 1955). The size of the ion and its hydration have a considerable impact on the distribution of cations between the Stern layer and the diffuse portion of the double layer (Shainberg and Kemper, 1966a, 1966b, 1967). This is true if the following assumptions are made:

1. Exchange spots in the surface are fixed charges.
2. Part of these charges are neutralized by ions in the Stern layer.
3. The remaining balancing cations are in the diffuse phase of the double layer.
4. The adsorbed ions may be fully hydrated or unhydrated with no water molecules between them and the adsorbing surface.

The total density of charge Q_t is equal to the sum of the charge densities of hydrated Q_h and unhydrated Q_u adsorbed cations:

$$Q_t = Q_h + Q_u \qquad\qquad (1\text{-}7)$$

The unhydrated ion is in the Stern layer directly on the surface and the hydrated ions in the diffuse phase. In other words, the hydrated ions correspond to the fraction active of the adsorbed ion that is dissociated, according to the concept of bonding energies previously discussed. The hydrated hull around a sodium ion, for example, would increase the distance of the center of the ion to the plane of negative charge in the octahedral sheet of the crystal lattice of montmorillonite by over 50 percent above that of the dehydrated cation. Computations of the distribution of the charge densities of monovalent cations in the double layer in relation to their hydration and polarization energies as shown in Table 1-5. These results confirm the well-known lyotropic series on

TABLE 1-5

Distribution of Monovalent Cations between Stern and Diffuse Layers (after Shainberg and Kemper, 1967)

Ion	Radius (Å)	Charge density (esu/cm²)		Cation in stern layer (percent)
		Q_u	Q_h	
Li	0.60	5,000	23,900	17.3
Na	0.95	11,900	17,000	41.2
K	1.33	17,100	11,800	59.1
Cs	1.69	21,000	7,900	76.0

the ease of replacement of adsorbed cations: $Li > Na > K > Cs$. Only 17 percent of the highly hydrated Li ion is in the Stern layer as compared with 76 percent of the unhydrated Cs ion. If the relative degree of adsorption of the Na ion in the Stern layer is assumed equal to 100, then $Li = 42$, $K = 144$, and $Cs = 184$. The difference between Na and K ions is of the same order of magnitude as the variations in their maximum bonding energies calculated according to equation 1-6.

Calculation of equilibrium constants from ionic exchange with the K ion indicate that variations in the energies of hydration and polarization can account for the differences in adsorption of Li, Na, and K ions but not for Cs (Shainberg and Kemper, 1967). There seem to be other specific forces of adsorption affecting the higher energy with which the Cs ion is held.

Summarizing, cation exchange in clays is related to the nature of the clay mineral and the valence, size, and hydration of the ions involved. Their distribution within the electric double layer on the particle surface is a function of the bonding energies of the individual ions. The Stern-Gouy model appears to satisfy the conditions of the clay double layer and can accommodate the various concepts of ion exchange reactions.

ADSORPTION OF ANIONS. The curves in Figure 1-8 point out that the net negative charge of colloidal clay systems decreases with increasing acidity (Fieldes and Schofield, 1960). Certain clay minerals, such as members of the kaolinite group and allophane, exhibit positive charges at lower pH values. These positive charge sites are the seat of anion adsorption and exchange. They originate from broken bonds primarily in the alumina octahedral sheet that exposes OH groups on the edges of the clay minerals. Anion exchange may also occur with OH groups on the hydroxyl surface of kaolinite (McAuliffe et al., 1947). The 2:1 lattice-type clay minerals also have broken bonds that can participate in anion exchange under acid conditions. However, the cation exchange capacity of these minerals overshadows that of anion exchange. A comparison of cation and anion exchange of different clay minerals is given in Table 1-6. These data indicate that the average ratio of cation to anion exchange

TABLE 1-6
Cation- and Anion-Exchange Capacities of Various Clay Minerals
(after Schoen, 1953)

Mineral	Cation-exchange capacity (me/100 g)	Anion-exchange capacity (me/100 g)	c/a	Equilibrium pH
Nontronite	87	13	6.7	2.4
Bentonite	62	15	4.1	3.1
Illite	21	9	2.3	4.5
Kaolinite	27	43	0.63	4.7
Halloysite	4	7.7	0.52	4.3

for montmorillonite, illite, and kaolinite groups is approximately 6.7, 2.3, and 0.5, respectively. Clays with large cation exchange capacities exhibit anion adsorption at very low pH values. Chloride and sulfate ions are only weakly adsorbed at low pH (Mattson, 1931). Phosphate ions, on the other hand, are strongly adsorbed even at pH values above the neutral point. The adsorbed phosphate can be replaced by other anions, such as hydroxide, fluoride, arsenate, and silicate (Scarseth, 1935; Dickman and Bray, 1941; Dean and Rubins, 1947; Wiklander, 1964). At low pH, it forms insoluble compounds with hydrous iron and aluminum oxides in the soil. One of the major problems is to distinguish between purely anion exchange effects and the precipitation of phosphate ions by iron and aluminum.

ION ADSORPTION AND FLOCCULATION. Early investigations pointed to the hydration of the exchangeable cations as a dominant factor in the stability of clay suspensions (Wiegner, 1925; Tuorila, 1928; Marshall, 1931; Jenny and Reitemeier, 1935). It was shown that the dispersity of the system followed the lyotropic series Li > Na > K > Rb > Cs. Clays saturated with divalent were less stable than those containing monovalent cations in the outer part of the double layer.

The flocculation of colloidal kaolinite suspensions by chlorides of monovalent and divalent cations followed the series H > Cs > K > Na > Li and Ba > Ca > Mg (Tuorila, 1928). The degree of flocculation of homionic systems with KCl decreased according to the series H > Cs > K > Na > Li for the monovalent ions and Ba > Sr > Mg > Ca for the divalent. Only the Ca ion occupied a position that is not in accordance with its ionic size. The H ion is a special case. In light of the modern concept that a H-clay is partially an Al-clay, this could be an Al-ion effect.

Polyvalent cations have a high flocculating power (Jenny and Reitemeier, 1935; Beavers and Marshall, 1951). This is in accordance with the Schulze-Hardy rule which states that the flocculating power of active ions of opposite charge to that of the surface of the particle increases with their valence. This rule is not universally applicable to the flocculaton of clays, because this is affected by the nature of the adsorbed cation, the hydrogen-ion concentration, and the concentration of the suspension (Bradfield, 1928). Polyvalent cations are tightly adsorbed and exhibit compressed double layers. There is a small percentage release of La and Th ions and a low flocculation value with respect to KCl. Adsorption of these polyvalent cations often results in a reversal of charge of the clay suspensions (Beavers and Marshall, 1951). This is dependent upon the type of clay mineral and the nature of the cation.

When a salt is added to a clay suspension that is saturated with the same cation, flocculation is the result of compressing the double layer and lowering the negative charge. This reduces the repulsion of the like-charged particles. If a salt containing a different cation is added, there will be both ionic exchange and compression of the double layer (Jenny, 1938; Kahn, 1958; van Olphen, 1963).

Anion adsorption on positive edge double layers of montmorillonite and kaolinite may produce flocculation (van Olphen, 1951, 1963; Schofield and Samson, 1954). It is obvious from previous discussions that the broken bonds on the edges of the clay minerals produce a structure that is not exactly the same as that of the planar surfaces. The broken bonds of the octahedral sheets produce a positive charge in acid solutions and a negative charge under alkaline conditions. This means that Al ions are on the inner layer at low pH values and OH ions above pH 7.0. In the case of broken bonds at sites in the tetrahedral sheet where Al substitutes for Si, there will be a positive charge. Similarly, soluble Al may be adsorbed on broken-bond silica surfaces to make them positive.

These positive edge charges can lead to flocculation of the suspension because of electrostatic attraction between positive edges and negative planar faces. This would be an edge-to-face association. Na-kaolinite exhibits this type of flocculation in salt-free suspensions (Schofield and Samson, 1954). In the presence of electrolytes, the thickness of both diffuse double layers is diminished and there can be face-to-face or edge-to edge association because of van der Waals forces. The significance of the positive edge charge to the dispersion of soils will be discussed in a later section.

ADSORPTION FROM SOLUTION. Soil particle surfaces are highly active and tend to adsorb a wide range of nonelectrolytes. Adsorption on surfaces is a complex subject which has received considerable attention and the reader is referred to the specialized reference of Adamson (1967) for details. A few of the more basic concepts of adsorption are discussed here. Adsorption of many substances on soil particle surfaces is usually pictured as occurring in a monolayer or at well-defined sites on the surface; the distribution more than a monomolecular layer away from the surface is the same as that in the bulk solution. The relation between the amount of substance adsorbed to the concentration of that substance in solution at any given temperature is known as the *adsorption* isotherm. It is an important characteristic of the adsorbent-adsorbate interaction. From a few simple considerations, it is possible to derive the well-known and widely applicable function called the Langmuir adsorption equation (Langmuir, 1918).

At equilibrium, the number of adsorbed molecules that leave the adsorption sites and move into the solution must be equal to the number that are adsorbed from solution during any given time. If it is assumed that the rate of desorption is proportional to the number of adsorbed ions, one can write

$$R_{des} = k_1 n \tag{1-8}$$

where n is the number of adsorbed molecules per unit mass of soil and k_1 is a rate constant, which should be proportional to $e^{Q/RT}$, where Q

is the heat absorption, R is the gas constant, and T is the absolute temperature. The constant k_1 is a measure of the probability that a given molecule will have enough energy to overcome the force of attraction of the surface. It is further aussmed that the rate at which molecules are adsorbed is proportional to the concentration of the solution C and the number of vacant adsorption sites. If Q is the total number of adsorption sites, or adsorption capacity, then the number of sites available for adsorption is $(Q - n)$; so the rate of sorption is

$$R_{sorp} = k_2(Q = n)C \qquad (1\text{-}9)$$

Setting $R_{des} = R_{sorp}$ gives

$$k_1 n = k_2(Q - n)C$$

which can be solved explicitly for n to give

$$n = \frac{aQC}{(1 + aC)} \qquad (1\text{-}10)$$

where $a = k_2/k_1$. The isotherm is linear at low values of the concentration and the amount of adsorption is proportional to the concentration. The adsorption at high concentrations approaches the adsorption capacity Q. It is more convenient when analyzing experimental data to write equation 1-10 in the form

$$\frac{1}{n} = \frac{1}{aQC} + \frac{1}{Q} \qquad (1\text{-}11)$$

The reciprocal of the adsorption, $1/n$, is plotted against the reciprocal of the concentration, $1/C$. A straight line results if the data obey the Langmuir isotherm. The intercept of this line with the $1/c$ axis gives the reciprocal of the adsorption capacity, $1/Q$. The constant a can then be evaluated from the slope of the line.

Equation 1-10 was derived upon the assumption that all adsorption sites had the same heat of adsorption. This is often not the case with soil minerals. For example, the adsorption isotherms for phosphorus and for boron in soils do not yield a single straight line when plotted according to equation 1-11. This implies that sites with two or more energies of adsorption are involved. The analysis of such an isotherm is relatively complicated unless only two site energies are involved. However, it can be shown (Adamson, 1960) that if the distribution of heats of adsorption is exponential $[f(E) \sim \exp -E/nRt]$, then the adsorption isotherm obeys an equation of the form

$$n = bC^{1/n} \qquad (1\text{-}12)$$

where b is a constant and n is a constant greater than unity. This isotherm does not show an adsorption maximum or limiting value. However, by plotting log n vs. log C, the intercept bn gives a measure of the adsorption capacity of the surface and the slope $1/n$ a measure of the intensity of adsorption.

The adsorption isotherm is stressed at this point because it is extremely important in the movement of substances through the soil. The mobility of any substance is a function of the fraction of the substance which is in solution at any given time. If the water content of the soil on a weight basis is denoted by w, then the total amount of a substance in solution associated with a unit mass of soil is wC, where C is the soil solution concentration. The amount of adsorbed substance is n, as given above, and the total amount of substance associated with a unit mass of soil is $(n + wC)$. The ratio of that in solution to the total is

$$R = wC(n + wC) \qquad (1\text{-}13)$$

This ratio varies from one to zero as n varies from zero to infinity.

The Adsorption of Gases

The adsorption of gases upon solid surfaces is perhaps even a broader subject than adsorption from liquids. The Langmuir adsorption isotherm has also been applied to this case even though only one shape of isotherm is involved—one in which there is a definite adsorption capacity that presumably is associated with adsorption of a monolayer. Brunauer, Emmett, and Teller (1938) extended the Langmuir approach to multilayer adsorption and developed an equation which is known as the BET equation. The principle assumption of their derivation, in addition to those already discussed, is that the heat of adsorption of the first layer has some value which is different from that for all succeeding layers, which are assumed to have a heat of adsorption equal to the heat of vaporization of the liquid adsorbate. This leads to an equation of the form

$$\frac{x}{v}(1 - x) = \frac{1}{cv_m} + \frac{(c - 1)x}{cv_m} \qquad (1\text{-}14)$$

where $x = P/P_0$, the ratio of the vapor pressure of the gas to the saturated vapor pressure of the liquid at the specified temperature, v is the volume of substance adsorbed at relative humidity P/P_0, v_m is the volume adsorbed in a monolayer, and c is a constant related to the heat of adsorption. Experimental adsorption data yield v as a function of x so that the function on the left-hand side of equation 1-14 can be plotted against x to give a straight line, if the isotherm obeys the BET equation. The intercept and the slope v_m and c can be evaluated. The BET equation appears to work reasonably well for the adsorption of nitrogen on soil minerals but is less satisfactory for the adsorption of water.

Measuring Specific Surface

Methodologies have been developed for measuring both the external and internal surfaces of clays. The BET equation has been used with nitrogen on soils (Nelson and Hendricks, 1943). If the samples are heated before adsorption to a sufficiently high temperature, the interlamellar layers of such minerals as montmorillonite and vermiculite collapse and do not permit the entry of the gas molecules. Only the external surface is measured in this case. If the layers are not completely collapsed, it is possible in some cases for a single layer of N molecules to be adsorbed in the interlamellar spaces. The surface area is calculated from a measurement of the quantity v_m necessary to occupy a monolayer (equation 1-14). The surface area can be calculated from a knowledge of the area of a single gaseous molecule.

A second method for calculating surface areas uses the principle of the adsorption of polar molecules on the surfaces of the clay minerals. This can be accomplished with a polar liquid that has a large number of hydroxyl groups in the molecule. Water, ethylene glycol, and cetyl pyridinium bromide have been employed for this purpose (Quirk, 1955; Dyal and Hendricks, 1950; Bower and Gschwend, 1952; Greenland and Quirk, 1964). Water and ethylene glycol can be either associated with the exchangeable cations present or hydrogen bonded with the oxygen atoms in the surface. The extent to which this occurs has not been established. Cetyl pyridinium bromide is considered to be adsorbed at first through cation exchange and subsequently through van der Waals forces between the cetyl chains. Ethylene glycol is adsorbed in two layers within the interlamellar spaces of montmorillonite. Only one such layer is adsorbed on the internal surfaces of vermiculite. It has been estimated that 3.1×10^{-4} g/m^2 of ethylene glycol are required to form a monolayer on the adsorbing surface.

The principle of this adsorption technique is to saturate a dry surface with ethylene glycol in a vacuum desiccator. The excess of this polar liquid is removed under vacuum over either anhydrous $CaCl_2$ or a $CaCl_2$-ethylene glycol mixture. When the weight of the clay mineral-glycol mixture reaches a constant value, the surface area is calculated from the equation

$$A = \frac{W_g}{W_s \times 0.00031} \tag{1-15}$$

where A is the specific surface in square meters per gram, W_g is the weight of the ethylene glycol in the sample, W_s is the weight of the air-dried sample and 0.00031 represents the grams of glycol that are required to form a monolayer per square meter of surface [see Mortland and Kemper (1965) for details].

The cetyl pyridinium bromide (CPB) technique is much simpler (Greenland and Quirk, 1964). It involves shaking the sample with solutions of CPB of varying concentrations, centrifuging, and determining the change in concentration of the supernatant liquid with ultraviolet absorption methods. The maximum adsorption value represents the formation of a double layer on the external surfaces. Unlike ethylene glycol, there is only one layer of CPB adsorbed in the interlamellar layer which is shared by both surfaces. It has been found that the mean external area per adsorbed molecule is equal to 27 Å. That of the internal surface is 54 Å. However, this method may not measure accurately the surfaces of expanding-lattice type clays or those with a low density of charge.

Comparisons of specific surfaces by various methods are shown in Table 1-3. It is seen that there is general agreement between the BET, CPB, and ethylene glycol methods for illites and kaolinites. It is obvious that the BET technique does not determine the interlamellar surfaces of montmorillonite.

When the expanding-lattice type minerals are heated to 600° C, the interlamellar layers collapse irreversibly and ethylene glycol is not adsorbed except on external surfaces. A comparison of the BET and ethylene glycol techniques for measuring specific surface under these conditions shows rather close agreement (Bower and Gschwend, 1952).

The internal surface of expanding-lattice-type clays can be determined either by comparing ethylene adsorption before and after heating the sample or by measuring external surface by the BET technique and subtracting this value from the total surface obtained by the ethylene glycol or CPB methods. For example, the external surfaces of the three montmorillonites shown in Table 1-3 varies between 47 and 101 m²/g; the internal surfaces range between 551 and 753 m²/g.

PARTICLE—SIZE ANALYSIS

Because of the correlation between specific surface and particle size, the percentage distribution of the various sizes of individual particles within a soil has served as an important soil characteristic. Determination of the particle-size distribution in order to characterize the soil texture is one of the most common soil physical analyses. The success of any such analysis depends first upon the preparation of the sample to ensure the complete dispersion of all aggregates into their individual primary particles without breaking up the particles themselves and second upon the accurate fractionation of the sample into the various separates.

Preparation of the Sample

The main objectives in the preliminary treatment of the soil sample preparatory to a mechanical analysis are to obtain maximum dispersion and to maintain this dispersion during the process of making the analysis. These can be accomplished by (1) the removal of cementing agents, such as organic matter and the oxides of iron and aluminum, (2) the rehydration of the clay particles by mechanical means, and (3) the physical and chemical dispersion of the particles.

Removal of the cementation effects of organic matter and other colloidal materials and the formation of a water film around each clay particle are not sufficient to ensure maximum dispersion. Particles may be separated physically in a suspension, but they may coalesce again into floccules or aggregates before an accurate particle-size analysis can be made. As previously discussed, clay particles possess a negative charge. The potential on the surface of the particles must be kept above a certain critical level in order to prevent flocculation. Highly hydrated monovalent cations must be substituted for strongly adsorbed H, Ca, and Mg ions on the exchange complex of the clay.

The positive edge charges must be changed to negative to eliminate the edge-to-edge and edge-to-face attractions of the clay minerals. This can be accomplished through anion adsorption of OH ions or complex phosphate anions (van Olphen, 1963; Schofield and Samson, 1954). Sodium hexametaphosphate achieves both objectives of increasing the negative charges on the planar surfaces and reversing the positive edge charges. The result is highly negatively charged particles that repel each other and remain thoroughly dispersed in suspension. Sodium hydroxide will accomplish this with H-saturated systems. Sodium carbonate, sodium hexametaphosphate, sodium oxalate, and sodium silicate may be used with soils containing exchangeable or carbonate calcium.

Fractionation of the Sample

Inasmuch as soil consists of particles of various sizes and since the fundamental objective of a particle-size analysis is to determine the percentage distribution of these groups in the soil mass, the question naturally arises as to the means of accomplishing such an analysis. If a nest of graded sieves were available, separation into different fractions might be accomplished. It is immediately obvious, however, that only the coarser fractions could be separated in this manner. A different principle must be used with the smaller particles. It has been shown that the rate of fall of particles in a viscous medium depends upon the size, density, and shape of the particle. In a given medium, such as water, larger particles fall more rapidly than smaller ones with the same density and consequently settle out of suspension more quickly. This principle serves as the basis of practically all mechanical analyses.

Fall of Particles in a Liquid

STOKES' LAW. Stokes (1851) was the first to suggest the relation between the radius of a particle and its rate of fall in a liquid. He stated that the resistance offered by the liquid to the fall of the particle varied with the radius of the sphere and not with the surface, according to the formula

$$V = \frac{2}{9} \frac{(d_p - d)gr^2}{\eta} \tag{1-16}$$

where V is the velocity of fall in centimeters per second, g the acceleration due to gravity, d_p the density of the particle, d the density of the liquid, r the radius of the particle in centimeters, and η the absolute viscosity of the liquid. It is obvious that the velocity of fall of particles with the same density, in a given liquid, will increase with the square of the radius.

Assumptions in Stokes' Law. There are several fundamental assumptions upon which the validity of Stokes' formula is based:

1. The particles must be large in comparison to liquid molecules so that Brownian movement will not affect the fall.
2. The extent of the liquid must be great in comparison with the size of the particles. The fall of the particle must not be affected by the proximity of the wall of the vessel or of adjacent particles. Odén (1925) has shown that particles fall independently of each other in suspensions of 1 percent or less of solid particles.
3. Particles must be rigid and smooth. This requirement is difficult to fulfill with soil particles. It is highly probable that the particles are not completely smooth over their entire surfaces. It is fairly well established that the particles are not spherical but are irregularly shaped with a

large number of plate-shaped particles present in the clay fractions. Since variously shaped particles fall with different velocities, the term "equivalent or effective radius" is used to overcome this difficulty in Stokes' law. Equivalent or effective radius is defined as the radius of a sphere of the same material which would fall with the same velocity as the particle in question.

4. There must be no slipping between the particle and the liquid. This requirement is easily fulfilled in the case of soils because of the water hull around the particles.

5. The velocity of fall must not exceed a certain critical value so that the viscosity of the liquid remains the only resistance to the fall of the particle. Particles larger than silt cannot be separated accurately by the use of Stokes' law.

Limitation of Stokes' Law. The effect of different particle shapes on the settling velocities of clay particles is one of the greatest limitations to the accurate application of Stokes' law. Kelley and Shaw (1942), using the electron microscope to measure the cross-sectional area of clay particles, shed considerable light on the hydrodynamic relationships in the fall of particles as determined by shape. They found that the fractionation of rod-shaped particles, based upon settling velocity, is not as accurate as that of disk-shaped ones.

In addition to the effect of the size and shape of the particles upon the applicability of Stokes' law to particle-size analyses, there are certain experimental limitations that must be considered in the use of this principle. Since the rate of fall varies inversely with the viscosity of the medium, it is necessary to maintain a known constant temperature during the analysis. For example, the velocity of fall is about 12 percent faster at 30° C than at 25° C.

A constant temperature also helps to prevent convection currents which might arise as a result of differences in temperature near the walls of the vessel and within the suspension. Such currents would prevent the uniform settling of the particles. Convection currents may also be set up when the soil suspension is stirred. Currents due to stirring are more difficult to eliminate than those caused by temperature variations.

The density of the soil particle is another factor that affects the accuracy of Stokes' law. Density depends upon the mineralogical and chemical constitution of the particles as well as upon their degree of hydration. It is questionable whether the actual density of a particle as it falls in water can be accurately determined. The pycnometer method is most commonly used for measuring density. The samples are dried at 110° C to expel the adsorbed water hull. It is obvious that a hydrated particle, falling in a suspension, should have a lower density than one that is completely dehydrated. Attempts have been made to investigate the hydration effect by using heavy aqueous liquids as a means of determining density. Marshall (1935a) used solutions of potassium mercuric iodide of different densities. Flocculation of the clay particles takes place, and it is easy to observe whether the liquid or the soil particle has the higher density. His results are shown in Table 1-7, along with other values for the specific gravity of different soil constituents.

TABLE 1-7
The Density of Various Soil Constituents

Orthoclase	2.50 to 2.60	Putnam clay	2.52 to 2.78
Mica	2.80 to 3.20	Cecil clay	3.35
Quartz	2.50 to 2.80	Rothamsted clay	2.74
Limonite	3.40 to 4.00	Putnam clay*	
Hematite	5.10 to 5.20	1–2 μ	2.56
Fe(OH)$_3$	3.73	0.5–1 μ	2.50
Kaolin	2.50	0.2–0.5 μ	2.42
Humus	1.37	0.1–0.2 μ	2.39

* Determined with KHgI$_3$ solutions (Marshall, 1935a).

It may readily be seen from these data that a specific gravity determination should be made for each soil for the best results to be obtained. The generally accepted value of 2.65 represents an average figure that is sufficiently exact for the majority of mechanical analyses. Accurate studies, however, require a more exact value. It is essential to note that the density of clay particles decreases with particle size when determined in aqueous suspension. This is due to the higher hydration of the smaller particles.

In light of the effects of particle size, shape, and density upon the applicability of Stokes' law to mechanical analysis, it should be remembered that any particle which is separated upon the basis of its settling velocity does not necessarily have the exact calculated size. Its effective or equivalent radius corresponds to a given size grouping, in that all particles within such a group have the same velocities of fall. For this reason Robinson (1936) defined particle sizes in terms of their settling velocities rather than by their equivalent radii or diameters.

Although these limitations to Stokes' law may become serious in refined analyses of the size distribution of various particles in the soil, for most purposes the results will not warrant extra precaution in technique, provided that extreme temperature fluctuations are prevented and care is exercised in the stirring of the suspension.

Analyses Based upon the Size Distribution of Soil Particles

THEORETICAL CONSIDERATIONS. The methods of particle-size analysis which have been discussed in the preceding sections permit the separation of the soil particles into a number of distinct fractions between certain size limits. It is obvious that if the number of separated fractions could be sufficiently increased, one would obtain a picture of the percentage distribution of the particles of different sizes in the sample. Such a complete analysis by any of the aforementioned methods, with the possible exception of the two-layer technique of Marshall, would be too laborious and inaccurate to be practical.

Odén (1915) developed a theory of the settling of particles in suspension, as based upon Stokes' law, and gave it a mathematical interpretation. Later Fisher and Odén (1924) presented a more general analysis of the theory, which has served as the basis for most of the present methods of particle-size analysis. A brief interpretation of their analysis is given here.

Let us consider a uniform suspension of soil particles in which all particles are settling independently of each other. At any time t and at a given depth h those particles with settling velocities greater than h/t cm/sec will have settled below the depth. In addition, particles with settling velocities smaller than h/t cm/sec, but falling through a shorter distance, will have settled below h. If the pan of a balance is placed at the depth h, the total weight of sedimented particles at the end of time t will be equal to the weight of that fraction with a settling velocity of or greater than h/t cm/sec, which has settled out completely, plus a portion of the material with a settling velocity less than h/t. The actual rate of increase in weight dp/dt at time t is due to this latter group of particles, since they have been falling at constant rate since the beginning of the experiment. The actual weight of these smaller particles on the pan of the balance is tdp/dt.

Perhaps the concept of the settling of particles in suspension will be clearer if a hypothetical case is discussed. Visualize a suspension containing only two different sizes of particles. The height of the suspension above the pan of a balance is 10 cm. The suspension contains 2 g of particles that will settle out completely in 20 min and 2 g of particles that will sediment in 40 min. The settling velocities therefore are $\frac{1}{120}$ and $\frac{1}{240}$ cm/sec (h/t), respectively. At the end of 20 min all the faster-sedimenting particles will settle on the pan. During sedimentation they will increase the weight at the rate of $\frac{1}{600}$ g/sec, since $\frac{1}{1200}$ of this fraction will settle out every second. During this same time, the smaller particles will settle at the constant rate of $\frac{1}{240}$ cm/sec. Since $\frac{1}{2400}$ of the 2 g will sediment every second, these particles will increase the weight at the rate of $\frac{1}{200}$ g/sec (dp/dt). At the end of the 20 min 1 g of this material will settle out:

$$\left(t\frac{dp}{dt} = 1200 \times \frac{1}{1200} \right)$$

Therefore at the end of time t (20 min) the rate of increase in weight will be due to the 1 g of these smaller particles that will continue to settle out for 20 more minutes at a constant rate of $\frac{1}{1200}$ g/sec. It is obvious that the same reasoning applies to an indefinite number of fractions with settling velocities of less than h/t cm/sec.

Instead of visualizing the pan of a balance at a given depth, onto which the falling particles settle out at different rates to cause a continuous increase in weight, one might consider the changes in concentration that take place in a thin plane of the suspension at this same depth. At the beginning of sedimentation, every plane in the suspension has the same number of particles of different sizes. As sedimentation proceeds, however, particles with settling velocities greater than h/t cm/sec will settle from the top through the plane at depth h. Smaller particles will settle out of this plane but will be replaced by similar particles that have fallen from above into this thin layer. As the time of settling increases, it may be seen that there is a progressive decrease in concentration or density of the layer at this given depth.

These changes in concentration may be calculated according to the following reasoning: Let W equal the weight of particles in a given volume V of suspension, D_p the density of the particles, D_s the density of the suspension, v the settling velocity of the particles, t the time of settling, h the depth of settling, and X the fraction or percentage by weight of particles having a velocity less than $v = h/t$.

At time t_0 in a completely dispersed uniform suspension, every cubic centimeter of the suspension will contain W/VD_p cm³ or W/V g of solid particles and $(1 - W/VD_p)$ cm³ or g of water. The specific gravity of the suspension at the beginning will be

$$D_{s0} = 1 + \frac{W}{VD_p}(D_p - 1) \tag{1-17}$$

If X equals the fraction of particles having settling velocities less than h/t, then at time t and depth h there will be left in suspension XW/V

g of particles per cubic centimeter immediately above h, because the faster-moving particles will have settled below this depth. The particles included in X will be in the same concentration as they were at the start. At time t therefore, the specific gravity of the suspension in a thin layer above h will be

$$D_s = 1 + \frac{XW}{VD_p} (D_p - 1) \qquad (1\text{-}18)$$

or

$$D_s = 1 + KX \qquad (1\text{-}19)$$

This equation is fundamental to the several methods that are based upon the Odén theory.

FUNDAMENTAL REQUIREMENTS. There are several fundamental requirements of methods based upon the change in concentration of a suspension at a given depth with time in order to permit an accurate measurement of this change.

1. It is necessary to have complete dispersion of the soil particles and to prevent any coagulation effects during settling. It is obvious that an incomplete dispersion will not permit a true measurement of the quantity of a given fraction. Unlike methods based upon repeated fractionations where dispersion may be effected as the separations proceed, the principle of determining changes in concentration requires that each particle fall independently of the others.

2. The concentration of the suspension must be dilute enough so that the particles do not interfere with each other during their fall through the liquid. Moreover, the density of the suspension should never vary greatly from that of the liquid. Odén advocated the use of a 1 percent suspension; Robinson used 2 percent; and Wiegner carried out most of his investigations in a concentration of 5 percent. Not only must the suspension be dilute but the sedimentation vessel also must have a diameter large enough to prevent errors arising from the attraction between the particles and the wall.

3. The suspension must be thoroughly stirred so that there is a uniform distribution of particles throughout at the beginning time t_0. All methods presuppose that every particle settles at a uniform rate from all portions of the suspension.

4. Measurements should be made at a constant temperature so that no convection currents occur during sedimentation. Differences in temperature between the wall of the vessel and the center of the suspension

may produce currents that will prevent the uniform settling of the finer particles. It is also essential to maintain a constant temperature in order not to change the viscosity of the dispersion medium.

Although these four requirements may seriously affect the results of size-distribution analyses if improper techniques are employed, they are so easily met under good experimental conditions that they are not a handicap to routine determinations.

METHODS USED. Although there are several principles that could be employed to determine the size-frequency distribution of particles, the most widely used is based upon changes in density of the suspension with time at a given depth. The pipette (Robinson, 1922) and hydrometer (Bouyoucos, 1927) methods are classical examples of this principle (Day, 1965).

GRAPHICAL PRESENTATION OF DATA. Summation percentage curves of the different separates from particle-size analyses can be plotted as a continuous function of log settling velocity or log diameter. Such curves for two subsurface samples of Miami silt loam are shown in Figure 1-12a.

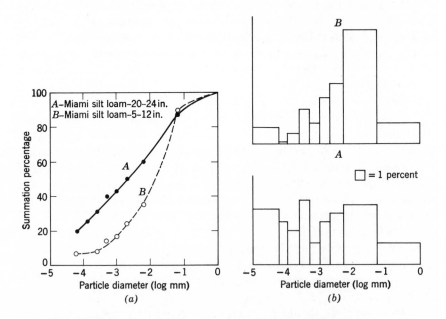

FIG. 1-12. Graphical presentation of particle-size analysis data. (a), summation curves; (b), size-distribution diagrams. (Data of Steele and Bradfield, 1934.)

The slope of these curves is a measure of the frequency of the fraction conforming to the corresponding diameter. The most abundant fraction is given by the steepest portion of the curve if the fractions are chosen at equal logarithmic intervals. The curves point out that there is a more uniform distribution of particles between the silt and clay fractions in sample *A* than in sample *B*. The latter indicates a high percentage of particles between 0.05 and 0.005 mm diameter. Although these curves are commonly employed to represent the particle-size analysis of a soil, the size-frequency distribution curves can be used to greater advantage from the standpoint of ready interpretation.

Once the weight or percentage of particles between chosen size limits is obtained by analysis, the data can be plotted in a histogram to get the size-frequency distribution. A small area is chosen to represent a given percentage and the total amount of particles within each size group is shown on the graph as an area. A smooth curve may be drawn across the tops of the histograms. Actually, this type of graph is equivalent to plotting the slope of the summation curve as a function of log diameter. The size-frequency histograms in Figure 1-12*b* depict the higher silt content of sample B over sample A and the higher percentages of clay in the latter.

References

Adamson, Arthur W. (1967). *Physical Chemistry of Surfaces*, 2nd ed. Interscience Publishers, New York.

Atterberg, A. (1912). Die mechanische Bodenanalyse und die Klassification der Mineralböden Schwedens. *Intern Mitt. Bodenk.*, 2:312–342.

Baver, L. D. (1928). The relation of exchangeable cations to the physical propreties of soils. *J. Am. Soc. Agron.*, 20:921–941.

Beavers, A. H., and C. E. Marshall (1951). The cataphoresis of clay minerals and factors affecting their separation. *Soil Sci. Soc. Am. Proc.*, 15:142–145.

Bolt, G. H. (1955a). Ion adsorption by clays. *Soil Sci.*, 79:267–276.

Bolt, G. H. (1955b). Analysis of the validity of the Gouy-Chapman theory of the electric double layer. *J. Colloid Sci.*, 10:206–218.

Bouyoucos, G. J. (1927). The hydrometer as a new method for the mechanical analysis of soils. *Soil Sci.*, 23:343–353.

Bower, C. A., and F. B. Gschwend (1952). Ethylene glycol retention by soils as a measure of surface area and interlayer swelling. *Soil Sci. Soc. Am. Proc.*, 16:342–345.

Bradfield, Richard (1928). Factors affecting the coagulation of colloidal clay. *Missouri Agr. Exp. Sta. Research Bull.* 60.

Bradley, W. F. (1940). The structural scheme of attapulgite. *Am. Mineralogist*, 25:405–410.

Brindley, G. W. (1951). The kaolin minerals. *X-ray Identification and Structures of Clay Minerals*. Mineralogical Society, London (1951), Chap. 11.

Brown, G. (1961). *The X-ray Identification and Crystal Structures of Clay Minerals*. Mineralogical Society, London, England.

Brunauer, S., P. H. Emmett, and E. Teller (1938). Adsorption of gases in multimolecular layers. *J. Am. Chem. Soc.*, 60:309–319.

Chatterjee, B., and C. E. Marshall (1950). Studies in the ionization of magnesium, calcium and barium clays. *J. Phys. Coll. Chem.*, 54:671–681.

Coleman, N. T., S. B. Weed, and R. J. McCracken (1959). Cation exchange capacities and exchangeable cations in Piedmont soils of North Carolina. *Soil Sci. Soc. Am. Proc.*, 23:146–149.

Day, Paul R. (1965). Particle fractionation and particle-size analysis. *Methods of Soil Analysis, Agronomy Monograph*, Part 1, pp. 545–567, Academic Press, New York.

Dean, L. A., and E. J. Rubins (1947). Anion exchange in soils. *Soil Sci.*, 63:377–406.

de Villiers, J. M., and M. L. Jackson (1967). Cation exchange capacity variations with pH in soil clays. *Soil Sci. Soc. Am. Proc.*, 31:473–476.

Dickman, S. R., and R. H. Bray (1941). Replacement of adsorbed phosphate from kaolinite by fluoride. *Soil Sci.*, 52:263–275.

Dyal, R. S., and S. B. Hendricks (1950). Total surface of clays in polar liquids as a characteristic index. *Soil Sci.*, 69:421–432.

Edelman, C. H., and J. C. L. Favejee (1940). On the crystal structure of montmorillonite and halloysite. *Z. Krist.*, 102:417–431.

Fieldes, M., and R. K. Schofield (1960). Mechanism of ion adsorption by inorganic soil colloids. *N. Z. J. Sci.*, 3:563–579.

Fieldes, M. (1962). The nature of the active fraction of soils. *Trans. Joint Meeting, Com. IV and V, Int. Soc. Soil Sci.*, New Zealand, pp. 62–78.

Fisher, R. A., and Sven Odén (1924). The theory of the mechanical analysis of sediments by means of the automatic balance. *Proc. Roy. Soc., Edinburgh*, 44:98–115.

Forslind, Erik (1950). Some remarks on the interaction between the exchangeable ions and the adsorbed water layers in montmorillonite. *Trans. 4th Int. Cong. Soil Sci., Amsterdam*, 1:110–113.

Fripiat, J. J. (1964). Surface chemistry and soil science. *Experimental Pedology, Proc. 11th Easter School in Agr. Sci.*, Univ. Nottingham, 1964, pp. 3–13.

Greene-Kelly, R. (1962). Charge densities and heats of immersion of some clay minerals. *Clay Minerals Bull.*, 5:1–8.

Greeland, D. J., and J. P. Quirk (1964). Determination of the total specific surface areas of soils by adsorption of cetyl pyridinium bromide. *J. Soil Sci.*, 15:178–191.

Grim, R. E., and R. H. Bray (1936). The mineral constitution of various ceramic clays. *J. Am. Ceram. Soc.*, 19:307–315.

Grim, R. E., R. H. Bray, and W. F. Bradley (1937). The mica in argillaceous sediments. *Am. Mineralogist*, 22:813–829.

Grim, Ralph E. (1953). *Clay Mineralogy*. McGraw-Hill Book Co., New York.

Grim, Ralph E. (1962). *Applied Clay Mineralogy*. McGraw-Hill Book Co., New York.

Gruner, J. W. (1934). The structures of vermiculites and their collapse by dehydration. *Am. Mineralogist*, 19:557–575.

Harmon, C. G., and F. Fraulini (1940). Properties of kaolinite as a function of its particle size. *J. Am. Ceram. Soc.*, 23:252–258.

Hendricks, S. B. (1942). Lattice structure of clay minerals and some properties of clays. *J. Geol.*, 50:276–290.

Hofmann, V., K. Endell, and D. Wilm (1933). Kristallstruktur und Quellung von Montmorillonit. *Z. Krist.*, 86A:304–348.

Humbert, R. P., and B. T. Shaw (1941). Studies of clay particles with the electron microscope. *Soil Sci.*, 52:481–487.

Jackson, M. L. (1963). Aluminum bonding in soils: A unifying principle in soil science. *Soil Sci. Soc. Am. Proc.*, 27:1–10.

Jackson, Marion L. (1964). Chemical composition of soils. *Chemistry of Soil* (Firman E. Bear, Editor). Reinhold Publishing Corp., New York, pp. 71–141.

Jenny, Hans (1932). Studies on the mechanism of ionic exchange in colloidal aluminum silicates. *J. Phys. Chem.*, 36:2217–2258.

Jenny, Hans (1938). *Properties of Colloids*. Stanford University Press, Stanford, Calif., p. 50.

Jenny, Hans, and R. F. Reitemeier (1935). Ionic exchange in relation to the stability of colloidal systems. *J. Phys. Chem.*, 39:593–604.

Kahn, Allan (1958). The flocculation of sodium montmorillonite by electrolytes. *J. Colloid Sci.*, 13:51–60.

Kelley, O. J., and B. T. Shaw (1942). Studies of clay particles with the electron microscope: III. Hydrodynamic considerations in relation to shape of particles. *Soil Sci. Soc. Am. Proc.*, 7:58–68.

Kelley, W. P., W. H. Dore, and S. M. Brown (1931). The nature of the base-exchange material for bentonite, soils and zeolites as revealed by chemical investigations and x-ray analysis. *Soil Sci.*, 41:259–274.

Kelley, W. P. (1948). *Cation Exchange in Soils*. Am. Chem. Soc. Monograph 109, Reinhold, New York.

Langmuir, Irving (1918). The adsorption of gases on plane surfaces of glass, mica and platinum. *J. Am. Chem. Soc.*, 40:1361–1403.

McAuliffe, C. D., N. S. Hall, L. A. Dean, and S. B. Hendricks (1947). Exchange reactions between phosphates and soils: Hydroxylic surfaces of soil minerals. *Soil Sci. Soc. Am. Proc.*, 12:119–123.

McLean, E. O. (1951). Interrelationships of potassium, sodium, and calcium as shown by their activities in a beidellite clay. *Soil Sci. Soc. Am. Proc.*, 15:102–106.

McNeal, B. L., W. A. Norwell, and N. T. Coleman (1965). Effect of solution composition on the swelling of extracted soil clays. *Soil Sci. Soc. Am. Proc.*, 30:313–317.

Marshall, C. E. (1935a). Mineralogical methods for the study of silts and clays. *Z. Krist. (A)*, 90:8–34.

Marshall, C. E. (1935b). Layer lattices and base exchange clays. Z. Krist. (A), 91:433–449.

Marshall, C. E., R. P. Humbert, B. T. Shaw, and O. G. Caldwell (1942). Studies of clay particles with the electron microscope: II. The fractionation of beidellite, nontronite, magnesium bentonite, and attapulgite. Soil Sci., 54:149–158.

Marshall, C. E. (1949). The Colloidal Chemistry of the Silicate Minerals. Agronomy, Vol. I. Academic Press, New York.

Marshall, C. E. (1950). The electrochemistry of the clay minerals in relation to pedology. Trans. 4th Int. Cong. Soil Sci., Amsterdam, 1:71–82.

Marshall, C. Edmund (1964). The Physical Chemistry and Mineralogy of Soils. John Wiley and Sons, New York.

Mattson, S. (1929). The laws of soil colloidal behavior I. Soil Sci., 28:179–220.

Mattson, Sante (1931). The laws of soil colloidal behavior: VI. Amphoteric behavior. Soil Sci., 32:343–365.

Mitscherlich, E. A. (1901). Untersuchungen über die physikalischen Bodeneigenschaften. Landw. Jahrb., 30:360–445.

Mortland, M. M. (1954). Specific surface and its relationship to some physical and chemical properties of soils. Soil Sci., 78:343–347.

Mortland, M. M., and W. D. Kemper (1965). Specific Surface Methods of Soil Analysis. Agronomy Monograph, Part I. Academic Press, New York, pp. 532–544.

Nelson, R. A., and S. B. Hendricks (1943). Specific surface of some clay minerals, soils, and soil colloids. Soil Sci., 56:285–296.

Odén, Sven (1915). Eine neue Methode zur mechanischen Bodenanalyse. Int. Mitt. Bodenk., 6:257–311.

Pauling, L. (1930). The structure of micas and related minerals. Nat. Acad. Sci. Proc., 16:123.

Portland Cement Association (1962). PCA Soil Primer, Chicago, Ill.

Quirk, J. P. (1955). Significance of surface areas calculated from water vapor sorption isotherms by use of the B.E.T. equations. Soil Sci., 80:423–430.

Radoslovich, E. W. (1960). The structure of muscovite, $KAl_2(Si_3Al)O_{10}(OH)_2$. Acta Cryst., 13:919–932.

Radoslovich, E. W., and K. Norrish (1962). The cell dimensions and symmetry of layer-lattice silicates. Am. Mineralogist, 47:599–616.

Robinson, G. W. (1922). A new method for the mechanical analysis of soils and other dispersions. J. Agr. Sci., 12:306–321.

Robinson, G. W. (1936). Soils, Their Origin, Constitution and Classification, 2nd ed. Thomas Murby and Company, London, p. 16.

Ross, C. S., and E. V. Shannon (1926). The minerals of bentonite and related clays and their physical properties. J. Am. Ceram. Soc., 9:77.

Scarseth, G. D. (1935). The mechanism of phosphate retention by natural aluminosilicate colloids. J. Am. Soc. Agron., 27:596–616.

Schoen, U. (1953). Kennzeichung von Tonen dürch Phosphatbindung und Kationenumtausch. Z. Pflanzenernähr. Düngung u. Bodenk., 63:1–17.

Schofield, R. K. (1949). The effect of pH on electric charges carried by clay particles. J. Soil Sci., 1:1–8.

Schofield, R. K., and H. R. Samson (1954). Flocculation of kaolinite due to the attraction of oppositely-charged crystal faces. *Dis. Faraday Soc.*, 18:135–145.

Shainberg, Isaac, and W. D. Kemper (1966a). Conductance of adsorbed alkali cations in aqueous and alcoholic bentonite pastes. *Soil Sci. Soc. Am. Proc.*, 30:700–706.

Shainberg, I., and W. D. Kemper (1966b). Hydration status of adsorbed cations. *Soil Sci. Soc. Am. Proc.*, 30:707–713.

Shainberg, I., and W. D. Kemper (1967). Ion exchange equilibria on montmorillonite. *Soil Sci.*, 103:4–9.

Shaw, B. T., and R. P. Humbert (1941). Electron micrographs of clay minerals. *Soil Sci. Soc. Am. Proc.*, 6:146–149.

Steele, J. G., and Richard Bradfield (1934). The significance of size distribution in the clay fraction. *Am. Soil Survey Assoc. Bull.* 15, pp. 88–93.

Stokes, G. G. (1851). On the effect of the internal friction of fluids on the motion of pendulums. *Trans. Cambridge Phil. Soc.*, 9:8–106.

Tuorila, P. (1928). Über Beziehungen zwischen Koagulation, elektrokinetischzen Wanderungsgeschwindigkeiten, Ionenhydration und chemischer Beeinflussung. *Kolloidchem. Beihefte*, 27:44–181.

United States Department of Agriculture (1951). *Soil Survey Manual.* Handbook 18, pp. 207–208.

van Olphen, H. (1951). Rheological phenomena of clay sols in connection with the charge distribution on the micelles. *Disc. Faraday Soc.*, 11:82–84.

van Olphen, H. (1963). *Clay Colloid Chemistry.* Interscience Publishers, New York.

Wiegner, G. (1925). Dispersität und Basenaustauch. *Rept. 2d Com. 4th Int. Conf. Soil Sci.*, Rome.

Wiklander, Lambert (1964). Cation and anion exchange phenomena. *Chemistry of the Soil* (Firman E. Bear, Editor). Reinhold Publishing Corp., New York, pp. 163–205.

Whitt, D. M., and L. D. Baver (1937). Particle size in relation to base exchange and hydration properties of Putnam clay. *J. Am. Soc. Agron.*, 29:905–916.

CHAPTER 2

The Viscosity and Swelling of Soil Colloids

The physical and chemical properties of soils and the associated processes are intimately related to the interaction between the soil particles and water molecules. The adsorption of water molecules by colloidal clay particle surfaces is a function of the nature of the mineral surface and the associated exchangeable cations. The adsorbed water molecules are oriented into ordered arrangements as a result of hydrogen bonding between the water molecules and the oxygen atoms in the surface of the crystal and the hydration of the adsorbed cations.

Hydrogen bonding between water molecules extends beyond the first layer of molecules adsorbed. At least one or two molecular layers of water are adsorbed by soil particles even in relatively dry air. Water is present within the soil profile in films on particle surfaces, in interstices between particles, and within soil pores. Particle surfaces strongly attract liquid water by virtue of two types of force: van der Waal-London forces and hydrogen bonding. The van der Waal-London forces are partly electrical and partly gravitational; they are large close to particle surfaces. They vary inversely with the sixth or seventh power of the distance from the surface. The surface has a strong influence on the water closest to the surface because of these forces. If it is assumed that a water molecule is about 3 Å thick, the force of attraction at the midpoint of the second molecular layer is from $\frac{1}{1000}$ to $\frac{1}{2000}$ that in the first layer. This force drops off very rapidly with distance from the particle surface and probably becomes relatively small within a few molecular layers. A polar liquid such as water is responsible for a longer range force that results from the interaction of the water dipoles with the electrostatic field that emanates from the charged particle surface. This force may extend as far as 100 Å and probably overshadows the very short-range

54

forces, except at very low water contents. The water molecule is composed of an oxygen atom with six electrons in its outer orbital shell and two electrons in an interior orbit. Two hydrogen atoms share electrons with the oxygen atom to fill its outer shell; an angle of about 105° results between the lines that join the oxygen proton with the two protons of hydrogen. The positive hydrogen protons give the water molecule two positive poles with the two interior electrons of oxygen forming negative poles so that the molecule is a tetrahedron with charged corners (see Chapter 8, Figure 8-1). This polar nature of water promotes strong binding both between water molecules and to charged surfaces. Beyond the first thin film of water, hydrogen bonding is an important force that holds water on particle surfaces and in the pore space.

There are a number of ways in which water molecules may bind together; consequently, the structure of water is not a constant entity. A most interesting consequence of the variation in the binding pattern is that which causes water to freeze with an increase in volume as it cools below 3.96° C. This phenomenon is uncommon for most liquids. The biologic consequences of a volume increase with freezing range all the way from frost damage in soils and plants to the floating of ice in lakes and streams while the aquatic life below is preserved.

Water close to solid surfaces and in the presence of electrical charges undergoes structural changes which have been found to increase water volume and decrease density.

VISCOSITY OF COLLOIDAL CLAYS

Nature of Viscosity of Colloidal Suspensions

Viscosity is a structure-sensitive property of a fluid. The viscosity of colloidal clay suspensions gives valuable information about the influence of the clay particles upon the structure of the adsorbed water. The viscosity of a liquid refers to the internal friction between the molecules of the liquid. In a colloidal suspension, friction takes place between the particles and the molecules of the dispersion medium. If the colloidal particles are hydrated, then the friction is between the water molecules on the surface of the dispersed phase and those in the medium. Moreover, interaction is possible between the colloidal particles. This means that the viscosity of the suspension represents the integrated effects of the molecules of the liquid, the influence of the particles on the liquid, and the impact of particle interaction.

Colloids are generally divided into two distinct groups, hydrophile (or lyophile) and hydrophobe (or lyophobe), with respect to their viscosity. Hydrophilic sols are characterized by a relatively high viscosity.

$$\phi = \frac{V(N_s - N_m)}{N_s + 1.5 N_m}$$

The hydrophobic colloids do not possess a viscosity appreciably different from that of their dispersion medium. Clay sols can be considered as occupying an intermediate position, being neither truly lyophilic or lyophobic. Colloidal clay possesses the properties of hydrophilic colloids because of its hydration. Its sensitivity to electrolytes is a hydrophobic characteristic.

Viscosity measurements have been used extensively for the characterization of lyophilic colloids, even though most of the work has been relative in nature. It has been established that the viscosity of colloids is a function of the volume occupied by the disperse phase. Einstein proposed a formula showing the relation between viscosity and the volume of the dispersed phase. His equation indicates that any increase in viscosity in a colloidal system depends on the total volume of the particles and is independent of the degree of dispersion. This formula is

$$\eta_s = \eta_m(1 + 2.5\phi) \tag{2-1}$$

where η_s is the viscosity of the colloidal system, η_m the viscosity of the dispersion medium, and ϕ the volume of the dispersed phase per unit volume of sol.

This equation assumes that the particles are spherical and rigid and that the suspension is dilute enough to offset particle interference. The volume ϕ includes the total volume associated with the particles. This means any water hull around individual particles or water enmeshed within aggregates. Since the viscosity of a colloidal system is a function of the "active volume" of the dispersed phase, a colloidal aggregate, containing water enmeshed between the particles constituting the aggregate, will have a larger volume than the total volume of the individual particles. In other words, the volume of the aggregate will be the sum of the volume of the particles and the volume of the occluded water. Therefore a suspension containing aggregates will possess a higher viscosity than one containing the same number of particles in a monodispersed state, provided the dispersed particles have no adsorbed water hull. On the other hand, if a highly hydrated ion is adsorbed by the particle when the aggregate is dispersed, the volume of the particle will be larger, causing an increase in viscosity.

The factor 2.5 applies to spherical particles but must be increased several times for rodlike or platelike particles that have major and minor axes. Increases in concentration will augment particle interaction. Such interactions will also be a function of the size, shape, and hydration of the particles. Shape affects not only the magnitude of the viscosity but also the concentration at which particle interaction increases the viscosity. This is clearly shown in Figure 2-1. The largest particles, R-2, have the smallest

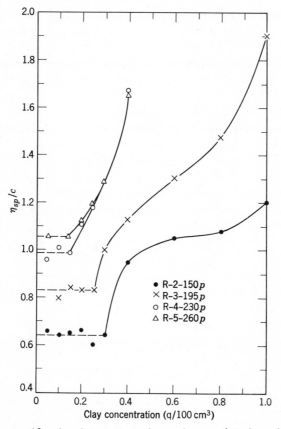

FIG. 2-1. The specific viscosity-concentration ratio as a function of concentration. (After Van Olphen and Waxman, 1958.)

major to minor axis ratio. The specific viscosity of this fraction with a ratio of 150 is about 0.62; that of the smallest particles with a ratio of 260 is approximately 1.05. Particle interaction is manifested at a concentration of about 0.15 percent for the smallest sizes and 0.30 percent for the largest. The rapid rises in the specific viscosity/concentration curve as a function of enlarged clay concentrations indicate conditions of flocculation brought about by edge-to-edge and edge-to-face structures of the interacting particles.

Effect of Exchangeable Cations and Clay Minerals on Viscosity

The amount and nature of exchangeable cations have different effects on the viscosity of beidellite as compared with montmorillonite and keolinite. The effects of saturating H-Putnam clay (beidellite) with various cations on its viscosity are shown in Figure 2-2 (Baver, 1929). A careful study of the viscosity curves (Figure 2-2) indicates that the effects of monovalent cations are manifested in three different ways as

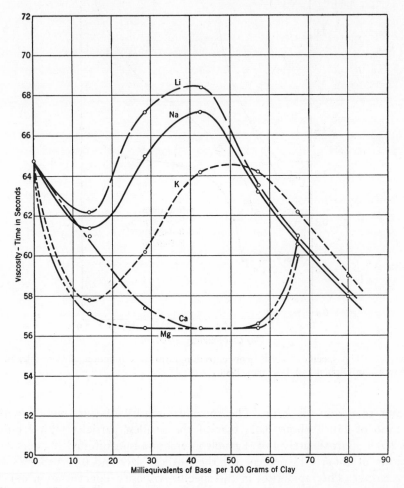

FIG. 2-2. The relation of viscosity of clays to the amount and nature of adsorbed cations.

increased amounts of base are added to H-clay. The H-saturated clay is not a monodispersed system. It is composed of loosely bound, hydrated aggregates. The relative viscosity of this system is 1.4232. The first effect of adding 14.25 me/100 g of a monovalent base is to markedly decrease the viscosity. This decrease is greatest with the K ion and least with the Li ion. The differences between the Li, Na, and K ions are undoubtedly lyotropic effects. These lower viscosities are due to the dispersion of the H-clay aggregates. The enmeshed water is released and there is a diminution of the total volume of the dispersed phase. Such a dispersion has been verified by ultramicroscopic counts. Edge-to-face associations in the H-clay that result from coulombic forces are broken up by the additions of the base (van Olphen, 1951).

The second influence on viscosity is associated with the hydration of the dispersed particles by the exchangeable cations. The water hulls formed around the particles cause an increase in total volume and viscosity. The K ion, with the lowest hydration of the monovalent ions used, produces the least hydration of the particles and decreases the viscosity the greatest amount. The high hydration of the Li ion nearly counterbalances the effect of dispersion, inasmuch as the viscosity decreases only slightly. Thus with only a slight addition of an exchangeable monovalent cation to this colloidal clay, a distinct effect upon the structure and viscosity of the system is observed.

As the concentration of the monovalent cations becomes greater, the viscosity increases to a maximum and then rapidly decreases to a minimum. Hence raising the viscosity to a maximum represents an augmentation in hydration as more ions with a water hull replace the H ions on the surfaces of the clay mineral. It is accompanied by a rapid rise in the negative charge of the particle and an increase in the dispersion and specific conductivity of the suspension. There can be little doubt about an enlargement of the thickness of the diffuse double layer. The maximum viscosity represents the point of highest hydration of the particle as produced by the hydrated ions on the exchange complex. In other words, the exchange complex contains the largest number of cations functioning as active ions in the physicochemical behavior of the colloid. The colloid contains about 40 me of exchangeable Li and Na ions and about 50 me of K ions per 100 g of clay at this point of maximum viscosity. The pH value of the suspension is approximately 6.5.

The variations between the three ions conform to the differences observed in their bonding energies (Marshall, 1950) or in their relative percentages in the Stern layer (Shainberg and Kemper, 1966).

The decrease in viscosity from this maximum point is accompanied by a rapid fall in migration velocity. The potentiometric and conductometric results showed that the number of ions in solution, free from the surface of the colloid, increases rapidly above 50 me/100 g of clay. This suggests that the higher concentration of ions in solution repressed the degree of dissociation of the cations from the surface and compressed the diffuse double layer. An increase in the concentration of electrolyte in solution causes the outer layer of the colloidal particle, containing the exchangeable cations, to become tighter, thereby decreasing the hydration of the particle (Wiegner, 1924).

The third effect on viscosity takes place at higher concentrations of added base. A point is reached where slow coagulation occurs, probably as a result of edge-to-edge associations through van der Waals forces. As aggregation proceeds, water is enmeshed between the particles in the floccule and this hydration brings about an increase in total volume and viscosity. The flocculation values of Li- and Na-clays were 536 me/100 g; that for K-clay was 218. The corresponding viscosities were equivalent to 280, 230, and 190 sec, respectively. Both the Li and Na suspensions exhibited thixotropy; they were solid gels when at rest and liquids when stirred.

The divalent ions exhibit a different type of curve. There is a continuous decrease in viscosity until about 60 me of Ca or Mg ions per 100 g of clay have been added. At this point there is an increase in viscosity due to slow coagulation of the sols. The flocculation values for Ca- and Mg-sols were 67 me/100 g, which is equivalent to the saturation capacity of the colloids. The continuous decrease in viscosity signifies a lowering of the hydration of the particles with respect to that of the original H-saturated aggregates. A diminution in the volume of the dispersed phase indicates that a replacement of the H ions from the particle by Ca ions caused at least a partial dispersion of the aggregates. This was confirmed by ultramicroscopic counts.

Somewhat different effects on viscosity are produced by adding monovalent bases to H-kaolinite (Johnson and Norton, 1941). The first few increments of base decrease the viscosity gradually until about 3.5 me/100 g are added. This is followed by a sharp decrease to about $\frac{1}{100}$ of the original viscosity. The pH jumps at the same point. Viscosity is constant up to 20 me/100 g. Flocculation occurs at higher concentrations, with a corresponding elevation of viscosity and the occurrence of gelation. There is an increase in zeta potential, which is undoubtedly due to the adsorption of OH ions on the positive broken bonds. The effects of Li, Na, K, and Cs ions are similar. The sharp decrease in viscosity is less with divalent than with monovalent ions.

The addition of NaCl to a 0.23 percent Na-montmorillonite suspension decreases the viscosity at low concentrations of salt (van Olphen, 1951). This is followed by an increase as flocculation occurs. These changes are explained on the basis that the original salt-free suspension is subject to internal mutual flocculation because of edge-to-face linkages of positive edges to negative faces. This edge-to-face electrostatic attraction is dominant over face-to-face repulsion. The addition of small amounts of

NaCl breaks down the edge-to-face structure, which results in more complete dispersion and a decrease in viscosity. Increased amounts of NaCl reestablish edge-to-face and edge-to-edge attraction with a raising of the relative viscosity. Na-kaolinite does not exhibit an augmentation of viscosity after the original decline because face-to-face associations dominate after the initial edge-to-face linkages are destroyed. Ca-montmorillonite does not show a decrease in viscosity upon the addition of $CaCl_2$. When NaCH is added to H-montmorillonite, there is a rapid increase in charge, a compression of the double layer with further additions, and flocculation due to crosslinking of edge-to-edge surfaces (M'Ewen and Pratt, 1957).

Summarizing, the viscosity and thixotropy of clay suspensions are determined in large measure by (1) the surface density of the exchangeable cations, (2) the bonding energies between the exchangeable cations and the surface, and (3) the hydration of both the cations and the surface. Increases in viscosity appear to be closely associated with the relative volume of the disperse phase as it reflects the total hydration of the system. This can be due to water hulls around the particles arising from an expansion of the diffuse double layer or to water occluded between edge-to-face or edge-to-edge associations in flocculated systems.

SWELLING OF COLLOIDAL CLAYS

A dried colloidal system adsorbs water molecules from the vapor phase. The amount of this adsorption depends upon the vapor pressure, the type of clay mineral, and the nature of the exchangeable cations (see Chapter 8). If such a colloidal system is placed in direct contact with liquid water, it will absorb water and enlarge its volume in a process known as swelling (Mattson, 1929, 1932). The principle of comparing the uptake of polar and nonpolar liquids by clays provides an index of swelling (Winterkorn and Baver, 1934).

The rate of water sorption (intake) varies considerably with the nature of the colloid. It is rapid at first and then becomes slower with time. Equilibrium for ordinary colloidal clays is obtained within 1 to 3 days; for bentonites, however, sorption continues for almost a week. On the other hand, the sorption of nonpolar liquids, such as benzene or carbon tetrachloride, is complete within 5 minutes. The sorption of nonpolar liquids increases with the dielectric constant of the liquid. This should be expected if swelling is associated with the electric field forces on the surface of the colloidal particles. Such nonpolar liquids as benzene and carbon tetrachloride do not cause significant swelling and consequently may be used to measure the total pore space of the system.

Part of the sorbed water fills the pores and part is oriented on the surface of the particles to produce the phenomenon of swelling. The difference between the sorption of water and the nonpolar liquid therefore represents the amount of water that is taken up in the swelling process, although it does not represent the total volume increase of the clay because of the contraction effects previously mentioned.

The principle of comparing absorption of benzene and water by clays can be used to measure the swelling of soil aggregates at different soil moisture tensions (pF values) (Quirk and Panabokke, 1962). Apparently, about 50 percent of the swelling of clay aggregates occurs before pF 4 is reached on the wetting cycle and 75 percent before pF 3 is attained.

Factors Affecting Swelling

The data in Table 2-1 point out that the swelling of colloidal clays varies with the type of clay mineral and the nature of the adsorbed cation. The expanding-lattice type of colloids (montmorillonite and beidellite) swell considerably more than the fixed-lattice types (kaolinite, halloysite). The swelling of montmorillonite is greater than that of beidellite. As will be discussed later, this suggests two types of swelling, interlayer and interparticle. Since the kind of clay mineral structure and the degree of isomorphous substitutions within the crystal lattice affect the exchange capacity of the colloid, one should expect swelling to increase with cation-exchange capacity.

The results in Table 2-1 demonstrate that the swelling of Putnam clay (beidellite) varies with the nature of the adsorbed cations as follows: Li > Na > Ca > Ba > H > K. The monovalent K ion causes the least amount of swelling. This same effect has been observed on all other clays of the beidellite type. On the other hand, the order of swelling for bentonite (montmorillonite) is Na > Li > K > Ca=Ba > H. In the montmorillonitic colloids the K ion occupies its normally expected place. Lateritic colloids* (halloysite) do not swell, irrespective of the nature of the adsorbed ion on the exchange complex (Lutz, 1934).

The swelling of Li- and Na-clays increases with the concentration of these ions on the complex (Baver, 1929). Maximum swelling is reached at about 60 percent saturation of the exchange complex with Li ions. On the other hand, there is a continuous decrease in swelling as the percentage of K ions in the system increases.

---

* The term "lateritic" or "laterization" as used in this text refers to those soils and soil-forming processes usually associated with such soil groups as ultisols, oxisols, and certain inceptisols.

TABLE 2-1

The Relation of Swelling to the Type of Clay Mineral and Nature of Exchangeable Cations

Type of colloid	Type of clay mineral	CEC (me/100 g)	Swelling (cm³/g colloid)					
			H^+	Li^+	Na^+	K^+	Ca^{++}	Ba^{++}
Bentonite	Montmorillonite	95	2.20	10.77	11.08	8.55	2.50	2.50
Lufkin clay	Montmorillonite-beidellite	82	1.18	—	—	—	—	—
Putnam clay	Beidellite	65	0.81	4.97	4.02	0.50	0.91	0.85
Susquehanna clay	Beidellite-halloysite	47	0.57	—	—	—	—	—
Cecil clay	Halloysite	13	0.05	—	—	—	—	—
			Swelling (cm³/me cation)					
Bentonite	Montmorillonite	95	2.44	11.3	11.6	9.0	2.63	2.63
Putnam clay	Beidellite	65	1.24	7.6	6.2	0.77	1.4	1.3
Ratio: $\dfrac{\text{bentonite}}{\text{Putnam clay}}$			1.97	1.49	1.87	11.68	1.88	2.02

It is significant that there is little difference in the swelling of H- and Ca-saturated bentonite or Putnam clay. In all cases, the H system swells slightly less than the Ca colloid. This fact will be emphasized further in discussions of soil structure in Chapter 4.

Concepts of Swelling

The experimental evidence on the adsorption of water vapor by colloidal systems as discussed in Chapter 8 point to the fact that both the attractive forces of the clay mineral surfaces and the exchangeable cations for water are responsible for the hydration of the colloid. The effect of hydration depends upon (1) the number of the exchangeable ions on the complex, (2) the tightness with which they are held, that is, their degree of dissociation from the surface, and (3) the hydration energy of each ion as determined by its charge and size. The preceding data show that the same factors are operative in the swelling of colloidal clays. The mechanisms involved in the interaction of these factors have been the subject of a great deal of research and theoretical discussions. Swelling has been considered an osmotic phenomenon that functions according to the basic principles of the Donnan equilibrium (Mattson, 1929, 1932). The colloidal particle is visualized as being surrounded by a swarm of diffusible ions that are dissociated from the surface. Electrostatic attraction prevents these ions from diffusing too far from the surface. Consequently, the colloidal particle in a dilute solution of electrolyte consists of a micelle in which there is water, dissociated cations from the surface, and ions from the electrolyte. According to the Donnan principle, at equilibrium the product of the ion concentrations in the micelle is equal to that of the ion concentrations in the medium:

$$x^2 = y(y + z) \tag{2-2}$$

where x is the concentration of each ion of the electrolyte in the medium, y is the concentration of each ion of the electrolyte in the micelle, and z is the concentration of the dissociated adsorbed cations. Osmotic pressure is higher at the surface than in the outside medium. It represents the inflow of water from the medium into the micelle to counteract the higher concentration of cations there that result from dissociation from the surface. This inflow of water causes swelling. Swelling increases with the ionic hydration of monovalent cations, which determines the degree of dissociation. Suppression of swelling is governed by the concentration and valence of the electrolyte. Additions of NaOH to a H-clay increase swelling to a maximum at the exchange capacity above which there is a decline with further increments of the base. This decline is attributed to the repression of the dissociation of the adsorbed Na ions by the excess of these ions in the medium.

The swelling process has been explained on the basis of two types of hydration of the colloid (Baver and Winterkorn, 1935). First, water molecules are oriented at the surface of the clay minerals as a result of the electrical properties of the liquid, the exchangeable cations, and the surface. Second, water may be adsorbed because of osmotic effects, according to the concept of Mattson. The first process is associated with the release of an appreciable amount of heat. The osmotic type of hydration takes place without the liberation of measurable quantities of heat. In the osmotic type of swelling, the hydrated ions surround the hydrated surface in a diffuse double layer. These ions keep within a distance from the surface in which the mean osmotic force is in equilibrium with the mean van der Waals and electrostatic attractions due to the different charges of the colloidal surface and the exchangeable ions. The apparent volume of the colloidal particle is then defined by the ionic atmosphere. Since the diffuseness of the double layer is a function of the osmotic pressure due to the dissociated cations, this type of swelling is called "osmotic." The diffuseness of the ionic atmosphere is also a function of the mineralogical structure of the colloidal surface. The kind of ion adsorbed on the surface has an impact upon both the hydration of the surface and osmotic effects.

The high swelling of bentonites (montmorillonite) in comparison with Putnam (beidellite) and other colloidal clays strongly suggests that bentonites attract large amounts of water as a result of forces associated with the surfaces of the inner layers of the clay mineral. It is interesting to compare the magnitude of swelling of Li-, Na-, and K-saturated bentonites and Putnam clay in relation to the amount of exchangeable cations present (Table 2-1). The relative swelling of bentonite over that of Putnam clay per milliequivalent of exchange capacity is 149, 187, and 1168 percent for the Li-, Na-, and K-saturated systems, respective. These large differences cannot be explained solely by ionic hydration and the bonding energies of these three cations. The attractive forces of the inner layer undoubtedly have a sizable impact upon both the hydration of the surface and the mobility of the ions that cause an osmotic type of swelling.

The peculiar behavior of the K ion is particularly significant. The K-saturated bentonite is similar to the Li and Na systems in its swelling, even though it causes about a 21 percent decrease compared with the other monovalent alkali cations. However, in the case of the Putnam clay, the K ion behaves similarly to the divalent cations and the H ion. In fact, it causes the least swelling of all the added cations. The difference between its action on montmorillonite and its specific influence on beidellite is clearly shown at the bottom of Table 2-1. Whereas the ratio of the swelling of bentonite to that of Putnam clay for the other ions varies between 1.49 and 2.02, that of the K ion is 11.68.

It has been suggested (Low, 1961) that the larger size of the K ion interferes with the ordered arrangement of water molecules near the surface of the clay particle. This would account for a lower heat of wetting and water adsorption. It seems to be a logical explanation of the depressed swelling of K-bentonite. However, the tremendous impact of the K ion on beidellite requires a different explanation. In light of the present concepts of the structure of the crystal lattice, it is possible that the exchangeable K ions are wedged into the holes in the silica sheet upon drying and hold the layers together as in the illite structure. This explanation is supported by the fact that K ions are fixed between the sheets of an expanding-lattice-type clay mineral upon drying (Page and Baver, 1939). Since beidellite is one of the end members of the montmorillonite group in which the isomorphous substitutions are in the tetrahedral sheet, its behavior on drying in the presence of K ions should be expected to be different from montmorillonite where the substitutions are in the octahedral sheet. Norrish (1954a, 1954b) visualizes the mechanism involved in the swelling of montmorillonite as follows. In the first region of swelling, the interlayer expansion is stepwise below a spacing of 22 Å. It is dependent on the nature of the exchangeable cations as their hydration energies overcome the electrostatic attraction forces. Spacings for the H and Li ions increase stepwise from 15.4 to 19.0 or 22.5 Å. Then there is a jump to 36 Å. Spacings for the Na ion enlarge to 19.0 Å where there is a jump to 40 Å. The larger monovalent cations, K, NH_4, and Cs do not exhibit spacings beyond 15 Å. Apparently there are insufficient hydration energies to overcome the close-range attractive forces. The Ca and Mg ions show a stepwise expansion from 15 to 19 Å but no jump thereafter.

After the jump to 35 Å or more, expansion of the Li- and Na-montmorillonites is continuous due to the development of a thick, diffuse double layer. This is the second region of swelling where osmotic repulsive forces are dominant. Interlayer spacings may reach 130 Å. To pass from region one to region two, the hydration energy of the cation must be sufficient to overcome the barrier due to electrostatic attractions.

In another version of this principle the swelling process is considered as particle-to-particle interactions in clay-water systems rather than particle-to-water phenomena (van Olphen, 1954, 1962). First, there are short-range particle interactions that involve not only the separation of unit layers in expanding-lattice clay minerals but also the separation of planar surfaces in stacks of individual mineral crystals. The potential energy of interaction between plates or particles when ions are midway between the sheets is a function of the distance between them as described by the equation

$$E = \frac{+4\pi\sigma^2 x}{\epsilon} \qquad (2\text{-}3)$$

where E is the potential energy of interaction, σ is the surface charge density, x is the half distance between the plates, and ϵ is the dielectric constant. These short-range interactions involve van der Waals attraction, electrostatic interaction between the negative charges of the surface and the exchangeable cation, and hydration energy. Considerable emphasis is placed upon hydration energy as a major contributing factor to swelling. It may be due to hydrogen bonding between the water molecules and the oxygen atoms in the surface of the tetrahedral sheets or to the water hull around the exchangeable cations, or a combination of both. The hydration energy of the clay is dependent upon the hydration energies of the cations. If ions in the dehydrated state were in the Stern layer, they would affect the hydration of the colloidal system by their influence on the formation of ordered water arrangements on the surfaces of the crystal. Such forces would be active at plate distances less than 10 Å, which is equivalent to four monomolecular layers of water (as suggested by Hendricks et al., 1940).

Osmotic swelling is due to long-range particle interactions. The energy of surface hydration is no longer operative and diffuse double-layer repulsion becomes the major force causing the separation of the plates. Osmotic swelling takes place from an interlayer spacing of 10 to about 120 Å at equilibrium. Osmotic swelling is responsible for considerable swelling pressure, which may exceed 10 atm. This swelling pressure is the force that must be applied to the colloidal system to prevent the layers from moving under the force of double-layer repulsion.

Equilibrium is reached when the double-layer repulsive forces are balanced by attractive forces between the layers of particles. These forces have usually been considered to be van der Waals attraction between the layers; van der Waals forces are inadequate to provide this balancing effect and it is proposed that positive edge to negative face attractions give a crosslinking force that compensates the double-layer repulsion (van Olphen, 1962). According to this concept, it would take a relatively small number of nonparallel crosslinking plates to limit the swelling of a sizable number of parallel plates. The effects of edge-to-face bonds in resisting swelling and compensating for double-layer repulsion are shown in Figure 2-3. Li-vermiculite has a higher swelling than any of the montmorillonite systems. This should be expected because of its larger

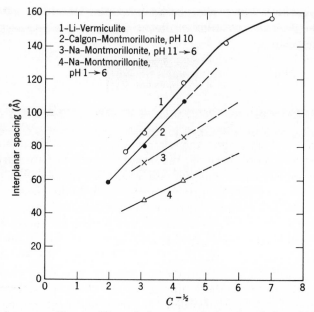

FIG. 2-3. Intercrystalline swelling of clay minerals in salt solution. (After Norrish and Rausell-Colom, 1963.)

surface charge density (vermiculite = 5.8×10^{-4} esu/cm²; montmorillonite = 3.05). The swelling of the Na-montmorillonite at pH 11 → 6 is greater than the system at pH 1 → 6. When the pH is raised above 6, OH ions are adsorbed on the permanent sites of positive charge, thereby increasing both the pH-dependent negative charges and the swelling. Calgon (sodium hexametaphosphate) is also adsorbed on the positive charge sites and converts them to negative sites. This treatment caused the swelling of the calgon-montmorillonite to approach that of Li-vermiculite.

The Gouy-Chapman theory of the diffuse double layer has been used successfully (Bolt and Miller, 1955; Warkentin, Bolt, and Miller, 1957; Warkentin, 1962) to predict the osmotic type of swelling, especially with monovalent cations. Since the repulsive forces between parallel plates is the result of active ions in the diffuse double layer, the addition of the strongly adsorbed tetravalent Th ion to bentonite greatly represses swelling (Hemwall and Low, 1956). The strong adsorption of the Th ion by the clay mineral surface so reduces the number of osmotically active cations that the double-layer repulsion is practically eliminated.

The applicability of the diffuse double-layer concept to the swelling of clays, especially those that are saturated with Ca ions, has been questioned (Aylmore and Quirk, 1959, 1960; Quirk and Aylmore, 1960). The concept of clay domains has been introduced to explain the swelling of colloids upon the uptake of water. When clay-water systems dry out, the particles are organized on the domain basis in which there is a parallel alignment of individual crystals to produce a smaller volume of oriented particles. Pore volume decreases and the coherence between the domains is weaker than that between the crystals within the domain. When these domains are rewetted, they swell as an entity. The pore volume swells proportionally to the overall volume of the clay mass. Domains of illite crystals are similar to the montmorillonite crystal in that swelling takes place between crystals as compared with intracrystalline swelling of the montmorillonite. Montmorillonite is considered to be a special type of domain with the single unit layers constituting the basic units instead of the entire crystal as in an illite domain. Ca-saturated montmorillonite exhibits limited swelling because coulombic attraction forces hold the crystals together. Consequently they are not subject to diffuse double-layer repulsion.

By way of summary, the swelling of colloidal clays is related to their mineralogy and the nature of the exchangeable cations on the surface. Swelling of the different minerals decreases with diminishing hydration according to the series vermiculite > montmorillonite > beidellite > illite > kaolinite (halloysite). There are two types of swelling involved. The first is the result of short-range interacting forces operating between the unit layers of expanding clay minerals or between the planar surfaces of domains of individual mineral crystals. The hydration energies of the clay mineral surfaces and the exchangeable cations determine the extent of this type of swelling. The second type of swelling is caused by long-range interacting forces that produce osmotic swelling which is an expression of double-layer repulsion. Equilibrium is brought about when the double-layer repulsion is balanced by crosslinking forces arising from edge-to-face attraction between positive edges and negative faces of the clay minerals. Considerable pressure is generated by this type of swelling.

The osmotic type of swelling is most highly expressed with Li- and Na-saturated systems. K ions repress swelling either as a result of their interference with the ordered arrangement of water molecules in the surface hydration of the mineral surface or by their tendency to hold expanding lattices together due to their entering the holes of the tetrahedral sheets upon dehydration of the crystal. The divalent cations reduce swelling because of coulombic attractions that prevent enough water from entering between the sheets to cause lattice expansion. The H-saturated clays act similarly to those containing divalent cations. This is probably due to the influence of the trivalent Al ion.

Potential Volume Change (PVC) of Soils

Volume changes in expansive clays pose many problems for soil engineers, including both swelling and shrinkage (shrinkage will be discussed under soil consistency). Physicochemical stresses involved in particle interactions are related to the stresses that the soil engineers use to forecast soil behavior (Lambe, 1958a, 1958b, 1960a, 1960b):

$$\bar{\sigma} - \sigma - u = R - A \qquad (2\text{-}4)$$

where $\bar{\sigma}$ is the effective stress, or the force transmitted between interacting particles per unit of soil, σ is the total external stress applied to the soil, u is the pressure in the free pore water of the soil (that water not associated with surface adsorptive properties), R is the repulsive pressure (as discussed in previous sections under the diffuse double layer), and A is the attractive pressure between clay particles. Therefore, $R - A$ becomes the net repulsive pressure. It cannot be measured even though it is responsible for soil behavior. Both σ and u can be measured to provide information to predict the behavior of the soil. Equation 2-4 states that under a given soil-water situation an effective stress equivalent to the net repulsive force must be applied to the soil in order to prevent swelling and maintain the volume constant.

The pore water in a partially dried soil is in a state of tension because of insufficient water to satisfy the demands of the diffuse double layer. When such a soil is placed in contact with water, water will be absorbed, the tension will decrease (u will increase) and the effective stress will diminish. Expansion of the diffuse double layers and the resultant swelling will take place until the net repulsive pressure $(R - A)$ between the particles is balanced by an added effective stress.

In accordance with the concepts of swelling previously discussed, plastic clays with a large specific surface absorb large amounts of water and exhibit considerable swelling (Lambe, 1960b). The amount of swelling depends upon the initial density and water content, the thickness of the sample as it affects the time to reach equilibrium, and the confining pressure to which the soil is subjected. Swelling increases with increasing density and decreasing water content and confining pressure. A special instrument has been designed for the Federal Housing Administration to measure the swell index of soils (Lambe, 1960b). It determines the swelling pressure caused by a sample of compacted soil that is being wetted. Pressure readings are registered on a dial gage. The value obtained after two hours of swelling is called the *swell index*. On the basis of correlations published in the literature that involve heave from dry and moist water contents, plasticity index, plasticity index activity, linear shrinkage, and water adsorption at 85 percent relative humidity, certain PVC categories have been established:

PVC category	Rating
Noncritical	2
Marginal	2–4
Critical	4–6
Very critical	6

These ratings are plotted against the swell index for dry (50 percent RH) and moist (100 percent RH) as well as wet (plastic limit) soils to produce the curves shown in Figure 2-4. The PVC category is obtained

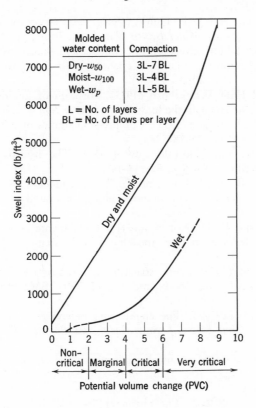

FIG. 2-4. Relation of potential volume change to swell index. (After Lambe, 1960b.)

graphically from the swell index values from the curve for the dry and moist soil. Swell indexes in excess of 4750 lb/ft^3 are associated with heavy clays with plasticity indexes above approximately 40 and fall in the very critical category. The noncritical category includes soils having swell indexes lower than 1700 lb/ft^3 and plasticity indexes below about 14. High degrees of correlation are obtained between the swell index and both the plasticity index and linear shrinkage (from about field capacity to the shrinkage limit), except for the highly plastic clays with plasticity indexes above 35. Consequently, the swell index measurements can be used to obtain graphically an approximte value of the plasticity and linear shrinkage for the less plastic soils.

References

Aylmore, L. A. G., and J. P. Quirk (1959). Swelling of clay-water systems. *Nature*, **183**:1752–3.

Aylmore, L. A. G., and J. P. Quirk (1960). The structure status of clay systems. *Clays and Clay Minerals, 9th Nat. Conf. Proc.*, pp. 104–130.

Baver, L. D. (1929). The effect of the amount and nature of exchangeable cations on the structure of a colloidal clay. *Missouri Agr. Exp. Sta. Research Bull. 129.* Also *Soil Sci.*, **29**:291 (1930).

Baver, L. D., and Hans Winterkorn (1935). Sorption of liquids by soil colloids: II. Surface behavior in the hydration of clays. *Soil Sci.*, **40**:403–419.

Bolt, G. H., and R. D. Miller (1955). Compression studies of illite suspensions. *Soil Sci. Soc. Am. Proc.*, **19**:285–288.

Hemwall, John B., and Philip F. Low (1956). The hydrostatic repulsive force in clay swelling. *Soil Sci.*, **82**:135–145.

Hendricks, S. B., R. A. Nelson, and L. T. Alexander (1940). Hydration mechanism of the clay mineral montmorillonite saturated with various cations. *J. Am. Chem. Soc.*, **62**:1457–1464.

Johnson, A. L., and F. H. Norton (1941). Fundamental study of clay: II, Mechanism of deflocculation in the clay-water system. *J. Am. Cer. Soc.*, **24**:189–203.

Lambe, T. William (1958a). The structure of compacted clay. *Proc. Am. Soc. Civil Engr., J. Soil Mech. and Found. Div.*, SM2 Paper 1654, pp. 1–34.

Lambe, T. William (1958b). The engineering behavior of compacted clay. *Proc. Am. Soc. Civil Engr., J. Soil Mech. and Found. Div.*, SM2, Paper 1655, pp. 1–35.

Lambe, T. William (1960a). A mechanistic picture of shear strength in clay. *Am. Soc. Civil Engr., Res. Conf. on Shear Strength of Cohesive Soils*, Colorado, pp. 555–580.

Lambe, T. William (1960b). The character and identification of expansive soils. FHA Soil PVC Meter. *Fed. Housing Adm., Technical Studies Report*, FHA-701.

Low, Philip F. (1961). Physical chemistry of clay-water interaction. *Advances in Agronomy*, 13:269–327, Academic Press, New York.

Lutz, J. F. (1934). The physicochemical properties of soils affecting soil erosion. *Missouri Agr. Exp. Sta. Research Bull.* 212.

Marshall, C. E. (1950). The electrochemistry of the clay minerals in relation to pedology. *Trans. 4th Int. Cong. Soil Sci., Amsterdam*, 1:71–82.

Mattson, S. (1929). The laws of soil colloidal behavior. I. *Soil Sci.*, 28:179–220.

Mattson, Sante (1932). The laws of soil colloidal behavior: VIII. Forms and functions of soil water. *Soil Sci.*, 33:301–322.

M'Ewen, Marjorie B., and Margaret I. Pratt (1957). The gelation of montmorillonite. Part I. The formation of a structural framework in sols of Wyoming bentonite. *Trans. Faraday Soc.*, 53:535–547.

Norrish, K. (1954a). Manner of swelling of montmorillonite. *Nature*, 173:256–257.

Norrish, K. (1954b). The swelling of montmorillonite. *Disc. Faraday Soc.*, 18:120–134.

Norrish, K., and J. A. Rausell-Colom (1963). Low-angle X-ray diffraction studies of the swelling of montmorillonite and vermiculite. *Clays and Clay Minerals, 10th Nat. Conf. Proc.*, pp. 123–149.

Page, J. B., and L. D. Baver (1939). Ionic size in relation to fixation of cations by colloidal clay. *Soil Sci. Soc. Am. Proc.*, 4:150–155.

Quirk, J. P., and L. A. G. Aylmore (1960). Swelling and shrinkage of clay-water systems. *Trans. 7th Int. Cong. Soil Sci., Madison*, 11:378–385.

Quirk, J. P., and C. R. Panabokke (1962). Pore volume-size distribution and swelling of natural soil aggregates. *J. Soil Sci.*, 13:71–81.

Shainberg, I., and W. D. Kemper (1966). Hydration status of adsorbed cations. *Soil Sci. Soc. Am. Proc.*, 30:707–713.

van Olphen, H. (1951). Rheological phenomena of clay sols in connection with the charge distribution of the micelles. *Disc. Faraday Soc.*, 11:82–84.

van Olphen, H. (1954). Interlayer forces in bentonite. *Clays and Clay Minerals. 2nd Nat. Conf.*, pp. 418–438.

van Olphen, H., and M. H. Waxman (1958). Surface conductance of sodium bentonite in water. *Clays and Clay Minerals, Proc. Fifth Nat. Conf.*, pp. 61–80.

van Olphen, H. (1962). Unit layer interaction in hydrous montmorillonite systems. *J. Coll. Sci.*, 17:660–667.

Warkentin, B. P. (1962). Water retention and swelling pressure of clay soils. *Can. J. Soil Sci.*, 42:189–196.

Warkentin, B. P., G. H. Bolt, and R. D. Miller (1957). Swelling pressure of montmorillonite. *Soil Sci. Soc. Am. Proc.*, 21:495–497.

Wiegner, G. (1924). Dispersität und Basenaustausch. *Proc. 4th Intern. Soc. Soil Sci.*, 2d Com. (Rome), pp. 390–424.

Winterkorn, Hans, and L. D. Baver (1934). Sorption of liquids by soil colloids: I. Liquid intake and swelling by soil colloidal materials. *Soil Sci.*, 38:291–298.

CHAPTER 3

The Dynamic Properties of Soils

The dynamic properties of soils refer to the behavior of the soil to an applied stress. They are properties that are expressed through soil movement resulting from externally applied forces.

SOIL CONSISTENCY

As the concentration of soil in the soil-water system becomes large enough so that the mass is no longer free flowing, the forces of cohesion and adhesion come into play. The soil is then said to have a certain "consistency."

Soil consistency is usually defined as a term "to designate the manifestations of the physical forces of cohesion and adhesion acting within the soil at various moisture contants. These manifestations include, first, the behavior toward gravity, pressure, thrust and pull; second, the tendency of the soil mass to adhere to foreign bodies or substances; and third, the sensations which are evidenced as feel by the fingers of the observer" (Russell, 1928). This definition implies that the concept of soil consistency includes such properties of the soil as resistance to compression, shear, friability, plasticity, stickness. All these properties are manifested differently as the forces of cohesion and adhesion within the soil mass vary.

Forms of Soil Consistency

The suggestions of Atterberg (1912) can be condensed to give four essential forms (not including the viscous state) of consistency that most soils may be expected to exhibit:

1. The sticky consistency, as evidenced by the property of stickiness or adherence to various objects.

2. The plastic consistency, as manifested by the properties of toughness and the capacity to be molded.

3. The soft consistency, as characterized by friability.

4. The harsh consistency, which has the pronounced characteristics of hardness.

At low moisture contents the soil is hard and very coherent because of a cementation effect between the dried particles. Clods will be produced if the soil is tilled in this condition. As the moisture content increases, however, water molecules are adsorbed on the surface of the particles and decrease the coherence and impart friability to the soil mass. The zone of friable consistency represents the range of soil moisture in which conditions for tillage are at an optimum. As the amount of water in the soil is augmented further, the cohesion of water films around the particles causes the soil to stick together and be plastic. Soils are easily puddled in this moisture range. Some soils exhibit stickiness within the plastic range; others do not become sticky until they approach a viscous consistency.

Consistency of Moist and Wet Soils

Friability

Friability characterizes the ease of crumbling of soils. That moisture range in which soils are friable is also the range in which conditions are optimum for tillage. Soils are usually in good tilth when they are friable and mellow. The individual granules are soft; cohesion is at a minimum. There is sufficient moisture between the individual particles to minimize the cementation effects that are dominant in the zone of harsh consistency. On the other hand, there is not enough water present to cause the formation of distinct films around particle contacts to produce the cohesion that exists in the plastic range.

The crumbs are probably held together, in part at least, by the orientation of water molecules between the individual particles and the exchangeable cations present. Such a linking system consists of particle-oriented water molecule-exchangeable cation-oriented water molecule-particle (Russell, 1934). As will be discussed later, the colloidal particles at this moisture content are undoubtedly arranged haphazardly, which will contribute to friability at the lower moisture contents.

Plasticity

It has been shown that soils (except those that are nonplastic, such as sands) become more plastic as the moisture content increases. They are tough and exhibit considerable cohesion; they can be molded like putty. Plasticity has been defined as "the property which enables a clay to change its shape without cracking when it is subjected to a deforming stress" (Mellor, 1922). It has also been visualized as that property of material which enables it to be deformed without rupture when the material is subjected to a force in excess of the yield value (Wilson, 1927).

There are other similar definitions of plasticity. However, all these definitions imply that plasticity is a characteristic of clay to take up water, to form a mass that can be deformed into any desirable shape after the force applied exceeds a certain yield value, and to maintain this shape after the deformation pressure is removed. Moreover, the shape will remain unchanged after the water is removed. Sands can be molded when they are wet, but the molded form falls into pieces when it is dried; therefore they are not plastic.

Plasticity thus is the resultant effect of a stress and a deformation. The extent of the deformation for a given system is determined by the distance the particles can move without losing their cohesion. The pressure that is required to produce a specific deformation is an index of the magnitude of the cohesive forces that hold the particles together. These forces vary with the thickness of the water films between particles. Since the amount of deformation that can be produced varies with the size and shape of the particles, it is evident that the amount of surface present determines the number of water films contributing to cohesion. Thus it appears that plasticity is a property which expresses the magnitude of the film forces within the soil and the effects of these forces in determining the extent to which the shape of the soil mass can be permanently changed without breaking.

LAWS OF PLASTIC FLOW. The essential difference between viscous and plastic flow is that a certain amount of stress must be added to plastic soils before flow is produced. The volume of flow, as a function of the force applied, is characterized by the familiar Bingham equation:

$$V = k\mu(F - f) \tag{3-1}$$

where V is the volume of flow, μ is the coefficient of mobility, F the force applied, f the force necessary to overcome the cohesive forces of the system and just enough to start the flow (this force is the so-called "yield value"), and k is a constant. It is obvious that equation 3-1 may be used to characterize viscous flow when $f = 0$. Then the volume of flow is proportional to the force applied and the coefficient of viscosity of the liquid.

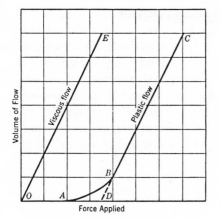

Volume of Flow

Viscous flow

Plastic flow

Force Applied

FIG. 3-1. A comparison of viscous and plastic flow.

The difference between viscous and plastic flow is illustrated in Figure 3-1. Curve OE shows how viscous flow increases directly with the applied pressure. Curve $OABC$ indicates that a certain pressure must be applied to plastic bodies before flow is started. Finally, flow is proportional to the applied force, as indicated by the segment BC. The yield value is obtained by extrapolating the segment BC to the point D on the abscissa. The magnitude of the yield value is correlated with the extent of the cohesive forces of the water films between particles.

COHESION AND ADHESION. It is necessary to distinguish between cohesion and adhesion in an analysis of the causes of plasticity. Adhesion refers to the attraction of the liquid phase on the surface of the solid phase. The water molecules may adhere either to the surface of the soil particles or to objects brought into contact with the soil. Cohesion in soils is a bonding of the particles due to attractive forces between the particles arising from physicochemical mechanisms (Seed, Mitchell, and Chan, 1960). These bonding forces may be (1) the van der Waals forces which vary inversely as the cube of the distance between particles, (2) electrostatic attraction between negatively charged clay surfaces and positively charged clay edges, (3) linking of particles together through cationic bridges, (4) the cementation effects of organic matter, aluminum and iron oxides, carbonates, and so on, and (5) surface tension of the curved menisci at the air-water interfaces that are always present in an unsaturated clay. Cohesion in wet soils takes place between the molecules of the liquid phase as bridges or films between adjacent particles.

Water Films and Cohesion. Haines (1925) developed a theoretical concept of cohesion in an ideal soil on the basis of the surface-tension forces which arise from the water films between particles. The ideal soil is visualized as consisting of uniform spheres, with radius a, that are arranged either in an open- or a close-packed state. In open packing or cubical arrangement there are six points of contact per particle. In the close-packed or tetrahedral grouping there are twelve contacts per sphere.

At low moisture contents, most of the film water is found as annular rings around the various points of contact (Figure 3-2). Each of these films tends to draw the particles together. The total cohesive force produced by all the films is equivalent to the summation of the individual forces exerted at each point of contact.

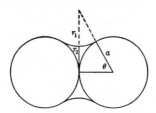

Fig. 3-2. Schematic representation of the nature of water rings between spheres. (After Haines, 1925.)

If r_1 and r_2 are the radii of curvature of the moisture film in each of its two principal directions and T is the surface tension of the liquid, then the pressure deficiency inside the meniscus is

$$p = T\left(\frac{1}{r_1} - \frac{1}{r_2}\right) \quad \text{or} \quad \frac{T(r_2 - r_1)}{r_1 r_2} \tag{3-2}$$

where $1/r_1 - 1/r_2$ is equal to the sum of the two curvatures at right angles to each other. (This value is sometimes erroneously called the total curvature.) The tensile strength of the film due to the pressure deficiency inside the liquid is equal to $\pi r_2^2 p$, or $\pi r_2^2 T(1/r_1 - 1/r_2)$.

The total tensile force of the water films is equal to the force due to pressure inside the meniscus plus the tension exerted by the air-water interface; the latter is equal to the circumference of this interface times the surface tension, or $2\pi r_2 T$ (Fisher, 1926). The total tensile force is given by the formula

$$F = \pi r_2^2 T\left(\frac{1}{r_1} - \frac{1}{r_2}\right) + 2\pi r_2 T = \pi r_2 T\left(\frac{r_2 + r_1}{r_1}\right) \tag{3-3}$$

One can reduce equation 3-3 to the following (see *Soil Physics*, third edition):

$$F = \frac{2\pi a T}{1 + \tan\left(\frac{1}{2}\right)\theta} \tag{3-4}$$

It may be seen from this equation that the total tensile strength becomes $2\pi aT$ as Θ approaches zero. This limit cannot be reached, however, since equation 3-3 breaks down as r_1 and r_2 approach molecular dimensions. Cohesion is the sum of these individual film forces over unit cross-sectional area.

When equation 3-4 is transformed to give the tensile stress per unit cross-sectional area within the soil mass, it is found that the total pull per unit cross section varies as T/a. Therefore cohesion varies directly as the surface tension and inversely as the radius of the particle. The validity of this formula breaks down when water fills more than one-fourth of the pore space, because of the coalescence of the films. High values for maximum cohesion within a soil depend upon small particles. This means that cohesion is a function of the number of films.

This concept of the cohesion of moisture films can be applied to plate-shaped particles (Nichols, 1931). The cohesive force of a water film between two particles is given by the formula

$$F = \frac{k4\pi rT \cos \alpha}{d} \tag{3-5}$$

where k is a constant, r the radius of the particles, T the surface tension, α the angle of contact between the liquid and the particle (generally assumed to be zero), and d the distance between particles.

The cohesive force should vary inversely with the moisture content for a given size and number of particles. Moreover, as long as no excess water is present, the product of the cohesive force and moisture content should be a constant. Nichols verified this fact experimentally with synthetic soils, as shown by the data in Table 3-1. It may be noted that

TABLE 3-1
Relation of Cohesion to the Moisture Content of
Soils (Nichols, 1931)

Soil	Moisture content D (%)	Cohesion F (g/in.²)	$F \times D$
Sand ⅔, clay ⅓	10.90	17.25	188
	12.90	15.00	193
Sand ⅓, clay ⅔	12.73	26.40	336
	13.10	22.50	294
Pure Cecil clay	13.55	56.00	759
	17.50	49.00	857

cohesion increases with the clay content but decreases with the amount of moisture, only because the lower moisture values in this table represent the points of maximum cohesion on the moisture content-cohesion curves. Cohesion increases up to a maximum and then decreases rapidly as the moisture content of the soil is raised.

These facts are readily explained on the basis of equations 3-4 and 3-5. Both these formulas suggest that cohesion is a function of the number and thickness of the films. In equation 3-4, cohesion decreases as Θ increases; in equation 3-5, cohesion becomes less as d enlarges. For a given value of Θ or of d, the magnitude of the cohesion will depend upon the summation of the individual film forces, that is, the total number of films.

The ordinary laws of surface tension are operative even if the water films on the surface of particles are only a few molecules thick (Schofield, 1938). The yield value of plastic clays corroborate the surface-tension concept of cohesion in moist soils (Norton, 1948; Schwartz, 1952; Kingery and Francl, 1954). The data in curve B of Figure 3-3 indicate that the workability of the clay (the yield value $\times$ maximum extension) increases linearly with the surface tension of the wetting liquid. The same linear

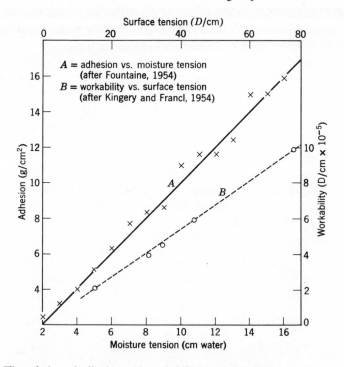

Fig. 3-3. The relation of adhesion and workability to tension forces.

relationship was obtained between yield value and surface tension. Michaels (1959) stated that the capillary tension in the pore fluid is an important contributive factor to soil cohesion. In a partially saturated soil, the air-water menisci retreat into the smaller pores where the water tension increases.

Although Schofield was undoubtedly correct in that the forces which are responsible for surface tension effects operate even in very thin films, it does not follow that such forces provide a complete and adequate description of soil cohesion. The experiments of Vomocil and Waldron (1962) on unsaturated glass bead systems showed that even in a relatively simple system surface films play an increasing role as the water content decreases.

Variation of Cohesion with Moisture. Cohesion increases with decreasing moisture content due to the decrease in thickness of the moisture films Clay has a much greater cohesion than a fine sandy loam; this is the result of more moisture films and greater surface contacts. There are two distinct portions to curves *A* and *B* in Figure 3-4. The break points represent the limit of shrinkage of the soil and the entrance of air into pores which were originally filled with water.

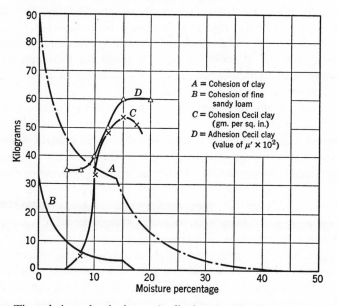

Fig. 3-4. The relation of cohesion and adhesion to soil moisture. Curves *A* and *B* are taken from Johannsen (1914) using puddled soils; curves *C* and *D* are from the data of Nichols (1929, 1931) using nonpuddled soils.

Cohesion at moisture contents above this point is due primarily to film forces; below this point, interparticle attractions due to van der Waals forces is the dominant factor. Thus the cohesion due to interparticle attraction of the fine sandy loam increases only slightly with decreasing moisture below the break point; that of the clay, however, increases rapidly at low moisture contents. It should be kept in mind in the interpretation of these curves that both types of cohesion were measured on puddled soils that were permitted to dry. This technique produces maximum contact between particles, which causes high cohesion due to interparticle attractions.

Curve C in Figure 3-4 shows the variation in the cohesion of nonpuddled Cecil clay as the moisture content is increased. The loose, dry soil has no cohesion. As water films are formed, however, cohesion increases rapidly to a maximum and then decreases. The decrease represents the loss in cohesion that results from a thickening of the water films between particles. Maximum cohesion increases with the clay content of the system.

Curves A and C in Figure 3-4 might seem contradictory at first glance. For example, curve A shows an increase in the cohesive forces below about 15 percent moisture, whereas curve C shows a decrease. This difference is due to the method of preparing the clay. The former has been puddled and dried; the latter has not been puddled but the moisture has been added to loose, dry soil. It should be noted that cohesion in curve C begins to decrease with moisture after reaching a maximum. It becomes puddled at about 15 percent moisture. Consequently both curves agree beyond this point. There is a decrease in cohesion as the moisture content increases in both cases.

Whatever the various mechanisms involved in soil cohesion, it is almost invariably observed that even a modest soil suction tends to provide a certain stability to soil of any texture. As the water content is increased so that the suction is reduced from only a few millibars to zero, there is a pronounced increase in the tendency of a soil to disperse or slake.

Water Films and Adhesion. The adhesion of a foreign object to a soil should take place only at moisture contents above that of maximum cohesion. At these higher moisture contents, the water is held less tightly by the particles and is attracted on the surface of the object to form connecting films between it and the soil. The adhesion of the soil to the object is through the medium of these films. The moisture content at which maximum adhesion occurs depends upon the amount of water required to satisfy the films between the individual particles and the attractive forces in the surface of the foreign object. The force of adhesion of soil to metal is a linear function of the colloid content (Nichols, 1929).

The moisture content of maximum adhesion is uniformly higher than that for maximum cohesion in the same group of soils. This is illustrated by curve D in Figure 3-4. The adhesion and cohesion curves are S-shaped; the former is located slightly higher on the moisture scale, as should be expected from the film theory of cohesion and adhesion.

Adhesion is directly proportional to the surface tension of the dispersion liquid. This linear relationship for a clay loam soil is depicted in curve A, Figure 3-3. The straight line is the theoretical relationship that would be obtained if the adhesion were due entirely to the moisture tension within a water film that was perfectly continuous. The experimental values agree closely with the theoretical values. The value of adhesion therefore is equivalent to the surface area of the film and the tension within it. This means that the force of adhesion of soil to metal is dependent upon the clay content, which determines the number of films and the amount of water that regulates the thickness of the films.

Consistency of Dry Soils

Shrinkage

Atterberg's (1911) original classification of soil consistency included a semisolid form or state. The moisture content below which the soil ceased to shrink was called the shrinkage limit. It represents the lower moisture limit of the semisolid state of consistency, which is approximately equivalent to the soft-friable consistency.

The force that produces shrinkage arises from the tensions formed at the air-water interfaces at the surface of the soil-water system. Evaporation from the soil surface withdraws water from within the soil, thereby causing the soil particles to be drawn closer together. Shrinkage takes place in proportion to the volume of water withdrawn. Finally, a point is reached where there is an interaction between particles. Further shrinkage is caused by a compression and further orientation of particles as a result of the increased surface tension forces of the air-water interfaces that have penetrated into the smaller pores of the soil mass as air replaced the increments of water removed.

These effects are seen in Figure 3-5. The volume of a puddled block of soil is measured as it dries. When the volume of the soil is plotted as a function of the volume of water removed, several very significant facts concerning shrinkage become obvious. Curve A represents a clay separate containing 90.5 percent clay, curve B is for kaolin containing

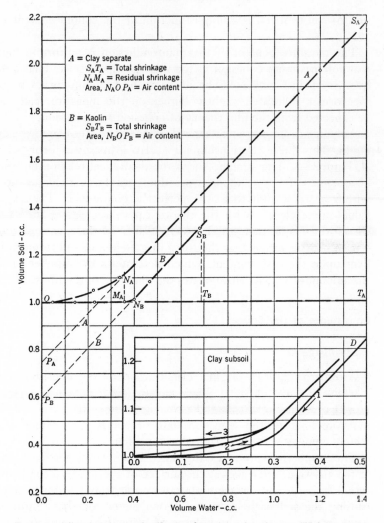

Fig. 3-5.   The shrinkage of soils as a function of moisture. (Haines, 1923.)

52.8 percent clay, and curve D represents a clay subsoil that has been alternately dried and rewetted.

It is noted that as the thoroughly puddled samples are dried, the decrease in the volume of the soil is equal to the volume of the water lost. All the curves are parallel in the wet region and have a slope of 1. As dehydration progresses, there is a distinct break in the curve, and the change in soil volume becomes much less than the volume of water removed. This break signifies the point at which air enters the soil. The shrinkage over the lower portion of the curve varies with the nature of the soil. This portion of the curve is called "residual shrinkage," as distinguished from the total shrinkage that takes place from complete saturation to dryness. If the straight portion of the curve, where the volume change is proportional to the water loss, is extrapolated to intersect the y-axis, then the values of OP_A and OP_B will represent the volume of the pore space that is occupied by air in the dried soils.

The total porosity of A and B is 22.5 and 37.8 percent, respectively. The clay separate has a total shrinkage of 130 percent $(S_A T_A)$ and a residual shrinkage of 8.6 percent $(N_A M_A)$, calculated as a percentage of the dry volume. Kaolin has a total shrinkage of 30 percent $(S_B T_B)$ and no residual shrinkage. Total shrinkage appears to be dependent on the clay contents. Residual shrinkage varies with the hydration of the soil colloidal material, both organic and inorganic.

The shrinkage curves of clods upon the removal of water are different from those of puddled blocks of soils (Lauritzen, 1948). The bulk density of the natural clod determines to a great extent whether the shrinkage curve is similar to that of the blocks. Generally speaking, natural clods exhibit only a small amount of normal shrinkage. The volume-change ratio, (change in volume of soil mass)/(volume of water lost), denotes the extent to which shrinkage is associated with the loss of water. The ratio of the bulk density of dried clods to that of dried blocks is considered a measure of soil porosity resulting from structural development.

Total shrinkage is determined by (1) filling a cylindrical ring of known volume with the soil molded at a moisture content above the plastic limit, (2) drying the soil, and (3) measuring the volume of the dried soil by mercury displacement.

Linear shrinkage is measured by noting the decrease in length of a strip of molded clay after drying at 105° C. It is expressed as a percentage of the dried length. Shrinkage increases with decreasing particle size and increasing plasticity index. It is affected by the nature of the clay mineral and the type of exchangeable cations, just as these factors influence soil plasticity.

The coefficient of linear extensibility, COLE, has been adopted as a standard measurement in soil survey investigations in the United States (United States Department of Agriculture, 1967). It is calculated from the bulk densities of the moist soil Db_m (usually at $\frac{1}{3}$ bar tension) and of the air-dried soil Db_d. The measurements are made on natural clods that have been covered with a special plastic coating.

$$\text{COLE} = \frac{Db_d}{Db_m} - 1 \qquad\qquad (3\cdot6)$$

Shrinkage characteristics of soils are important in ceramics and in engineering problems associated with the construction of buildings, dams, and highways. Alternate shrinkage and swelling play a significant role in the aggregation of clay soils (see Chapter 4).

Modulus of Rupture

Thoroughly dried soils with normal compaction exhibit a decided hardness or coherence in the field. The extent of this coherence naturally varies with the structure of the soil, since porosity determines the number of particles per unit volume. This, in turn, correlates with the amount of surface contacts. Measurements of the cohesion of dry soils are usually based upon the breaking strength of dried briquettes. Atterberg called it the *Festigkeitzahl*. It is now referred to as tensile strength, breaking strength, dry strength, or modulus of rupture. There are several techniques used to measure the modulus of rupture. Soil engineers use the method in which the soil is wetted, kneaded, and molded into rectangular briquettes. These briquettes are dried, supported on both ends as a simple beam, and ruptured by applying a pressure at the center of the beam. This procedure has been modified by soil physicists in order to simulate crust formation in soils. The air-dried soil is placed in a rectangular mold, soaked 1 hour in water, and dried at 50° C. The resulting beam is ruptured by applying pressure at the middle (Reeve, 1965).

For a rectangular briquette, the modulus of rupture S (dynes/cm^2) is calculated from the equation

$$S = \frac{3FL}{2bd^2} \qquad (3\text{-}7)$$

where F (dynes) is the breaking force, L is the distance (cm) between the bars supporting the briquette, b is the width, and d is the thickness of the briquette (cm). The final results may also be expressed in bars (1 bar = 10^6 dynes/cm^2).

The modulus of rupture has also been used as an index of soil crusting (Richards, 1953; Lemos and Lutz, 1957). Its use is based upon two assumptions: (1) the physical properties of the prepared briquette simulate those of natural crusts; and (2) the modulus of rupture represents the force of germinating seedlings in breaking these crusts. Apparently no single soil characteristic correlates with the values obtained. The hardness of the crust on a given soil is the result of interactions of complex physical and physical-chemical processes. The modulus of rupture is higher for montmorillonitic clays than for kaolinitic, provided the particle-size compositions are the same. Raindrop impact and puddling produce high values.

The tensile strength varies with the nature of the soil, as illustrated by the values for three H-saturated soils: Cecil clay, with 77 percent 5-μ particles, 61 lb/in.[2]; Hagerstown silt loam, with 42 percent 5-μ clay, 105 lb/in.[2]; and Putnam silt loam with 31 percent 5-μ clay, 165 lb/in.[2]. In spite of the high clay content of the Cecil soil, its lateritic nature is responsible for the lowest coherence.

The foregoing observations may be explained by the differences in the surface properties of colloidal clays as discussed in Chapter 1. The lateritic colloids, which are composed primarily of kaolinitic clay minerals, possess a low surface activity. Since only weak attractive forces exist in the surface, a lower cohesion between particles should be expected. Undoubtedly, the particle shape is an important factor in the cohesion of dried briquettes. Plate-shaped particles can be oriented to give a close packing during the kneading of the briquettes. These particles, with their high specific surface and greater contact per unit surface, should cohere tenaciously, as experimental observations indicate.

The cohesion of dried soils therefore may be visualized as depending upon the amount of surface contacts per unit volume of the soil mass and the magnitude of the attractive forces between solid particles. This is evidenced by the fact that the addition of small amounts of water, to form a thin layer of water molecules on the surface of the individual particles, causes a decrease in cohesion and imparts friability to the soil mass.

SOIL PLASTICITY

Methods for Determining

Many techniques have been proposed for measuring soil plasticity (Bodman, 1949; Bloor, 1957). Basically, the various methods can be grouped into the following categories:

1. Determining the amount of water required to produce plasticity in the soil and the moisture range over which this plasticity occurs. This involves causing plastic flow by impact or pressure at the desired moisture content. The Atterberg limits and the point of equal stiffness (Bodman and Tamachi, 1930) fall in this category.

2. Resistance to compression. There are two methods for measuring this resistance. One is to determine the point of failure of unconfined cores of soil under load. The other is the decrease in height of a confined soil under stress.

3. The shear strength of soils involving either uniaxial or triaxial techniques.

4. The deformation of soil by penetration of an object. The soil penetrometer belongs to this grouping.

5. Torque tests to measure the stress required to deform the clay (yield point) and the maximum extension that will take place before the clay ruptures. This technique, which employs bars of molded clay, is used in the ceramic industry. Both the amount and the rate of applying torque are important in making determinations.

The Atterberg limits are used extensively by soil scientists and soil engineers for measuring plasticity. Atterberg (1911, 1912) studied plasticity from the point of view of the moisture range over which plasticity was manifested. He suggested three values that have attained wide usage among soil investigators. These are, first, the upper plastic limit (liquid limit), or the moisture content at which the soil will barely flow under an applied force, second, the lower plastic limit, or that moisture content at which the soil can barely be rolled out into a wire; and third, the plasticity number (index), or the difference between the liquid and plastic limits. The last is taken as an index of plasticity.

The liquid limit was determined originally by placing a small amount of soil in a round-bottomed dish, working it into a stiff paste, pressing it tightly against the bottom, cutting a V-shaped groove in the plastic mass, and jarring the dish to make the two segments flow together. If flow was not produced, additional water was added and the process repeated. If too much flow was obtained, dry soil was mixed with the plastic mass. This cut-and-try process was repeated until the correct flow was obtained. The moisture content of the plastic soil was then determined.

Casagrande (1932) designed a special apparatus to minimize the personnel errors in the determination of the liquid limit and to make it possible for various individuals to reproduce the same results. It has become the standard procedure for the American Society for Testing and Materials (Sowers, 1965).

The plastic limit is determined now in the same manner as it was originally. Dry soil is mixed with water in a round-bottomed dish until it begins to lose its crumbly feel and show a tendency to become plastic. The mass is then kneaded in the hands. A small portion is rolled between the fingers and a glass plate, or piece of glazed paper, until a wire is formed. The process of adding water or soil is repeated until that moisture content is reached when the plastic mass will just barely roll out into a wire that breaks into pieces about 1/4 to 3/8 in. long. Although this technique seems crude, it is possible to duplicate results with considerable accuracy. Tough clays are the most difficult to study.

Film Theories of Plasticity

Baver (1930b) suggested that the colloidal clay particles in the soil act as a lubricant between coarser particles and diminish their friction. It is highly probable that the plate-shaped particles are oriented so that their flat surfaces are in contact. This orientation increases the amount of contact between the colloidal particles. The increased contact, together with the raising of the ratio of water-film surface to the particle mass, may be considered as producing the plastic effects. In other words, within a certain moisture range, the tension effects of the water films between the oriented platelike colloidal particles, which impart to the soil its cohesive properties, enable the soil to be permanently molded into any desired shape or form. This moisture range corresponds to the range of plasticity of a soil. Orientation of particles and their subsequent sliding over each other takes place when sufficient water has been added to provide a film around each particle. The amount of water necessary to produce these films corresponds to the moisture content at which the soil ceases to be friable. With an excess of water, the water films become so thick that the cohesion between particles decreases and the soil mass becomes viscous and flows.

The colloidal particles in a soil at low moisture contents are visualized to be in a haphazard arrangement. Such an arrangement has been proposed by other investigators (Rosenquist, 1959; Lambe, 1953).

When the amount of adsorbed water in the system reaches a moisture content corresponding to that of the lower limit of plastic consistency, the particles become oriented when pressure is applied. The tension of the adsorbed water films holds the adjacent oriented particles together. As the pressure is increased above that of the tension of the films, the particles slide over each other. After the pressure is removed, the particles do not return to their original positions because they are held in place in their new positions by the tension of the moisture films.

A slightly different version of the impact of the tension of water films on soil plasticity stresses the importance of the capillary force of the water films in compressing particles together and overcoming the repulsive forces that tend to separate particles (Norton, 1948; Schwartz, 1952; Kingery and Francl, 1954). The surface of the soil mass within the plastic range is visualized as being covered by a continuous water film that stretches down into the capillaries. The radius of curvature at the air-water interface depends upon the amount of water available for the film. The yield value of a clay-water system is inversely proportional to the distance between the particles. Curve B in Figure 3-3 shows that plasticity is directly related to surface tension.

The physical state of the adsorbed water that constitutes the water films around the particles has been suggested as the basic mechanism involved in soil plasticity (Grim, 1948). H-bonding, as discussed in Chapter 1, causes an ordered, oriented arrangement of water molecules on the surface of the particles. A strong water bond between particles is formed by the overlapping of layers of oriented water molecules that extend outward from adjacent colloidal surfaces. When these layers of oriented water molecules are thin, the H bonds are strong and the water has a rigid structure. As the thickness of the water layers increases, there will be less and less orientation of the water molecules at the junction of the adjacent adsorbed layers. Optimum plasticity is obtained when the adsorbed water layers are oriented enough to produce a strong bond between adjacent particles but sufficiently thick to reduce the rigidity of the water at the boundary of the adjacent adsorbed layers. At this point, the application of a small stress causes movement of the particles at the juncture of the adsorbed layers. In other words, plasticity is developed after two conditions are met. First, enough water must be added to provide for the formation of rigid layers of water on the adjacent colloidal surfaces. Second, there must be enough extra water to serve as a lubricant between the rigid water layers when the system is subjected to a small deforming stress.

When one considers that H-bonding is responsible for surface tension effects, then the aforementioned mechanism helps to explain the tension forces at the air-water interfaces in the plastic zone. Irrespective of the exact mechanisms involved, plasticity is caused by forces associated with water films around and between particles. Plasticity is a function of the number and thickness of the films in a given system. The number of films depends upon the size of particles and their specific surfaces. The thickness of the films is a function of the nature of the clay mineral, which determines the quantity of water that is adsorbed before a distinct film around each point of contact is formed, and the amount of water added to the system.

The Atterberg Concepts

Significance of the Atterberg Limits

The plastic limit represents the moisture content of the change from the friable to the plastic consistency. It represents the minimum moisture percentage at which the soil can be puddled. Orientation of particles and their subsequent sliding over each other take place at this point, since sufficient water has been added to provide a film around each particle; or there is sufficient water to satisfy the requirements for the development of the highly rigid adsorbed layers plus a slight excess for lubricating purposes. The moisture content of this limit depends upon the amount and nature of the colloidal material present. Cohesion is a maximum slightly above the plastic limit (Figure 3-11).

Moisture at the plastic limit is held at a tension equivalent to pF 2.8 to 3.3 (666 to 2000 cm of water) (Croney and Coleman, 1954; Greacen, 1960). The fact that the soil at the plastic limit is in an unsaturated state lends credence to the importance of the tension effect in determining the plasticity just as surface tension is considered important in affecting soil cohesion.

The liquid limit signifies the moisture content at which the moisture films become so thick that cohesion is decreased and the soil mass flows under an applied force. It also depends upon the number of films that are present; essentially it is the moisture content at which most of the films coalesce to fill the majority of the pore space; or, in terms of the physical state of the water present, the ratio of the so-called rigid water to that which is unoriented becomes so small that there is little bonding force between surfaces. The tension of the water at the liquid limit is in the vicinity of pF 0.5 (3.2 cm of water) (Croney and Coleman, 1954).

The plasticity index is an indirect measure of the force required to mold the soil. It is a function of the number of the films and represents the amount of water that must be added to the soil system to increase the value of d in equation 3-5 from a film thickness at which maximum

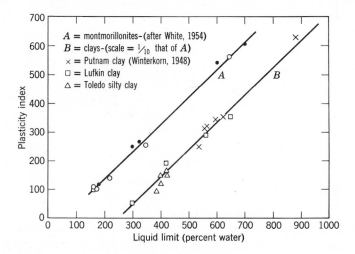

FIG. 3-6. Relation of plasticity index to liquid limit.

tension occurs to a thickness at which flow is produced. This means that there should be a direct relationship between the plasticity index and the liquid limit. This is verified by the data in Figure 3-6. The points on the montmorillonite curve represent two different samples that were saturated with different cations. The points for Putnam clay and Toledo silty clay relate to homionic systems; those for Lufkin clay are for different clay percentages. It is significant to note that the two curves are parallel, which means that curve B would be a continuance of curve A if both were plotted on the same scale.

Factors Affecting the Atterberg Limits

CLAY CONTENT. Since plasticity is a function of the finer soil fractions, various soils possess different plasticities according to the amount of clay they contain. Atterberg (1911, 1912) showed that an increase in the percentage of clay causes plastic limits to be higher on the moisture scale and increases the plasticity number. The results in Figure 3-7b point out the same effects of clay. The clay content of the Lufkin soil was varied by additions of silt. These results are typical of a large number of samples. It is seen that the moisture content of the plastic limit decreases slightly as the clay content is decreased. This should be expected since it takes more water to satisfy the surface forces of clay than of silt. The outstanding effect of a decreased clay content is the rapid lower-

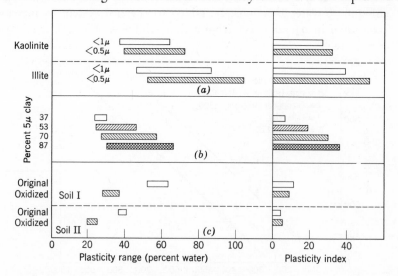

FIG. 3-7. Factors affecting Atterberg limits. (a), effect of particle size (after White, 1949); (b), effect of clay content; (c), effect of organic matter.

ing of the liquid limit and the consequent decrease in the plasticity index. Since the liquid limit is dependent upon the number of films that are present and inasmuch as the nature of the colloidal surface is the same, there should be a direct relationship between the decrease in the percentage of clay and the water content at this limit. Such a relationship is readily apparent from the data since the ratio of the observed to calculated liquid limit varies only between 1.00 and 1.08.

The data in Figure 3-7a indicate that the plasticity of the <0.5-μ fractions of both illite and kaolinite is higher than that of the <1-μ particles; the plasticity of illite is greater than that of kaolinite. Like the results in Figure 3-7b, all the Atterberg limits are raised as the surface is increased. Skempton (1953) showed that the plasticity index was related to the percentage of <2-μ clay in different clay systems.

The clay content therefore determines the amount of surface that is available for water adsorption. In other words, it more or less regulates the number of films contributing to cohesion and plasticity. For a given clay mineral, the amount of adsorbed water required at the plastic limit will increase with the amount and size of the particles present. The plasticity index then becomes an indirect measure of the clay content as it represents the amount of water in a given number of films from the thickness of the plastic limit to that of the liquid limit.

NATURE OF CLAY MINERALS. Atterberg (1911, 1912), in his original investigations on soil plasticity, was interested in finding out to what extent the different minerals from which soils are derived affect plasticity. Consequently, he ground various minerals to give particles the size of clay and measured their plasticities. The results of this experiment are outstanding in light of more recent knowledge on the nature of clay minerals. His studies showed that only those minerals that have a platy or sheetlike structure exhibit plasticity when ground. Quartz and feldspar, whose crystals are made up of linked tetrahedra, are nonplastic. On the other hand, kaolinite, talc, muscovite, biotite, and others whose crystal lattices are built up in sheets are plastic. These differences are attributed to a greater surface and increased contact in the case of plate-shaped particles. It is realized that few soils contain sufficient amounts of these primary minerals to affect plasticity very markedly, nevertheless, the fact that the secondary clay minerals have sheetlike structures similar to the aforementioned plastic primary minerals helps to explain the plasticity of clays.

It has been shown in previous discussions that the type of clay mineral has a tremendous influence upon the adsorption of water by the colloidal system. There can be little doubt that this adsorbed water adjacent to the particle surfaces possesses different physical properties from that in the bulk of the liquid water. It is only the nature of these properties and their relation to clay-water interactions that have not been clearly enough defined. The adsorption of water in the vapor form increases according to the clay mineral series montmorillonite > beidellite > illite > kaolinite. There is an ordered arrangement of water molecules near the surface. Data in Chapter 2 clearly indicated that the swelling of different clay minerals followed the order vermiculite > montmorillonite > beidellite > illite > kaolinite.

There is a wide variability in the Atterberg limits between different samples of the same clay mineral. These variations may be due to differences in the isomorphous substitutions within the crystal, the structure of the mineral, the nature of the exchangeable cation, and perhaps to the effect of larger particles.

The activity values for kaolinite range from 0.33 to 0.46; for illite 0.90; for Ca-montmorillonite 1.5; and for Na-montmorillonite 7.2 (Skempton, 1953). Although the plasticity index of all clay minerals increases with specific surface, the increase in plasticity per unit of specific surface is much greater for the montmorillonites (Platen and Winkler, 1958). When plasticity is determined in $3.6N$ NaCl, there are no differences in the index. Plasticity in the salt solution is entirely a function of the amount of specific surface. There is little difference in the plasticity indexes of illite between water and the salt solution. However, that of the kaolinite is slightly higher in the salt solution, which is apparently due to an increase in the specific surface and the activity per unit surface. The plasticity index of the montmorillonite is greatly repressed in the salt solution because of a prevention of interlayer swelling and the extreme shrinking of the diffuse double layer.

The type of clay mineral therefore influences plasticity because of the effect on the ability of the clay surfaces to adsorb and orient water molecules. The plasticity of the montmorillonite group is high because of greater surface hydration, which results from more orientation of water and the interlattice swelling as the thickness of the water films increases.

NATURE OF EXCHANGEABLE CATIONS. The exchangeable cations have considerable influence upon soil plasticity (Baver, 1928). The data in Figure 3-8 point out typical effects of exchangeable cations on different soils. The following tendencies occur consistently:

1. Na-saturated soils exhibit the lowest plastic limit, the second lowest liquid limit of the basic cations, and the highest plasticity index in three out of five cases.

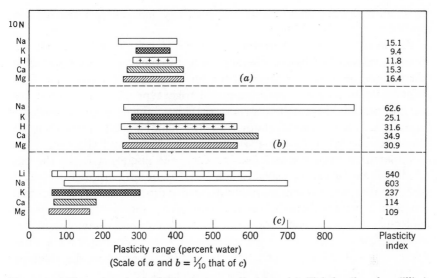

FIG. 3-8. Effect of exchangeable cations on plasticity. (a) Toledo silty clay (illite) (After Baver, 1928); (b), Putnam clay (beidellite) (After Winterkorn, 1948); (c), montmorillonite (After White, 1954).

2. K-saturated soils show the lowest plasticity index and the lowest liquid limit of the basic cations.

3. Ca- and Mg-saturated soils behave similarly. The plastic and liquid limits are generally higher than those of K- and Na-systems. Mg-saturated soils tend to have slightly higher plasticity indexes than those saturated with Ca.

4. H-saturated soils fluctuate considerably.

Plasticity values on beidellite (Putnam clay) are presented in Figure 3-8b. There are three distinct tendencies in evidence with this expanding-lattice type clay. First and foremost is the high liquid limit of the Na-clay. Second is the low liquid limit and plasticity index of the K-saturated soil. Third is the similar plastic limits of the clays saturated with the different cations. The Na- and K-saturated systems do not have lower plastic limits than those saturated with divalent cations.

The effect of exchangeable cations on the plasticity of montmorillonite is illustrated in Figure 3-8c. The Na-saturated montmorillonite has the highest plastic limit. The close but variable behavior of the K-, Ca-, and Mg-saturated montmorillonites with respect to all three Atterberg values indicates that the monovalent K ion acts as a divalent cation in affecting the plastic properties. Li- and Na-saturated montmorillonites have much higher liquid limits and plasticity indexes than the K-, Ca-, and Mg-saturated systems.

Previous discussions (Chapter 2) on the hydration of clays indicated that the valence and hydration energies of the exchangeable cations affect the amount of water adsorbed on the surface of the clay particles. Their effects differ between the expanding-lattice type clays and those that have a more rigid crystalline structure. These influences have a profound impact upon soil plasticity both from the standpoint of determining the moisture content at which plasticity is developed and the excess that is required to cause the soil mass to enter the viscous state. In the case of the expanding-lattice clays, Li and Na ions are responsible for appreciable osmotic swelling and a greater dispersity of the system. Na-montmorillonite becomes plastic when just enough of the unit layers have 10 molecular layers of water between them to produce a continuous film (White, 1954). The greater the amount of water taken up by the interlayer spaces, the more water will be needed to increase the thickness of the film to give a lubrication action (Michaels, 1959).

The repressive effect of the K ion on the Atterberg limits of montmorillonite and beidellite has long been recognized. There appears to be two explanations for its behavior. As previously pointed out in the discussion on swelling, the large size of the K ion causes it to have not only a low energy of hydration but also an interfering effect upon the ordered arrangement of water molecules on the hydrating colloidal surfaces. The K ions do not have enough energy of hydration to overcome the electrostatic forces of attraction when the clay surface is hydrated (Norrish, 1954a; Baver and Winterkorn, 1935). The unit spacing of montmorillonite decreases from 20.1 to 16.35 Å when the K ion replaces Na on the exchange compex (Moum and Rosenquist, 1961). This represents a loss of one molecule of water, which is changed from internal to external water.

As previously mentioned, the K ions are fixed between the inner layers when the expanding-lattice clays are dried. They hold the unit layers together, resulting in the formation of an illitic-type structure which decreases the water-adsorbing properties of the clay mineral. The plasticities of K-, Rb-, and Cs-saturated montmorillonites are analogous to those of illites (Rosenquist, 1959).

The polyvalent cations tend to hold the expanding lattices together and there is little development of water layers between them. Such an action explains the nonexpansion of Ca-montmorillonite even though the Ca ion has a much higher hydration energy than that of Na (Grim, 1942). If this be true, then the liquid limit of such systems should be approximtely the same. Ca- and Mg-montmorillonites generally have liquid limits of the same order of magnitude; similar effects are exhibited by Al-, Fe-, and Th-montmorillonites. If the plasticity of clay systems is caused by the sliding of water layers over each other, then the water held between the unit layers in montmorillonite and beidellite is not directly involved in the mechanisms involved. However, the requirements of the diffuse double layer between the expanding layers must be satisfied in order to provide the excess water for increasing the thickness of the films between individual colloidal particles, either for decreased tensions or for increased lubrication purposes.

The effect of these ions on nonexpanding clays is quite different from the effect on the montmorillonite group, The Li and Na ions tend to lower the plastic limit and the plasticity index as compared with the divalent cations. The differences are not large, probably because of the lower exchange capacities of illite and kaolinite. If the monovalent cations are not hydrated when they are adsorbed (Hendricks et al., 1940), their presence on the surface will interfere with the normal ordered arrangement of water molecules and consequently reduce the plasticity. The higher hydration energies of the divalent cations should cause a raising of the Atterberg limits. This is suggested in Figure 3-8a for a soil that contains an illitic-type clay.

ORGANIC MATTER CONTENT. Organic matter exerts an interesting effect upon soil plasticity. Measurements of the plasticity limits of different soil profiles usually show that these limits of the surface horizon are higher on the moisture scale than those of the other layers. This effect is apparently associated with the presence of organic matter in the surface horizon. Oxidation of the organic matter with hydrogen peroxide causes a decided lowering of both plasticity limits (Baver, 1930a). This is illustrated in Figure 3-7c. Soil II, with an organic matter content of 3.5 percent, became plastic at 36.5 percent moisture. Removal of the organic matter lowered this limit to 19.8 percent moisture. Moreover, the oxidized soil flowed at 25.1 percent moisture, whereas the soil with the organic matter was friable up to a moisture content of 36.5 percent. The plasticity index was not materially changed by oxidation of the organic matter. Soil I, which contained 7 percent organic matter, became plastic at 52.2 percent moisture. Removal of the organic matter caused the soil to exhibit plasticity at 27.7 percent moisture.

Plasticity measurements on virgin and cultivated Putnam silt loam in Missouri exhibited the same effects of organic matter. A decrease in organic matter from 3.9 percent in the virgin area to 2.6 percent for the soil under cultivation for 60 years lowered the plastic limit from 27 to 22 percent moisture. Thus oxidation of organic matter in the field under natural conditions produced results similar to artificial oxidation in the laboratory.

The causes of this decided lowering of the plasticity limits on the moisture scale, without a really significant effect upon the plasticity index, are readily understood on the basis of the film theory of plasticity. Organic matter has a high absorptive capacity for water. Hydration of the organic matter must be fairly complete before sufficient water is available for film formation around the mineral particles. Consequently, the plastic limit occurs at relatively high moisture contents. After the films are formed, however, practically all the additional moisture functions only to enlarge the films until flow is produced. The presence of organic matter has little effect on this type of water and therefore does not influence the plasticity index to any significant extent.

The results in Figure 3-7c point out the importance of considering the range of plasticity on the moisture scale in discussing soil consistency. It is evident from these data that two soils may possess the same plasticity index but exhibit plasticity at entirely different moisture contents. The practical significance of the plasticity range is well illustrated by soil I, Figure 3-7c. The presence of organic matter makes it possible to cultivate up to 52.2 percent moisture without puddling the soil. When the organic matter is removed, the soil puddles at 27.7 percent moisture. Thus the addition of organic matter to soils may be expected to extend the zone of friability to fairly high moisture contents.

The Sticky Point of Soils

Atterberg suggested that the scouring point (*Klebegrenze*) is an important character of soil consistency. This point represents that moisture content at which the soil no longer sticks to a foreign object. It is usually determined by drawing a nickel spatula across the face of a moist, kneaded mass of soil. The moisture content is regulated until that point is reached where no soil adheres to the spatula. The scouring point and the liquid limit vary between the same moisture contents. With highly plastic soils, the scouring point lies slightly below the liquid limit; with slightly plastic soils it occurs above the liquid limit.

The moisture percentage of the sticky point corresponds to that moisture content at which the attractive power of the soil for water is satisfied (Keen and Coutts, 1928). The sticky point is due to water films between the surface of a foreign object and the soil (Fountaine, 1954). This film is connected to the bulk of the soil water at the same tension that exists throughout the soil. According to the film theories of cohesion and adhesion, the sticky point should be approximately the same as the moisture content at which maximum adhesion occurs. There is a highly significant correlation between the sticky point and the plasticity index (Gill and Reaves, 1957). The effects of clay and organic matter on the sticky point are very similar to their influences on both the plastic and liquid limits. This is in accordance with the water film relationships involved. From a practical point of view, the sticky point indicates the maximum moisture content at which normal soils will scour during tillage.

SHEAR STRENGTH OF SOILS

Basic Principles

The shear strength of a soil is the maximum internal resistance of a soil to the movement of its particles; that is, resistance to slipping or sliding of soil over soil. The forces that resist shear are internal or intergranular friction and cohesion. According to Coulomb's law,

$$S = C + \tan \phi P \tag{3-8}$$

where S is the shear strength, C is the cohesion, P is the effective pressure normal to the shear plane, and $\tan \phi$ is the coefficient of friction where ϕ is the angle of friction.

The curves in Figure 3-9 illustrate the application of equation 3-8 to the shear value of soils from the Rothamsted plots. These shear strengths were measured *in situ*. The cohesion factor at zero normal stress is slightly less than 2 lb/in.[2] Then, according to theory, the shear strength increases linearly with the applied normal stress.

According to equation 3-8, the components of shear resistance are cohesion and friction. These two components are expressed through a combination of both physical and physicochemical factors (Lambe, 1960; Seed, Mitchell, and Chan, 1960). The physical factors affect primarily the friction component ($\tan \phi$) in equation 3-8. Two processes are involved, the resistance to sliding of one soil particle over another and the interlocking of particles (referred to as *dilatancy*). Movement of interlocked particles requires that the particle must move vertically under the applied stress before it can move horizontally over an adjacent particle. This means that there is a volume increase in dilatancy. It requires a larger horizontal stress than if the particle had to move horizontally only. The physical component of shear strength is directly proportional to the effective normal stress and is of greater importance with granular than with clay particles.

The physicochemical factors are expressed through the cohesion factor in equation 3-8. Cohesion is a function of the attractive and repulsive

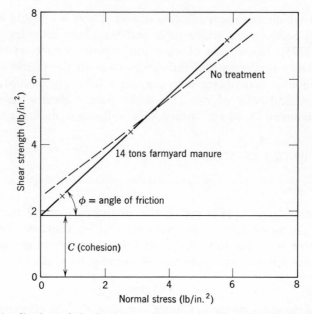

FIG. 3-9. Application of Coulomb's equation to plots from Broadbalk Field. (After Fountaine, Brown, and Payne, 1956.)

forces in clay-particle interactions. As previously stated in the discussion on soil cohesion, there are several interparticle bonding forces. In addition to the forces that tend to hold the particles together, there are the repulsive forces resulting from the diffuse double layer around the particles which are dependent upon the hydration of the adsorbed cations, and so on, as discussed in the sections on viscosity and swelling.

Water plays an important role in determining the magnitude of the cohesive component as it affects the distance between particles and the attractive forces associated with air-water menisci.

The application of compressive forces to the soil will increase cohesion by bringing about particle orientation, thereby decreasing the spacing between particles and affecting the attractive and repulsive forces.

The cohesion component may express itself through the formation of larger particles out of smaller ones, which will increase the angle of friction rather than raising the cohesion intercept.

Methods for Determining Shear Strength

There are two widely accepted methods for determining the shear strength of soils, the direct and triaxial shear tests. There are two varia-

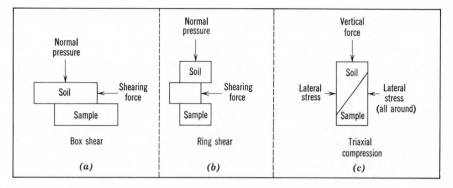

Fɪɢ. 3-10.   Schematic presentation of force components in shear tests.

tions of the direct method, the direct or box-shear and the double-direct or ring-shear tests. The direct shear test usually employs a rectangular box that is split in the middle so that the upper part can slide over the bottom half. The lid on the box can move vertically. The box is filled with soil, a pressure is applied to the lid normal to the surface, and a shearing force is applied to the side of the upper half, thereby shearing the sample in the middle. The pressure-shearing force relationships are shown in Figure 3-10a.

In the ring-shear test, a sample of soil is placed in a metal cylindrical container, the center of which is a ring that can be forced through the soil to give a double-shearing action. Pressure is applied normal to the shear planes by a piston at the top of the container. The pressure-shearing pattern is illustrated in Figure 3-10b. Both the direct and double-direct shear tests are relatively simple and lend themselves readily to routine determinations. They have the weakness of forcing shear to take place in a specified plane, which may or may not be the weakest in a specified sample. Moreover, there is not an equal distribution of stresses over the shearing surface.

The triaxial-shear test (also called triaxial compression) is more complicated but more reliable. A cylindrical sample of soil is surrounded by a rubber membrane with a rigid cap on one end and a piston on the other. It is then placed in a special closed cell to which a lateral stress is applied to all sides of the sample by regulating the air pressure within the cell. An axial load (vertical stress) is applied to the upper end of the cylinder through the piston. This stress is increased until shear takes place, usually along a diagonal plane through the sample. The stress-shearing pattern is shown in Figure 3-10c. See Sallberg (1965) for details for making direct- and triaxial-shear test.

Relationship to Plasticity

In discussing the shear strength of cohesive or plastic soils, one is interested in the shear value at a given moisture content as a function of the applied stress and the relation of strength at a given stress to the moisture content. The increased pressure applied to a soil should have the effect of diminishing the distance between particles and thereby increasing the attractive forces which would result in higher shear strength in direct proportion to the stress. The shear value of plastic soils increases proportionately to the stress applied normal to the plane of shear (Nichols, 1932). The shear value of unconsolidated soil increases linearly with moisture to a maximum near the plastic limit and then decreases to a very low amount at the liquid limit (Figure 3-11). In light of the film theories of plasticity, this should be expected, since maximum film tension and cohesion occur near the moisture content of the plastic limit. The film-tension forces at the liquid limit are very small and flow is easily produced. Once the soil becomes plastic, there is little internal friction; consequently shear is a function of the cohesion of the moisture films beyond the plastic limits. As the moisture content approaches that of the liquid limit, shear assumes the properties of viscous flow. The maximum shear value is proportional to the plasticity index,

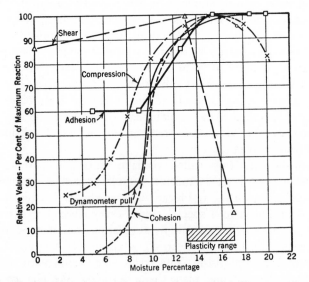

FIG. 3-11. The relation of dynamic factors involved in tillage to soil moisture, with special reference to the plasticity range. (The maximum value for each of these factors was taken as 100.)

as should be expected if shear is a function of cohesion due to moisture films. The higher the activity ($PI/2$-μ clay) of the clay, the greater is the contribution of the cohesion to shear strength.

The plastic limits can help to characterize the shear strength of soils in terms of the normal stress applied and the moisture content of the soil. Equation 3-7 expresses these relationships for unconsolidated plastic soils (Nichols, 1932).

$$F_s = \frac{(PL - W)(0.66PI + P + 1.8)}{PI} \qquad (3\text{-}9)$$

where F_s is the shear strength, P is the normal stress. W is the moisture content, PL is the plastic limit, and PI is the plasticity index.

Clay soils can develop suctions of about 666 cm of water (pF 2.82) when sheared at their maximum shear strength, which takes place at the plastic limit (Greacen, 1960). When the shear stress is first applied there is a rapid decrease in the voids ratio, which levels off to almost a constant value at the higher strains. This means that shearing produces a compaction effect. At the same time, there is a sharp rise in both moisture tension and shear strength, which also approaches a constant value at the higher strains. The straight line relationship between the moisture tension and moisture content between the plastic and liquid limits is expressed by the equations

$$W = -A \log P' + C \qquad (3\text{-}10)$$

where W is the water content, P' is the soil-moisture tension in g/cm^2, and A and C are constants.

$$A = -\frac{dW}{d \log P'} = \frac{(LL - PL)}{(pF_{LL} - pF_{PL})}$$

$$C = A + LL$$

Therefore

$$\log P' = 1 + 1.82 \frac{(LL - W)}{(LL - PL)} \qquad (3\text{-}11)$$

The equivalent clay strength, $\log S$, is equivalent to $\log P'$ since P' may be assumed to be the suction in the straining clay at zero applied mechanical load. After assuming that P' is the moisture tension of the clay under strain at zero applied normal stress, P' then becomes the effective compression load and equation 3-11 gives the relation of equivalent clay strength, $\log S$, to the moisture content. Equivalent strength increases with the soil suction. At the liquid limit ($pF = 1$), strength values as high as 1000 g/cm^2 can be obtained with heavy clays.

COMPRESSION-COMPACTION

Resistance to Compression

General Principles

Compression may be defined as the change in volume of a soil under an applied stress. Since the volume of the soil consists of both solid phase and the voids between it, compression denotes the decreases in the voids ratio per increment of applied load or pressure. The voids ratio e is the ratio of the volume of the voids V_v to the volume of the solid phase V_s. A soil with 60 percent porosity would have a voids ratio of 1.5.

The voids ratio-pressure relationship is given in equation 3-12:

$$e = A \log P - C \qquad (3\text{-}12)$$

where A is the compression index ($de/d\log P$), P is the applied load or pressure, and C is a constant equivalent to the voids ratio for unit load P.

Soils may be compressed under low or high pressures. Shearing stresses may be present or absent during compression. Particle orientation and changes in the size of the colloidal micelles are the major causes of compression (Lambe, 1958b). The former prevails primarily in low-pressure compression. Both are important in high-pressure consolidation. Under high-pressure compression, the soil expands or rebounds when the load is removed. This effect is due principally to the diffuse double layer and the swelling of the micelle.

The differences between low- and high-pressure compression are illustrated in Figure 3-12. It is seen in Figure 3-12A that an increase in pressure on the dry, compressed sample produces a linear decrease in the voids ratio. This is brought about by an orientation of particles parallel to each other to make a lower volume per unit mass. The decrease in the voids ratio of the wet-compacted sample is caused by a closer spacing of particles that have already been partially oriented. Since work is required to orient particles, there is greater compression per unit of pressure in the wet-compacted sample. A slightly different story is told by the curves in Figure 3-12B. Pressures are plotted on a logarithmic scale. Particles are oriented and brought closer together by pressure in the dry-compacted samples. In the wet-compacted sample, the spacings of the oriented particles are decreased by increasing pressures. Both systems come to about the same voids ratio. When the load is released, both samples rebound or swell because of an increase in the spacings between particles; depending upong the nature of the clay mineral, the type of exchangeable

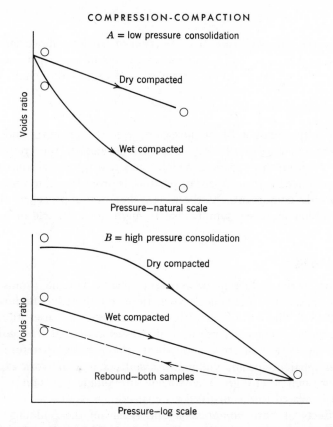

FIG. 3-12. The effect of compression on the voids ratio. (After Lambe, 1958b).

cations, or the presence of salts, which determine the thickness of the diffuse double layer. The repulsive forces resulting from the diffuse double layer could extend over a long range that may be equal to a change in the voids ratio of 1 to 20 (Bolt, 1956). The nature of the exchangeable cation influences the effect of the diffuse double layer on compressibility. The compressibility of kaolinite has been found to decrease according to the order $Li > Na > K > Ca > Ba$ (Salas and Serratosa, 1953). For bentonite, however, the order is $Li > Na > Ca > K > Ba$. The low compressibility of K-bentonite is due to the fixation of the K ion in the lattice which prevents the hydration of the inner surfaces. Compression increases with the degree of polarity of organic liquids, according to the series $CH_3CH_2OH > CH_3OH > CCl_4$.

Compression increases with moisture content to a maximum and then decreases as the amount of moisture becomes larger (Nichols and Baver, 1930). This maximum occurs within the plastic range, undoubtedly as a result of the enhanced ease of orientation of particles above the plastic limit (see Figure 3-11). The increase in the compression per increment of pressure can be expressed by the equation (Reaves and Nichols, 1955)

$$y = ae^{-bx} \qquad (3\text{-}13)$$

where y is the amount of compression, x is the pressure, and a and b are constants. Reaves and Nichols developed the relationship $dy/dx = ky$ from this equation, which states that the increment of compression produced per increment of pressure is proportional to the amount of soil already compressed. As the plastic limit was approached, the rate of compression increased rapidly due to augmented particle orientation.

Soil Puddling

Puddling is the reduction in apparent specific volume (voids ratio) of a soil caused by mechanical work done on the soil (Bodman and Rubin, 1948). The term "puddlability" expresses the susceptibility of soils to puddling. These are the normal stresses associated with compression and the tangential stresses of shear. Puddlability therefore is the change in apparent specific volume of a soil per unit of work expended in causing such a change. The change in volume per unit of work (dv/dw) is related to the air-filled pore space.

The effects of both compression and shear on the puddling of the Yolo silty clay loam are shown in Figure 3-13. There are several important features in these results. First, and probably most significant, is the increased destruction of air-filled pores with an increase in moisture content. At the high moisture content of 27.0 percent, which is practically equal to the moisture equivalent, very few air-filled pores remain at the applied pressure of 1.5 kg/cm² (21.3 lb/in.²). Second, compression is the major force responsible for "puddling" within the moisture range that was used. Third, the effect of shear becomes more pronounced as the moisture content increases. These data are in line with those shown in Figure 3-11.

One should expect compression to be most effective below the liquid limit. Saturated and unsaturated soils compress under a natural load to a voids ratio that appears to be a function of the strength of the aggregates (Greacen, 1959). When a shear force is applied, the soil compresses further to a voids ratio that appears to be constant. This has been called the "ultimate voids ratio" (UVR), which is dependent upon aggregate strength and the water-holding properties of the soil. The voids ratio decreases with increasing moisture at the same applied load. Puddling

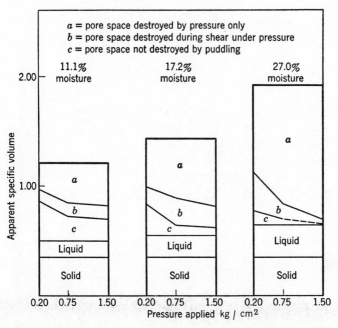

Fɪɢ. 3-13. Puddling of Yolo silty clay loam in relation to moisture and pressure. (Bodman and Rubin, 1948.)

therefore is a structural change associated with the consistency of the soil. The importance of both compressive and shearing forces in soil puddling as related to moisture content is clearly defined.

Compaction

Compaction refers to the increase in density of a soil as a result of applied loads or pressure. This means that the soil has a certain density or state of compaction before the added force is applied. In other words, "soil compaction is a dynamic soil behavior by which the state of compaction is increased" (Gill and Vanden Berg, 1967). Although drying and shrinkage may cause soil compaction, our discussion is limited to compaction from mechanically applied forces. Soil density increase is a function of both the compactive effort and the water content. These relationships are well illustrated in the compaction curves of Hawaiian soils in Figure 3-14. The force required to compact a soil to a given density decreases exponentially with the moisture content (curve A). The density of a soil at a given moisture content increases exponentially with the force applied (curve B). Both of these effects are related to the orientation

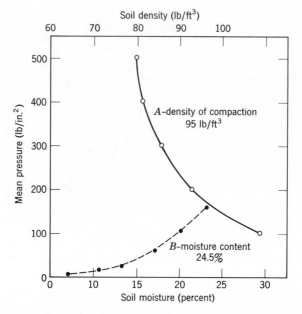

FIG. 3-14. The effect of pressure on soil compaction. (After Trouse, 1954.)

of particles, as will be clarified later. The moisture content of the soil in curve B is approximately that of the plastic limit.

The density of a soil under a constant compactive effort increases progressively with the water content to a maximum and then decreases with further additions of water. This maximum is known as the optimum water content for compaction. These effects are depicted in Figure 3-15. The curves point out that the maximum density levels become higher as the compactive effort is greater. Moreover, the optimum moisture content decreases as the force of compaction increases. In other words, the loci of the maximum density values are shifted toward the dry side as the compactive effort becomes greater.

The shape of the density-moisture curve is explained on the basis of the development of the diffuse double layer and the orientation of particles (Lambe, 1958a). In the 5-25-40 curve in Figure 3-15, there is insufficient water present at w_1 to form a double layer. The soil is flocculated and the particles are randomly organized. From w_1 to w_2, the double layer expands, the water films become thicker to produce a lubrication effect between the particles which become oriented so that they slide over each other and form a denser mass. From w_2 to w_3, there is further expansion of the thickness of the water films and the density decreases because of the dilution effect of the water on the concentration of parti-

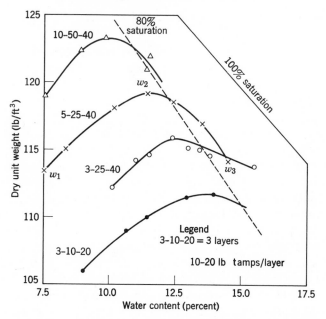

FIG. 3-15. Density-water relationships during soil compaction. (After Wilson, 1952.)

cles per unit volume. There is some displacement of air in the system but not as much as between w_1 and w_2. The higher densities obtained with increased compactive effort are due to the greater orientation of particles under the augmented forces of compaction. Greater orientation of particles and higher densities are obtained from kneading compaction than from a static application of force.

In the discussions on shear strength, compression, and compaction, facts have been presented to show that the application of compressive and shear forces to a cohesive soil results in a decrease in the void ratio and an increase in soil density and soil strength. Compaction destroys the larger pores, partially filling them with solid particles. Discussions in Chapter 5 on soil structure emphasize how compaction strikes at the core of root proliferation as it increases soil strength and decreases aeration.

The Compaction Problem in Soil Tillage

The use of agricultural machinery and transportation vehicles in the preparation of the seedbed-rootbed and the production and harvesting of crops is accompanied by applications of pressures to the soil. The distribution of these pressures in relation to soil compaction and plastic flow is of major importance in analyzing the impact of machines and vehicles on soil properties, both from the standpoint of the growth of the plant and of machine design to minimize these effects (Söhne, 1956).

Arch Action

A body resting upon the soil or pushed into it exerts a pressure that is distributed over a sizable area in an archlike pattern. Nichols (1929) defined arch action as the tendency of a soil to vector out a compressive force. This vectoring was explained as being due to friction of soil on soil, the interlocking of particles, and the cohesion of water films. Although arch action is an important soil reaction that occurs in most tillage operations, it is discussed here because of its role in soil compaction. Nichols observed that a plunger forced into the soil drove a cone-shaped mass of soil ahead of it. Soil movement occurred along the sides of the cone. The arch action was caused by soil movement in front of the advancing surface. The penetration of the plunger was directly proportional to the pressure applied, as soon as a cone was formed in the soil mass. Arch action before a 1-in. plunger in unconfined Vaiden clay is

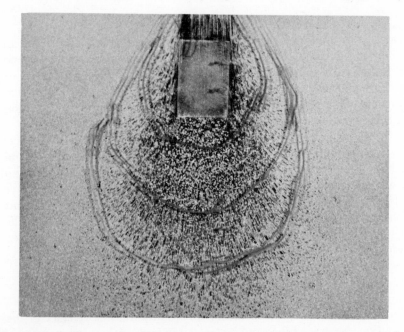

Fig. 3-16. Arch action from a 1-meter plunger in unconfined Vaiden clay. (After Reaves and Nichols, 1955.)

shown in Figure 3-16 (Reaves and Nichols, 1955). The crayon marks show lines of approximately equal pressure for ½-, 1-, and 1½-in. advancements of the plunger. The width of the arch is not affected by field bulk densities but is dependent primarily upon friction and particle interlocking; cohesion is a secondary factor.

Arch action produced by a piston caused greater compaction a few inches below the surface of the plunger than adjacent to it. This is illustrated in Figure 3-17 (Chancellor, Schmidt, and Söhne, 1962). Maximum

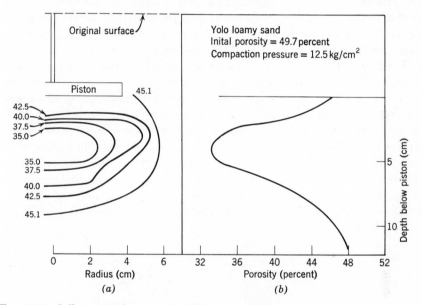

FIG. 3-17. Soil compaction test. (*a*) Lines of equal porosity; (*b*) porosity versus depth. (After Chancellor, Schmidt, and Soehne, 1962.)

compaction occurred at about 4 cm below the plunger surface. The porosity of the soil immediately below the plunger was equal to that at a depth of about 9.1 cm. This same porosity, representing a decrease of more than 9 percent from the original pore space, extended approximately 2 cm beyond the edges of the piston. These arch-action effects are significant in the impact of tires and tracks on soil compaction.

PRESSURE DISTRIBUTION UNDER TIRES AND TRACKS. The pressure distribution in the soil under tires depends on (1) the amount of load, which determines the total pressure exerted, (2) the size of the contact area between the tire and soil, which determines the amount of pressure per square inch (psi), (3) the distribution of this pressure within the contact area, and (4) the moisture content and density of the soil (Söhne, 1958). The pressure-distribution curves with depth are bulb-shaped. The lines of equal stress are circular in hard, dry soils; they are elliptical in soft soil due to soil flow on the sides. The depth of lines of equal stress increases with soil moisture content. Deep tracks in dense, wet soils result more from plastic flow than from soil compaction. This takes place when shear stresses exceed the yield strength. When there is slippage of tires, a thin layer of soil is compacted as a result of soil smearing; this orients the particles in the immediate surface. Soil compaction results from horizontal forces caused by thrust as well as from vertical forces produced by loading (Gill and Reaves, 1956). About 10 percent of the compacting pressures from tires comes from horizontal stresses.

Track-type tractors produce vibrating stresses that make the total stress considerably higher than the average for the same ground pressure per square inch. Reaves and Cooper (1960) studied the distribution of stresses under a 12-in. track and a 13-38 tire when both were loaded with 3600 lb and operated at a drawbar pull of 1500 lb. The ground pressure per square inch was 12.3 for the track and 25.4 for the tire. This difference was due to the greater length of contact of the track with the soil. The contact length for the track was 5 ft; that for the tire was only 2 ft. Maximum stresses occurred under the center of both at about 3 in. and then decreased both laterally and vertically (Figure 3-18). Stresses

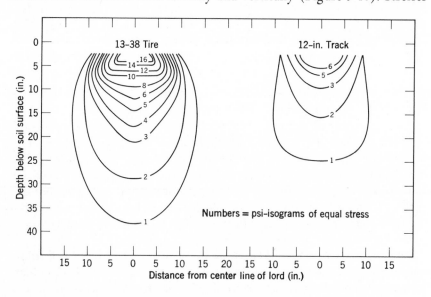

FIG. 3-18. Isograms of mean normal stress under a tire and a track. (After Reaves and Cooper, 1960.)

under the tire were at least twice those under the track. Note that the 6-psi isogram under the track reached a depth of about $5\frac{1}{2}$ in.; that under the tire extended to 12 in. The problem of removing harvested sugar cane with minimum compaction from the wet fields in Hawaii was solved by designing a cane buggy with long, wide tracks to give a ground pressure of about 6 lb/in.[2] when loaded.

"SOLE" FORMATION. It is common experience with many soils to find a compacted layer at the bottom of the zone of plowing. This layer has been termed the "plow sole." It is assumed to originate from a combination of tillage and other farm operations. Since the zone above this layer is plowed and tilled regularly, the compaction is observed only below the loosened layer. A thorough study of a number of tillage operations in Hawaii (Trouse and Baver, 1965b) pointed out that almost every agricultural implement created a sole of some kind under moist soil conditions. There were plow soles, subsoiler soles, disk-harrow soles, and traffic soles. All such soles reduced the permeability of the soil to water and restricted root elongation and proliferation.

Plow Soles. The compacting action of the plow can be especially injurious when the depth of plowing is constant. In addition to the effect of the plow itself, there are the influences of tractor wheels (or animal hoof prints). Frese and Altemüller (1962) made a thorough study of the morphology of plow soles with the help of thin-section techniques. They observed that the normal action of the plow left behind a loose surface layer and a dense subsoil where the soil aggregates had been pressed together by the sole of the plow. Photomicrographs of the compaction by the landside showed a smearing action that formed a very thin clay film only a few millimeters thick at the top of the compacted zone. This compacted condition diminished the permeability of the zone to such an extent that there could be hindering of plant growth.

Disk-plow soles are formed where the disks cut the soil from the bottom of the furrow (see Figure 3-19) (Trouse and Baver, 1965b). Subsoiler soles are produced where the duckfoot or blade of the subsoiler passes through the soil.

Disk-Harrow Soles. The disk is a compacting implement as well as one that can loosen soils. The same forces that cause penetration of the disk also produce compaction. Disking is usually one of the last operations in seedbed preparation. Its compactive effect in the formation of disk-harrow soles is clearly depicted in Figure 3-19 (Trouse and Baver, 1965b).

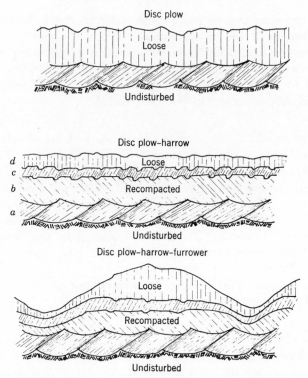

FIG. 3-19. Diagrammatic portrayal of seedbed profile in a tillage sequence.

One can observe in this perspective drawing (1) the disk-plow sole, (2) the recompacted zone, (3) the disk-harrow sole, and (4) the shallow loose layer remaining after disking. Observations on a number of seedbed profiles indicated that from 50 to 70 percent of the soil loosened by the disk-plowing operation was recompacted by subsequent disk harrowing. The Hawaiian studies showed that disk harrowing subsequent to subsoiling packed the soil into the subsoiler grooves to about the same density as the original untilled soil. These investigations caused important changes in the sequence of operations on many plantations. There was a 50 percent decrease from 1958 to 1964 in the number of plantations using disk harrowing as the final operation in seedbed-rootbed preparation.

Traffic Soles. Post-tillage operations, such as planting, weed spraying, and cultivation, create traffic soles. Although this type of compaction occurs between the rows, it has a restrictive action on water penetration and root development. Soil compaction by tillage implements in California orchards resulted in greatly decreased water infiltration. Although the detrimental effects of tractors were greater when the soil was tilled at mosture contents near the field capacity, compaction also occurred with dry soils (Parker and Jenny, 1945). Even disking caused appreciable compaction. There was a marked improvement in water infiltration when all cultivation was eliminated. Chemical weed control was found to be effective and is being practiced in many orchards with a minimum of machinery operations.

Tractor tires cause considerable compaction immediately below the tires (Jamison and co-workers, 1950; Weaver and Jamison, 1951). The depth of penetration of compaction effects increases with the initial looseness of the soil and the moisture content. Peak compaction occurs at moisture contents near the plastic limit, which is about the optimum condition for tillage. These detrimental effects can be minimized most effectively by designing traction equipment that impose the smallest unit pressures on the soil.

Compaction data from many situations involving vehicular traffic have been reported from Hawaii (Trouse and Baver, 1965a). The compacting effects of these vehicles are related to infiltraton rates and changes in soil porosity. A typical example of traffic soles caused by field machinery is illustrated in Figure 3-20. The side-mounted, cut-windrow harvester is propelled by a track-type tractor; it has a pneumatic tire on the end of the cutting mechanism. It cuts one row at a time in one direction only and must be backed to cut the next row. This means four passes of the track and two of the large tires in each interrow. The ground pressure under the track was estimated at 11 psi and that under the tire at 35 psi. The results in Figure 3-20 show that the infiltration rate was decreased from 3.1 in./hr in the no-traffic areas to 0.8 in./hr in the compacted interrow. This 74 percent decrease in water intake was due primarily to a 62 percent reduction in aeration porosity in the upper 5 to 6 in. of surface soil. Soil density was increased in the top 6 in. but not below this depth. However, the compacting effort upon the destruction of the large pores continued to a depth of 20 in. The major impact of sole formation was to change the size distribution of the soil pore space. There was an increase in the number of the smaller-sized pores. This effect was typical of other field equipment.

The reconditioning of compacted soils is a difficult operation. The use of subsoilers to break up the dense mass is almost mandatory. One of the major problems, however, is the cloddiness that results from this type of tillage. Hawaiian experience has shown that more than two-thirds of such tilled zones is composed of large, dense clods that are impervious to root penetration.

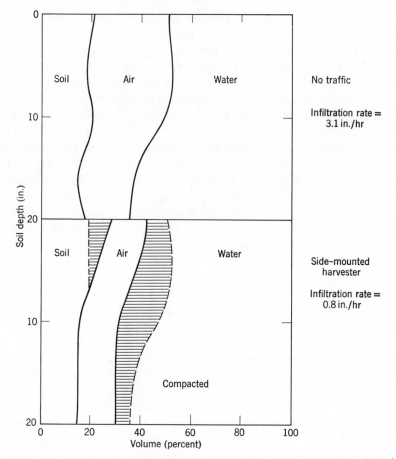

Fig. 3-20. Compacting effects of a side-mounted harvester on a humic latosol. (After Trouse and Baver, 1965a.)

RESISTANCE TO PENETRATION

The resistance of a soil to the penetration of a probing instrument is an integrated index of soil compaction, moisture content, texture, and type of clay mineral. In other words, it is an index of soil strength under the conditions of the measurement. It is a determination that involves both soil consistency and soil structure. The amount of penetration per unit force applied to a given soil will vary with the shape and kind of instrument used. As the penetrometer enters the soil, it encounters resistance to compression, friction between soil and metal, and the shear strength of the soil, which involves both internal friction and cohesion.

These soil conditions are manifested near the tip of the probe where there are localized failures.

The magnitude of the resistance to penetration is influenced by both compression and soil-metal friction (Farrell and Greacen, 1966; Greacen, Farrell, and Cockroft, 1968). There is a spherical compression of the soil at the point of a blunt probe. This compression consists of two zones as the soil adjusts to the volume of the probe. The radius of this zone can vary as much as six to ten times the radius of the probe, depending upon the index of compressibility of the soil. Outside this plastic area is a zone of elastic compression. The vertical component of the pressure required to produce this compression and the resulting frictional resistance constitute the total point resistance. The frictional component of a blunt probe with a coefficient of friction of 0.4 is about 42 percent of the total point resistance.

The gently tapered probe exhibits a cylindrical compression as it moves through the soil. Although there is not a great difference between the measured point resistances of the two types of probe, that of the gently tapered penetrometer is slightly higher. The bulk density of the soil is highest at the edge of the hole but falls off more rapidly with distance from the hole in the case of the blunt than the sharp probe.

Although there have been many types of penetrometer used for measuring soil compaction and soil strength (see Bodman, 1949; Vomocil, 1957), they fall into two categories: the impact and the recording types. The soil sampling tube has been used not only to obtain cores for bulk density but also to measure resistance to penetration by determining the energy needed to drive the tube 1 in. into the soil (number of blows) (Parker and Jenny, 1945).

The results shown in Figure 3-21 illustrate the use of the recording type. The pronounced effect of moisture content on penetrometer readings is clearly depicted in these curves. There is a rapid increase in resistance with decreasing moisture, indicating that soil strength becomes greater as the particles are brought closer together during the drying process. Soil moisture appears to be the dominant factor influencing the penetrometer readings, although there is no simple relationship between these readings and the amount of water present. Zones of compaction apparently move closer to the surface as the number of tillage operations increase.

The pattern of resistance to penetration is not affected by the type of instrument (Hénin, 1937). In a loose sandy soil, the resistance to penetration increases proportionally with the depth. In a silt loam soil with 16 percent clay that has been compacted in the moist state, the

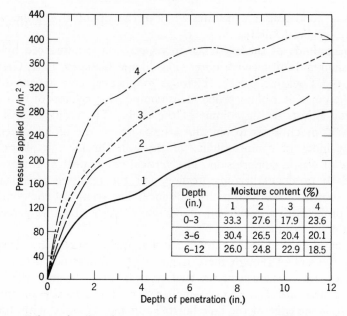

Depth	Moisture content (%)			
(in.)	1	2	3	4
0–3	33.3	27.6	17.9	23.6
3–6	30.4	26.5	20.4	20.1
6–12	26.0	24.8	22.9	18.5

FIG. 3-21. Effect of soil moisture on penetrometer readings. (Shaw et al., 1942.)

resistance increases rapidly with depth for several centimeters and then remains constant.

The penetrometer can be a useful tool to obtain information on soil strength and soil compaction if one keeps in mind the composite nature of the effects it measures. Davidson (1965) has given descriptions of several types of penetrometer that have proven satisfactory in soil studies.

DYNAMIC PROPERTIES OF SOILS INVOLVED IN TILLAGE

The dynamic properties of soils are expressed through soil movement that results from externally applied forces. Gill and Vanden Berg (1967) defined soil dynamics as "the relation between forces applied to the soil and the resultant soil reaction." The objectives of tillage can be achieved only by the application of forces to soils by agricultural implements. Therefore one must be familiar with the dynamic properties of soils that are expressed in tillage operations in order to understand the relationship between the physical properties of soils and tillage.

The pioneering investigations of Nichols (1929) led to a classification of variables that must be considered in relating soil dynamics to tillage operations.

The basic factors that affect the response of the soil to forces applied by tillage implements include the size distribution of soil particles, the amount and nature of the colloidal clay fraction, the density of packing of the particles, the amount of organic matter, and the moisture content.

Dynamic Properties of Soils

The shear value or coefficient of internal resistance of soils is generally accepted as the major dynamic property in soil-machine interactions. As previously discussed, it includes both soil cohesion and internal friction. The data in Figure 3-11 indicated that the shear value increases to a maximum at the moisture content of the plastic limit and then decreases sharply to a low amount at the liquid limit. The maximum value is proportional to the plasticity index. Almost all the soil variables previously listed contribute to the size of the shear component.

Friction between soil and metal is an important variable in tillage operations. Nichols observed that there are three phases in soil-metal friction interacton depending upon the moisture content and the weight and material of the metal slider. The first phase represents true friction between metal and dry soil. The second phase is governed by adhesion of soil to metal through water films as the moisture content increases. Adhesion increases to a maximum near the liquid limit. Maximum adhesion increases directly with the plasticity index (see Figure 3-11). Beyond the liquid limit, the moisture content is high enough to cause a lubrication effect and the coefficient of friction becomes constant or even decreases slightly. This is called the lubrication phase. The moisture content is the most important soil factor contributing to soil-metal friction.

Resistance to compression is of great importance in tillage operations since the application of pressure to the soil compresses it before any shear takes place. The data in Figure 3-11 show that resistance to compression increases to a maximum within the plastic range and then decreases. This is associated with the orientation of clay particles. Nichols observed that the degree of compressibility increases logarithmically with decreasing plasticity index.

The relationship of cohesion and adhesion to soil plasticity is also shown in Figure 3-11.

Dynamic Resultants

Nichols classified the resultant effects of forces applied by tillage implements and machinery as (1) shear as soil slips over soil, (2) fragmentation or the breaking up of soils, (3) arch action as a compressive force is vectored through the soil, and (4) compaction as the soil volume is changed by the compressive stress.

The Physics of Plow Action

Nichols (1929) and his associates (Nichols and Reed, 1934; Nichols, Reed, and Reaves, 1958) thoroughly studied the nature of plow action with the use of a glass-sided box that permitted careful visual observations. Their investigations showed that the first reaction to the point of the plow as it advances through the soil is to catch a part of the soil and drive it ahead in the form of a cone or wedge. This cone is compressed until the resistance of compression exceeds the shear value. Then a block of soil is sheared off. Figure 3-22 portrays soil movement before a plow share that has an exaggerated blunt point to make the soil reactions more perceptible. The different letters refer to the sites of different reactions.

A = The cone that is formed ahead of the blunt point by both forward and downward pressures. Part of the cone extends below the edge of the share.

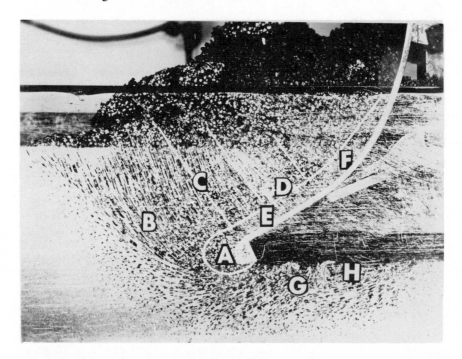

FIG. 3-22. Soil reaction to advancing share action. (A) Cone of soil; (B) compression area; (C) shear surfaces; (D) flow line up plow surface; (E) low-pressure area; (F) high-pressure area; (G) compacted area in plowsole; (H) vertical cracks above compacted plowsole. (After Nichols, Reed, and Reaves, 1958.)

B = The compression area ahead of the cone in the direction of movement of the plow. Pressure is the highest before the cone and
builds up to a peak at this point. The entire block of soil is
subjected to constantly increasing pressure upward as the inclined
plane of the plow advances.

C = The shear or rupture surfaces that form at angles of approximately 45° upward and forward as a result of pressures having
overcome the forces of cohesion of the soil. The entire block
moves up the inclined rupture surfaces, flowing over the adhering
mass of soil rather than the plow surface.

D = The flow line up the plow surface.

E = A low pressure area on the share back of the cone that is caused
by the soil being forced to flow around the cone and then
deflected upward.

F = Zone of adhesion of soil on the moldboard due to the flow of
deflected soil more nearly normal to the surface rather than
along the surface.

G = The compacted area in the plowsole caused by downward
pressures.

H = Vertical cracks and roughened surface at the bottom of the
furrow above a compacted plowsole.

Even though these reactions are magnified by the bluntness of the
plow point, these same types of behavior are observed under a wide
range of experimental conditions. The size of the cone, the amount of
adhesion of soil to the moldboard, and the extent of disturbance are
functions of soil physical properties. If the moldboard does not scour,
the functioning of the plow depends upon the soil properties rather than
the design of the plow.

Nichols and Reed (1934), in experiments under field conditions, observed that the primary shear planes (45° angles forward from the plane
of the shin) retain their same relative positions during their passage over
the moldboard. At the point where the furrow slice stands in the furrow
on its own furrow wall before it is inverted, the shear planes are parallel
to the undisturbed furrow wall ahead of the plow. From this point the
moldboard pushes the upper portion of the furrow slice forward and
to the right to make it lie smoothly against the preceding one. It is
common for these primary shear planes to retain their relative positions
after the slice has been inverted, especially in the case of the heavier
soils. The significant fact is that these shear planes move very little from
the plane in which they are originally formed. Inversion of the furrow
slice is accomplished by the rotation of the blocks of soil, formed by

the shear planes, forward and to the right at an angle 45° to the line of travel until these blocks stand on end. They are then pushed over more forward and to the right; they pivot on the upper edge of the original furrow wall.

The secondary shear planes are set up at an angle of 90° to the primary planes, which do not open until the furrow slice falls during the inversion process. These secondary planes are produced by the pressure of the advancing surface. The two types of shear plane are responsible for granulation of the furrow slice during the plowing operation.

A disk plow penetrates the soil and breaks it up by pressure (McCreery and Nichols, 1956). There is some cutting and pulverizing action. The extent of penetration depends primarily on the weight. The effectiveness of disk action is related to the angle of travel and the diameter, radius of curvature, thickness, and sharpness of the disks. Once the disk penetrates the soil, pressure is exerted as it moves forward. The soil is compressed until the pressure exceeds the shear value. Then there is a breaking up of the soil as shear planes are formed at an angle of about 45° with vertical. The weight of the disk plow provides penetration and the forward pull of the concave, angled disks cause compaction of the soil. The disk is the most widely used plow in the Hawaiian sugar industry primarily because of the lack of scouring of moldboard plows. The diameters of the disks range from 27 to 52 in. Disk plows do a good job of mixing the soil but are not very efficient in inverting the soil and burying surface residues.

Scouring

Scouring is defined as the self-cleaning flow of soil over tillage tools through a sliding action. When the soil adheres to the moldboard of a plow, soil flows over soil. This is because the force of adhesion of soil to metal is greater than the cohesion within the soil itself. The adhering soil acts as if it is part of the plow that is being pulled through the soil mass. The soil mass is pushed away from the moldboard; such soils are called nonscouring or "push soils." Fountaine and Payne (1954) analyzed the causes of nonscouring and concluded that five factors control the mechanisms involved:

1. The angle the implement makes with the direction of travel; it should be kept at a minimum.
2. Soil cohesion; dry soils have the greatest cohesion and scour the best.
3. The coefficient of soil-soil friction; aggregated soils have coefficients higher than nonaggregated.

4. The coefficient of soil-metal friction; metals that produce low co-efficients should be used.

5. The adhesion of soil to metal; low adhesion occurs at either high moisture tensions where there is insufficient water to cause a film between soil and metal or at low tensions where the film is too thick for strong adhesion and acts as a lubricant.

These principles have been thoroughly discussed by Gill and Vanden Berg (1967).

The disk plow is used in Hawaii because the moldboard plow does not scour in these latosolic soils unless the moldboard is covered with plastic. It is significant to note that scouring does take place in hydrol humic latosols where the moisture tension is always low.

Draft Requirements in Plowing

Nichols (1929) approached the problem of measuring the power requirements of a plow to overcome the forces of cohesion, adhesion, resistance to compression, shear, and soil-metal friction by pulling chisels through soils. These soils were compacted at varying moisture contents in a small glass-sided box so that the reactions of the soil to the chisels could be observed. A small dynamometer was attached to each chisel so that the actual pull in pounds could be recorded. The granulation of the soil ahead of the chisel was observed through the walls of the glass box. Nichols found that the primary shear planes that were present with the small plow also appeared ahead of the chisels. The normal shear angle of soil reaction to the plow was about 45°. This same angle was obtained with the chisels regardless of the angle or curvature of the chisel; both 45 and 90° chisels were used.

An analysis of the effect of moisture on the resistance to passage of the chisels through the soil showed that maximum pull occurred within the plastic range, slightly above the plastic limit. Since the primary factors affecting the resistance of the soil to the passage of tillage tools are cohesion and adhesion, it is clear why the maximum resistance should be within the plastic range. Maximum pull increased logarithmically with the plasticity number. From the slope of the curve below the plastic limit (Figure 3-11), one can readily see why it is essential to plow within the zone of friability if draft is to be decreased.

Even though these basic principles of draft requirements for pulling chisels through soils are applicable to plows in the field, it should be obvious that the manifold functions of the moldboard plow require additional applications of energy. Söhne (1956) established five categories to describe the work required in plowing:

1. The cutting work accomplished by the coulter and the plowshare.

2. The work necessary to overcome cohesion and shear forces involved in the compressing, shearing, and turning of soils to break up the soil mass into crumbs and clods.

3. The work of lifting and accelerating the furrow slice.

4. The work necessary to overcome friction between soil and moldboard.

5. The work necessary to overcome friction between soil and the land-side and sole of the plow.

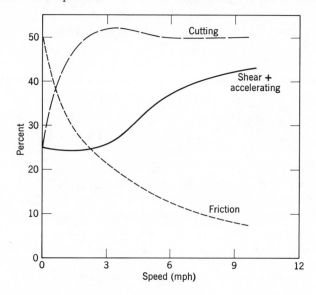

Fɪɢ. 3-23. Relative contributions of cutting, sliding friction, and shear plus accelera- tion to draft of an inclined tool. (After Gill and Vanden Berg, 1967, p. 318.)

Gill and Vanden Berg (1967) presented evidence to indicate the rela- tive contribution of these categories to the draft of the plow. They showed that the kind and position of the coulter with respect to the plow affects the draft considerably. The relative contributions of cutting, sliding friction, and shearing plus acceleration to total draft in relation to speeds of operation are indicated in Figure 3-23. It is seen that cutting and shear plus acceleration components increase with the speed of opera- tion; friction decreases. Cutting represents a high percentage of the re- quired draft. Gill and Vanden Berg called attention to attempts to reduce the friction factor by decreasing surface roughness, using plastic coatings, controlling the moisture content, and changing the character of the moldboard.

The draft of plows under field conditions may be measured by means of specially designed dynamometers. The most notable example of such dynamometer studies has been the work of Keen and Haines (1925) at the Rothamsted Experiment Station. They perfected a special dynamometer for soil work and then studied soil uniformity in relation to draft, as well as the effect of soil treatments on dynamometer pull. In their analysis of the variation in draft throughout a field, points of equal dynamometer pull were connected on a graph to form isodynes. Their isodyne maps therefore contained contour lines of equal dynamometer pull just as a topographic map contains lines of equal elevation.

Perhaps the most significant feature of their findings at Rothamsted was the rather obvious heterogeneity of fields that should have been uniform, in light of their past treatment. The highest resistance was over 40 percent higher than the lowest recorded. Interestingly enough, the location of the isodynes did not change with season. They were definite characteristics of the different portions of the field.

Only small differences were found between the various plot treatments. Annual applications of 14 tons of manure per acre showed some decrease in draft. The most significant decrease in soil resistance was accomplished by an exceptionally heavy application of chalk. The percentage reduction varied between 6 and 13 percent. Applications of 5 tons per acre had little effect.

The roots of live vegetation play an important part in changing the dynamometer pull during plowing. Work is required to cut or tear loose the numerous roots within the plow layer. This fact was evidenced in dynamometer trials at the Ohio Agricultural Experiment Station, when the limed end of a series of plots required more draft to plow than the unlimed. This increased draft was not due to the direct effect of lime on the soil but to the fact that liming had made possible a luxurious growth of a legume-grass mixture.

References

Atterberg, A. (1911). Die Plastizität der Tone. *Int. Mitt. Bodenk.* 1:10–43.
Atterberg, A. (1912). Die Konsistenz und die Bindigkeit der Boden. *Int. Mitt. Bodenk.*, 2:148–189.
Baver, L. D. (1928). The relation of exchangeable cations to the physical properties of soils. *J. Am. Soc. Agron.*, 20:921–941.
Baver, L. D. (1930). The Atterberg consistency constants: Factors affecting their values and a new concept of their significance. *J. Am. Soc. Agron.*, 22:935–948.

Baver, L. D., and Hans Winterkorn (1935). Sorption of liquids by soil colloids: II. Surface behavior in the hydration of clays. *Soil Sci.*, 40:403–419.

Bloor, E. C. (1957). Plasticity: A critical survey. *Trans. Br. Cer. Soc.*, 56:423–481.

Bodman, G. B. (1949). Methods of measuring soil consistency. *Soil Sci.*, 68:37–56.

Bodman, G. B., and J. Rubin (1948). Soil puddling. *Soil Sci. Soc. Am. Proc.*, 13:27–36.

Bodman, G. B., and M. Tamachi (1930). Studies of soils in the plastic state. *Soil Sci.*, 30:175–195.

Bolt, G. H. (1956). Physico-chemical analysis of the compressibility of pure clays. *Geotechnique*, 6:86–93.

Casagrande, A. (1932). Research on the Atterberg limits of soil. *Pub. Roads*, 13:121–130.

Chancellor, W. J., R. H. Schmidt, and W. H. Söhne (1962). Laboratory measurement of soil compaction and plastic flow. *Trans. Am. Soc. Agr. Engr.*, 5:235–239.

Croney, D., and J. D. Coleman (1954). Soil structure in relation to soil suction (pF). *J. Soil Sci.*, 5:75–84.

Davidson, Donald T. (1965). Penetrometer measurements. *Methods of Analysis, Agronomy Monograph* No. 9, Part 1, Academic Press, New York, pp. 472–484.

Farrell, D. A., and E. L. Greacen (1966). Resistance to penetration of fine probes in compressible soil. *Aust. J. Soil Res.*, 4:1–17.

Fisher, R. A. (1926). On the capillary forces in an ideal soil; correction of formulas by W. B. Haines. *J. Agr. Sci.*, 16:492–505.

Fountaine, E. R. (1954). Investigations into the mechanism of soil adhesion. *J. Soil Sci.*, 5:251–263.

Fountaine, E. R., N. J. Brown, and P. C. J. Payne (1956). The measurement of soil workability. *Trans. 6th Int. Cong. Soil Sci.*, 6:495–504.

Frese, H., and H. J. Altemüller (1962). Über einige morphologische Boebachtungen an Pflugsohlen. *Grundlagen der Landtechnik*, 15:10–14.

Gill, William R., and Carl A. Reaves (1956). Compaction patterns of smooth rubber tires. *Agr. Eng.*, 37:677–680.

Gill, William R., and Carl A. Reaves (1957). Relationships of Atterberg limits and cation-exchange capacity to some physical properties of soils. *Soil Sci. Soc. Am. Proc.*, 21:491–494.

Gill, W. R., and Glen E. Vanden Berg (1967). Soil dynamics in tillage and traction. *USDA, Agr. Res. Ser., Agriculture Handbook* No. 316.

Greacen, E. L. (1959). Water content and soil strength. *J. Soil Sci.*, 11:313–333.

Greacen, E. L. (1960). Aggregate strength and soil consistence. *Trans. 7th Int. Cong. Soil Sci.*, Madison, 1:256–264.

Greacen, E. L., D. A. Farrell, and B. Cockroft (1968). Soil resistance to metal probes and plant roots. *Trans. 9th Int. Cong. Soil Sci.*, Adelaide, 1:769–779.

Grim, Ralph E. (1942). Modern concepts of clay materials. *J. Geol.*, 50:225–275.

Grim, Ralph E. (1948). Some fundamental factors influencing the properties of soil materials. *Proc. 2nd Int. Conf. Soil Mech. and Found. Eng.*, 3:8–12.

Haines, W. B. (1923). The volume changes associated with variations of water content in soil. *J. Agr. Sci.*, 13:296–310.

Haines, W. B. (1925). Studies in the physical properties of soils. II. A note on the cohesion developed by capillary forces in an ideal soil. *J. Agr. Sci.*, 15:529–535.

Hendricks, S. B., R. A. Nelson, and L. T. Alexander (1940). Hydration mechanisms of the clay mineral montmorillonite saturated with various cations. *J. Am. Chem. Soc.*, 62:1457–1464.

Hénin, M. S. (1937). Signification des résultats obtenus avec les sondes dynamométriques. *Trans. 6th Com. Int. Soc. Soil Sci. Zurich*, B:461–467.

Jamison, V. C., H. A. Weaver, and I. F. Reed (1950). The distribution of tractor tire compaction effects in Cecil clay. *Soil Sci. Soc. Am. Proc.*, 15:34–37.

Johannsen, S. (1914). Die Konsistenzkurven der Mineralböden. *Int. Mitt. Bodenk.*, 4:418–431.

Keen, B. A., and J. R. H. Coutts (1928). Single value soil properties: A study of the significance of certain soil constants. *J. Agr. Sci.*, 18:740–765.

Keen, B. A., and W. B. Haines (1925). Studies in soil cultivation. *J. Agr. Sci.*, 15:375–406.

Kingery, W. D., and J. Francl (1954). Fundamental study of clay: XIII. Drying behavior and plastic properties. *J. Am. Cer. Soc.*, 37:596–602.

Lambe, T. William (1953). The structure of inorganic soil. *Proc. Am. Soc. Civil Engr.*, Vol. 79, Separate 315.

Lambe, T. William (1958a). The structure of compacted clay. *Proc. Am. Soc. Civil Engr. J. Soil Mech. and Found. Div.*, SM2, Paper 1654, pp. 1–34.

Lambe, T. William (1958b). The engineering behavior of compacted clay. *Proc. Am. Soc. Civil Engr. J. Soil Mech. and Found. Div.*, SM2, Paper 1655, pp. 1–35.

Lambe, T. William (1960). A mechanistic picture of shear strength in clay. *Am. Soc. Civil Engr., Res. Conf. on Shear Strength of Cohesive Soils*, Colorado, pp. 555–580.

Lauritzen, C. W. (1948). Apparent specific volume and shrinkage characteristics of soil material. *Soil Sci.*, 65:155–179.

Lemos, Petezval, and J. F. Lutz (1957). Soil crusting and some factors affecting it. *Soil Sci. Soc. Am. Proc.*, 21:485–491.

McCreery, W. F., and M. L. Nichols (1956). The geometry of disks and soil relationships. *Agr. Eng.*, 37:808–812.

Mellor, J. W. (1922). On the plasticity of clays. *Trans. Faraday Soc.*, 17:354–365.

Michaels, Alan S. (1959). Physico-chemical properties of soils: Soil-water systems. (Discussion). *Proc. Am. Soc. Civil Eng., J. Soil Mechanics and Found. Div.*, SM2, 85:91–102.

Moum, J., and I. Th. Rosenquist (1961). The mechanical properties of montmorillonitic and illitic clays related to the electrolytes of the pore water. *Proc. 5th Int. Conf. Soil Mech. and Found. Eng.*, 1:263–267.

Nichols, M. L. (1929). Methods of research in soil dynamics. *Alabama Agr. Exp. Sta. Bull.* 229.

Nichols, M. L. (1931). The dynamic properties of soil. I. An explanation of the dynamic properties of soils by means of colloidal films. *Agr. Eng.*, 12:259–264.

Nichols, M. L. (1932). The dynamic properties of soil. III. Shear values of uncemented soils. *Agr. Engr.*, 13:201–204.

Nichols, M. L., and L. D. Baver (1930). An interpretation of the physical properties of soil affecting tillage and implement design by means of the Atterberg consistency constants. *Trans. 2nd Int. Cong. Soil Sci.*, 6:175–188.

Nichols, M. L., and I. F. Reed (1934). Soil dynamics. VI. Physical reactions of soils to moldboard surfaces. *Agr. Eng.*, 15:187–190.

Nichols, M. L., I. F. Reed, and C. A. Reaves (1958). Soil reaction: To plowshare design. *Agr. Eng.*, 39:336–339.

Norrish, K. (1954). The swelling of montmorillonite. *Disc. Faraday Soc.*, 18:120–134.

Norton, F. H. (1948). Fundamental study of clay. VIII. A new theory for the plasticity of clay-water masses. *J. Am. Cer. Soc.*, 31:236–241.

Parker, E. R., and Hans Jenny (1945). Water infiltration and related soil properties as affected by cultivation and organic fertilization. *Soil Sci.*, 60:353–376.

Platen, H. V., and H. G. F. Winkler (1958). Plastizität und Thixotropie von fraktionerter Tonmineralen. *Koll. Z.*, 158:3–22.

Reaves, C. A., and M. L. Nichols (1955). Surface soil reaction to pressure. *Agr. Eng.*, 36:813–816.

Reaves, C. A., and A. W. Cooper (1960). Stress distribution in soils under tractor loads. *Agr. Eng.*, 41:20–21.

Reeve, R. C. (1965). Modulus of rupture. *Methods of Soil Analysis. Agronomy Monograph* No. 9, Part I. Academic Press, New York, pp. 466–471.

Richards, L. A. (1953). Modulus of rupture as an index of crusting of soils. *Soil Sci. Soc. Am. Proc.*, 17:321–323.

Russell, J. C. (1928). Report of Committee on Soil Consistency. *Am. Soil Survey Assoc. Bull.* 9, pp. 10–22.

Rosenquist, I. Th. (1959). Physico-chemical properties of soils: Soil-water systems. *Proc. Am. Soc. Civil Eng., J. Soil Mech. and Found. Div.*, SM2, 85:31–53.

Salas, J. A. Jiminiz, and J. M. Serratosa (1953). Compressibility of clays. *Proc. 3rd Int. Conf. Soil Mech. and Found. Eng.*, 1:192–198.

Sallberg, John R. (1965). Shear strength. *Methods of Analysis. Agronomy Monograph* No. 9, Part I, Academic Press, New York, pp. 431–447.

Schwartz, Bernard (1952). Fundamental study of clay: XII. A note on the effect of the surface tension of water on the plasticity of clay. *J. Am. Cer. Soc.*, 35:41–43.

Seed, H. B., J. K. Mitchell, and C. K. Chan (1960). The strength of compacted soils. *Am. Soc. Civil Engr., Res. Conf. on Shear Strength of Cohesive Soils.* Colorado, pp. 877–964.

Shaw, B. T., H. R. Haise, and R. B. Farnsworth (1942). Four years' experience with a soil penetrometer. *Soil Sci. Soc. Am. Proc.,* **7**:48–55.

Skempton, A. W. (1953). The colloidal "activity" of clays. *Proc. 3rd Int. Conf. Soil Mech. and Found. Eng.,* **1**:57–61.

Söhne, Walter (1956). Einige Grundlagen für eine landtechnische Bodenmechanik. *Grundlagen der Landtechnik.* **7**:11–27.

Söhne, Walter (1958). Fundamentals of pressure distribution and soil compaction under tractor tires. *Agr. Eng.,* **39**:276–281.

Sowers, George F. (1965). Consistency. *Methods of Soil Analysis, Agronomy Monograph* No. 9, Part I. Academic Press, New York, pp. 391–399.

Trouse, Albert C., Jr. (1954). Progress in the understanding of the soil compaction problem of our cane lands. *Report 13th Ann. Meeting Haw. Sugar Technologists,* pp. 5–7.

Trouse, A. C., Jr., and L. D. Baver (1965a). Tillage problems in the Hawaiian sugar industry. III. Vehicular traffic and soil compaction. *Tech. Suppl. to Soils Rpt. No. II. Exp. Sta. Hawaiian Sugar Planters' Assn.*

Trouse, A. C., Jr., and L. D. Baver (1965b). Tillage problems in the Hawaiian sugar industry. IV. Seedbed preparation and cultivation. *Tech. Suppl. to Soil Rpt. No. 12, Exp. Sta., Hawaiian Sugar Planter's Assn.*

United States Department of Agriculture (1967). Soil survey laboratory methods and procedures for collecting soil samples. *Soil Conservation Service, Soil Survey Investigations Report* No. 1, pp. 14–15.

Vomocil, J. A., and L. J. Waldron (1962). The effect of moisture content on tensile strength of unsaturated glass bead systems. *Soil Sci. Soc. Am. Proc.,* **26**:409–412.

Weaver, H. A., and V. C. Jamison (1951). Effects of moisture on tractor tire compaction. *Soil Sci.,* **71**:15–23.

White, W. Arthur (1949). The Atterberg plastic limits of clay minerals. *Am. Mineralogist,* **34**:508–512.

White, W. Arthur (1954). Water sorption properties of homionic montmorillonite. *Clays and Clay Minerals, 3rd Nat. Conf.,* pp. 186–204.

Wilson, Stanley D. (1952). Effect of compaction on soil properties. *Proc. Conf. on Soil Stabilization,* MIT, pp. 148–158.

Winterkorn, H. F. (1936). Surface-chemical factors influencing the engineering properties of soils. *Proc. 16th Ann. Meeting Highway Research Board,* pp. 299–301.

Winterkorn, Hans F. (1948). Physico-chemical properties of soils. *Proc. 2nd Int. Conf. Soil Mech. and Found. Eng.,* **1**:23–29.

CHAPTER 4

Soil Structure— Classification and Genesis

DEFINITION

The soil is essentially a three-phase system that contains an almost limitless number of components. These components, broadly categorized into solid, liquid, and gaseous phases, are not distributed at random thrughout the soil profile. Their arrangement is constrained by the sizes and arrangements of the solid phase. The very large number of components of even the solid phase makes an exact detailed characterization of the arrangement of the solid phase impossible. Thus any description of the arrangement of the soil particles is at best a gross oversimplificaton of the actual situation. The clay mineral crystals are often found to be arranged into layered particles of several unit layers in thickness and with some degree of stability. Blackmore and Miller (1961) found that the calcium montmorillonite in their system was organized into packets or "tactoids" of from four to five particles each. The number of particles in such a packet undoubtedly depends upon the past history of the system and the strength of the binding forces; this force is a strong function of the exchangeable cation.

Hence in considering structure, soil particles refer not only to the individual mechanical separates such as sand, silt, and clay but also to the aggregates or structural elements, which have been formed by the aggregation of smaller mechanical fractions. "Particle" therefore refers to any unit that is a part of the makeup of the soil, whether a primary (sand, silt, or clay fraction) or a secondary (aggregate, or tactoid) particle. Consequently, the structure of a soil implies an arrangement of these primary and secondary particles into a certain structural pattern. It necessarily follows that such patterns include the accompanying pore space.

Since many soils are made up of masses of secondary particles (crumbs, granules, aggregates), they have two types of pore space. These are the micropores within the granules and the macropores between the granules. The macropores are often orders of magnitude larger than pores within the granules (see Figure 6-1 and the accompanying discussion).

This concept of soil structure concurs with that of Marshall (1962). It is similar to soil fabric, a term adapted from petrology by both Kubiena (1938, p. 126) and Brewer and Sleeman (1960; also Brewer, 1964, p. 132) to characterize the arrangement of the soil components.

Soil structure implicitly includes some stability factors that impart characteristic structural breakdown patterns to soils that are subject to mechanical disturbance. Characteristic breakdown patterns can result from natural forces in the soil to organize soil particles into blocks or plates or into aggregates or granules. In this regard, Zakharov (1927) defined soil structure as "the very fragments or clods, into which the soil breaks up." Later, the Soil Survey Staff, United States Department of Agriculture (1951, p. 225) defined structure as referring "to the aggregation of primary soil particles, which are separated from adjoining aggregates by surfaces of weakness." This staff has called the individual aggregate (which is separated from adjoining aggregates by surfaces of weakness) a *ped*. Field descriptions are based upon the shape, size, arrangement, distinctness, and durability of the visible peds.

Both the gaseous and liquid phases are dependent upon the volume and geometrical arrangement of the voids in the solid matrix of the soil. This dependence, particularly that of the liquid phase, may be used to characterize the void structure that is formed by the solid phase and the surface area. This is discussed in later sections.

Many sandy soils are referred to as structureless, or as having *single-grain structure*, because few particles adhere to each other when the soil is dry. However, particles do adhere to each other under natural conditions where such soils are moist; such soils exhibit many physical properties that are similar to those of aggregated soils of finer texture. The term single-grain structure is appropriate for some purposes; but wherever such soils are moist, the individual particles do interact with each other similarly to the interacton between soil granules; they should not be considered structureless.

It will be seen from the discussions that follow that the characterization of soil structure at this time is largely qualitative. One can usually obtain only rather crude correlations between a structural classification and any given soil physical property. Quantitatively, about the best we can do at present is to conceive of the soil as a collection of primary particles that are organized into definable aggregates which vary in their size and stability. The size distribution of the aggregates as well as the amount and size distribution of the pore space both between and within the aggregates serve as our only direct quantitative measures of structure.

CLASSIFICATION OF STRUCTURE

Classification Based upon the Size, Shape, and Character of the Aggregates

Zakharov suggested a classification of soil structure based upon the size, shape, and character of the surface of soil aggregates, fragments or clods. The type of structure was distinguished by the main shape of the structural particles. Three principal types were recognized:

1. Cubelike structure, in which the secondary particles are equally developed along the three axes.
2. Prismlike structure, in which the secondary particles are elongated in the direction of the vertical axis.
3. Platelike structure, in which the structural units are shorter in the vertical direction and are developed more in the direction of both horizontal axes.

The structure within each of these principal types was distinguished on the basis of the character of the edges and faces of the aggregates. Two kinds of structural units were recognized: those in which the faces and edges were indistinctly manifested and those in which the faces and edges were clearly defined. For example, the main difference between cloddy and nutty structure was simply that the former had indistinct faces and edges. The fact that structural units generally broke out along certain planes of cleavage added emphasis to the use of the character of the surface as an important criterion for classification purposes.

The Soil Survey Staff has used the basic concepts of Nikiforoff (1941) to develop a classification of soil structure for field descriptions. This classification is based on (1) the type of soil structure, as determined by the general shapes and arrangements of the peds, (2) the class of soil structure, as differentiated by the size of the peds, and (3) the grade of soil strucutre, as determined by the distinctness and durability of the peds. Important details of this classification are given in Table 4-1.

Classifications Based upon Soil Fabric

Kubiena (Table 4-2) and Brewer and Sleeman (Table 4-3) have classified soil structure on the basis of petrographic studies. Several new terminologies have been introduced to develop their concepts of structure.

TABLE 4-1

Classification of Soil Structure according to Soil Survey Staff (1951)

A-Type: Shape and arrangement of peds

	Platelike. Horizontal axes longer than vertical. Arranged around a horizontal plane.	Prismlike. Horizontal axes shorter than vertical. Arranged around vertical line. Vertices angular.		Blocklike—Polyhedral—Spheroidal. Three approximately equal dimensions arranged around a point.			
				Blocklike—Polyhedral. Plane or curved surfaces accommodated to faces of surrounding peds.		Spheroidal—Polyhedral. Plane or curved surfaces not accommodated to faces of surrounding peds.	
		Without rounded caps	With rounded caps	Faces flattened; vertices sharply angular	Mixed rounded, flattened faces; many rounded vertices	Relatively nonporous peds	Porous peds
B-Class: Size of peds	Platy	Prismatic	Columnar	Blocky	Subangular blocky	Granular	Crumb
1. Very fine or very thin	<1 mm	<10 mm	<10 mm	<5 mm	<5 mm	<1 mm	1 mm
2. Fine or thin	1–2 mm	10–20 mm	10–20 mm	5–10 mm	5–10 mm	1–2 mm	1–2 mm
3. Medium	2–5 mm	20–50 mm	10–20 mm	10–20 mm	10–20 mm	2–5 mm	2–5 mm
4. Coarse or thick	5–10 mm	50–100 mm	50–100 mm	20–50 mm	20–50 mm	5–10 mm	
5. Very coarse or very thick	>10 mm	>100 mm	>100 mm	>50 mm	>50 mm	>10 mm	

C-Grade: Durability of peds

0. Structureless: No aggregation or orderly arrangement.

1. Weak: Poorly formed, nondurable, indistinct peds that break into a mixture of a few entire and many broken peds and much unaggregated material.

2. Moderate: Well-formed, moderately durable peds, indistinct in undisturbed soil, that break into many entire and some broken peds but little unaggregated material.

3. Strong: Well-formed, durable, distinct peds, weakly attached to each other, that break almost completely into entire peds.

TABLE 4-2

Classification of Microstructure based upon Arrangement of Fabric (after Kubiena, 1938)

Coating of mineral grains	Arrangement of the "fabric"	Occurrence
Grains not coated	1. Grains embedded loosely in a dense ground mass	Lateritic soils
	2. Grains united by intergranular braces	Chernozems, brown earths, lateritic soils
	3. Intergranular spaces containing loose deposits of flocculent material	Sandy prairie soils
Grains coated	1. Grains cemented in a dense ground mass	Desert crusts Podsol B-horizons
	2. Grains united by intergranular braces	B-horizon of podsolized, brown forest soils
	3. Intergranular spaces empty	B-horizon of iron and humus podsols

TABLE 4-3

Classification of Levels of Soil Structure (adapted from Brewer, 1964)

Structure level	Components described
A. Organization within the basic unit	
1. Plasmic structure	Size, shape, and arrangement of plasma, grains, and associated voids
2. Basic structure	Size, shape, and arrangement of plasma, grains, and voids in primary peds or apedal soil material, excluding pedological features*
3. Primary structure	An integration of the size, shape, and arrangement of all pedological features enclosed in S-matrix† and the basic structure
4. Elementary structure	An integration of a characteristic size, shape, and arrangement of specific pedological features and the basic structure
B. Organization between peds	
1. Secondary structure	Size, shape, and arrangement of the primary peds, their interpedal voids, and associated pedological features in a soil material
2. Tertiary structure	Size, shape, and arrangement of the secondary peds (compound peds resulting from the packing of primary peds), their interpedal voids and associated interpedal pedological features

* These include clay skins, concretions and nodules, krotovinas.
† Same as "ground mass" of Kubiena.

Soil fabric refers to the arrangement of the primary and secondary particles and the voids between them. The *skeleton* includes the individual mineral grains and resistant organic material that are larger than colloidal size. *Plasma* is that soil material that is or has been mobile in the soil. It is highly active and consists primarily of mineral and organic colloids. The arrangement of peds (Table 4-3) exhibits levels of organization based upon the size, shape, and arrangement of the simplest peds. These primary peds, which cannot be divided into smaller peds, become the basic units of description. In the case of soils with no peds (apedal), the whole soil material is the basic unit of description.

Structure in Relation to the Great Soil Groups

Aggregation of Surface Soils. The aggregation of soils formed under different climatic conditions is shown in Figure 4-1 (Baver, 1934). Let

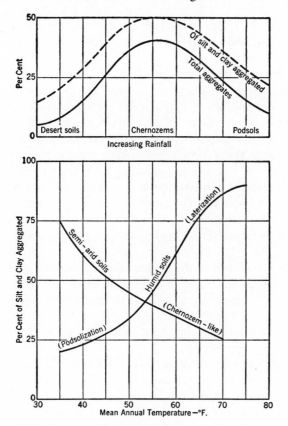

FIG. 4-1. The climatic aspect of soil aggregation.

us discuss the shape of the percentage of aggregates-rainfall curve. The percentage of aggregates is low in desert soils because of a small clay content. Under arid conditions chemical weathering does not proceed very far; consequently a small amount of clay is formed from the clay-forming minerals. This small clay content, even though it may be in the form of secondary particles, causes a relatively small number of aggregates to be present. As the rainfall becomes greater, chemical weathering is intensified and clay formation increases (Jenny and Leonard, 1934). This increase in the clay content of the surface soil obtains until the rainfall becomes great enough to cause eluviation of the clay from the A- to the B-horizon. The decrease in the clay content of the A-horizon diminishes the possibilities for aggregate formation. Not only is the clay content highest in the chernozem and dark humid-prairie regions but the percentage of organic matter in these soils is also largest. As a result of these two factors, the percentage of aggregates in the soil is greatest in the chernozem and dark humid-prairie soils. For example, the amount of aggregates >0.05 mm in podsol, dark humid-prairie, chernozemlike, chestnut-colored, and desert soils is approximately 10, 40, 40, 25, and 10 percent, respectively.

If one considers the percentage of silt and clay which is in the form of stable aggregates, the effect of texture becomes less important. It is seen that podsol soils have a relatively low percentage of the silt and clay aggregated. Only about 25 percent of the silt and clay are aggregated. The dark humid-prairie, the chernozemlike soils, and the chestnut-colored soils have about 50 percent of the total silt and clay aggregated. The low extent of aggregation in the podsol type of soils is probably due to the low organic matter content and the removal of alumina and iron from the A-horizons. The small amount of aggregation is the result of some organic matter which is present in the A_1-horizon. The effect of clay and organic matter on aggregation will be discussed in subsequent paragraphs.

The larger percentage of the silt and clay aggregated in the chernozem soils is due to organic matter and divalent bases (see section on genesis of structure). The presence of sodium in desert soils, together with low amounts of organic matter, causes only a slight aggregation of the small quantities of silt and clay in these soils. Less than 25 percent of the finer fractions are aggregated.

When rainfall is kept constant and temperature is increased, the percentage aggregation of the silt and clay is affected differently in humid and semiarid regions. In the chernozemlike soils of the semiarid regions, the percentage aggregation of the silt and clay decreases from Canada to Texas. Since the calcium content is practically constant, this decrease in aggregation is due to a lowering of the organic matter content with the increasing temperature. This curve is similar to the organic matter-temperature relationship (Jenny, 1930). The aggregation of these soils varies from 75 percent in Canada to 25 percent in Texas.

Similar relationships have been obtained with Russian soils (Savvinov, cited by Kononova, 1961). Aggregation decreases from chernozems to forest steppes to sod-podzolic soils. This is in the direction of vegetative changes from grass to trees as a result of increasing rainfall. In moving from north to south, aggregation follows the order chernozems > southern chernozems > chestnut soils > serozems.

The humid soils present a somewhat different picture. The aggregation varies from about 25 percent in the podsols to 95 percent in the true laterites. The causes associated with the small percentage aggregation of the silt and clay in the podsols have been discussed previously. Aggregation in the laterites is caused by the cementation effects of dehydrated alumina and iron oxides. Thus in lateritic soils alumina and iron contribute to aggregation, whereas in the podsols any secondary-particle formation is due to small amounts of organic matter. Differences in the factors contributing to the building of secondary particles therefore are responsible for the S-shaped curve, which correlates the percentage aggregation of the silt and clay in humid soils with temperature.

Shape and Character of the Aggregates. The surface aggregates of the chernozems are usually spherical or cubelike, have rounded edges, are distinctly pervious, and break up by a crumbling of secondary particles into smaller particles with properties similar to the larger ones. Figure 4-2*a* shows the surface of a ped from a Canadian chernozem; magnification is about 20 times. It is seen that this ped consists of a large number of small, rounded crumbs that seemed to be made up of smaller crumbs of the same shape; that is, it is constituted of secondary particles. This arrangement gives to the eye an impression of softness and perviousness. This type of crumb structure is at a maximum in the surface layers of chernozems.

Crumb structure becomes less pronounced as one passes through the humid prairies to the primarily podzolic forest soils. The peds, especially below the surface layer, become prismlike or platelike. The prismlike aggregates have clearly defined angular edges. The surface of such a ped is depicted in Figure 4-2*b*. It is observed that the ped is dense, compact, and impervious. The angular shape of this type of ped is shown by the small fragment that is breaking off along a definite cleavage plane (lower left-hand corner). There appears to be a distinct single-grained arrangement of the particles within this secondary unit.

Fig. 4-2. Photomicrographs of granular (a) and fragmental (b) types of structure (magnification about 20 times.)

The surface layers of gray forest soils and desert soils tend to have a single-grained to platelike structure. The nature of the peds within a given soil group changes with increasing depth. Zakharov (1927) visualized the structure of the various soil groups as follows:

The upper horizon of a typical loamy chernozem is of a clearly exhibited powder-like and granular structure, owing to the high content of humates, while the lower horizon assumes a nutty structure, which is deeper or replaced by prism-like structural aggregates. The podzolized soils, in their upper horizons, bear a feebly expressed cloddy-pulverescent structure altering under the influence of a higher content of silicic acid in the underlying layer into a platy-pulverescent one; deeper the structure becomes platy, then flat-nutty, and still deeper acquires a prismatic character. In the upper horizon of a brown alkali-soil, we frequently meet with a squamose or foliated structure which, with increasing depth, is suddenly replaced by clearly shaped columnar fragments with rounded upper bases (tops), and changes further into a sharp-faced structure.

Similar changes in the morphology of the soil aggregates with depth occur in heavy clay soils under prairie in Belgium (DeBoodt and DeLeenheer, 1958). The surface layer consists of irregular crumbs that contain fine pores. Total porosity is a maximum; this is true crumb structure. The second layer is composed of irregular, rounded aggregates that are relatively compact and practically nonporous. The crumbs become grains and the structure is termed granular. This granular structure grades into clearly irregular aggregates with rounded edges and corners, which are called subpolyhedric (similar to subangular blocky in Table 4-1). Beneath this layer, the structural elements become larger and break easily into angular aggregates with sharp edges. Porosity is greatly reduced. This is called polyhedric structure (similar to blocky in Table 4-1). The more deteriorated the structure of the soil, the closer the polyhedric forms approach the surface.

Chestnut soils on alluvium in North Dakota (White, 1966, 1967) change from a very fine granular to crumby A-horizon to a fine granular and subangular structure in the upper B-layer. This structure grades into a mixture of medium subangular and prismatic and then to prismatic in the lower B-horizons. Solodized soils on shale have a very fine platy A-horizon that change to medium columnar and blocky in the B-horizons. Parallelepiped faces are present in the lower B-layer.

GENESIS OF SOIL STRUCTURE

The genesis of soil structure refers to the causes and methods of formation of the structural units or aggregates. It has already been shown that an aggregate consists of an intimate grouping of a number of primary particles into a secondary unit. The mechanism of the formation of these aggregates is one of the most important phases of the soil-structure problem.

Since many of the present-day concepts of granulation and aggregate formation are built on *a priori* reasoning from the phenomenon of flocculation in dilute suspension, it is necessary at this time to distinguish between flocculation from the purely colloidal point of view and aggregation from the standpoint of soil structure. Flocculation and stable-aggregate formation are not synonymous. The former is primarily electrokinetic in nature. Primary particles with a high electrokinetic (zeta) potential repel each other when they collide in a suspension. When the potential is lowered sufficiently, a collision between particles results in a mutual attraction and the formation of a floccule. The floccule is stable as long as the flocculating agent is present. This is a "salt-type" of flocculation. It takes place rapidly with divalent or trivalent cations.

As discussed in Chapter 1, flocculation can occur as a result of electrostatic attraction between the positive edges and negative faces of clay minerals (Schofield and Samson, 1954). This edge-to-face type of flocculation can produce a much more stable system than the "salt-type."

Stable-aggregate formation requires that the primary particles be so firmly held together that they do not disperse in water. In other words, from the point of view of soil structure, aggregate formation requires a cementation or binding together of flocculated particles. Thus flocculation may aid in the aggregation process but is not aggregation in itself. As Bradfield (1936) so aptly stated, "Granulation is flocculation plus!" It is necessary to keep this fact in mind to avoid erroneous interpretations of flocculation data in terms of stable-aggregate formation.

The earlier Russian ideas on the genesis of structure considered the activity of root systems and soil fauna as primary factors in aggregate formation (Zakharov, 1927). This concept was then supplemented by the idea that periodical variations in moisture and temperature caused fracturation and fragmentation of the soil mass into aggregates (Gedroiz, cited by Zakharov, 1927; Tiulin, 1933). Aggregate formation depends primarily on coagulation and pressure. Roots of vegetation, and to some extent small burrowing animals, contribute the major pressure effects. This pressure produces more intimate contact between particles, so that

shrinkage

the cementing influences of the water films are rendered more effective. The (swelling) of the soil colloids plays an important role. The effect of swelling is of particular significance in the development of aggregates in the subsurface horizons. Genesis of structure therefore involves the impacts of cations, clay particle interactions in relation to moisture and temperature, clay-organic matter interactions, as well as vegetation, soil fauna, and microorganisms.

Cation Effects

It is almost universally recognized that lime and organic matter improve the physical properties of the soil. Moreover, it has been demonstrated conclusively that the poor structural qualities of alkali soils can be changed into a favorable physical condition if the sodium is replaced by calcium. These facts, along with the laboratory observations that clay suspensions can be flocculated by calcium salts, have led to the widely accepted viewpoint that the beneficial effects of lime are due to its ability to flocculate the soil colloids. The influence of organic matter has been attributed to its cementation effects.

One must distinguish between Ca ion effects on aggregation in alkali soils, where Ca and Na relationships are involved, as compared to acid soils, where Ca and H interactions are dominant.

There is little doubt that the effect of calcium on the reclamation of alkali soils is primarily associated with the more favorable influences of this ion on the properties of the soil colloids, as compared with the effect of the sodium ion. Sufficient evidence has been presented previously to show that Na-saturated soils are more highly hydrated and dispersed and swell more than Ca-saturated soils. The former swell and become impervious; this is not true of Ca soils. Consequently, the reclamation of alkali soils is based upon the removal of sodium ions from the base-exchange complex by replacement with Ca ions. The replaced sodium must then be removed from the soil through leaching. Irrigation waters must not contain appreciable amounts of sodium, but ample quantities of calcium salts should be present. Thus the flocculating ability of calcium is the main factor in the betterment of alkali soils.

The properties of different aklali soils vary, dependent upon flocculation effects that are produced by the presence of excess salts. Even though an abundance of sodium ions is characteristic of all alkali soils, the influence of sodium on the physical properties depends upon the nature of the sodium compounds that are present. For example, white alkali soils contain sodium on the exchange complex and an excess of sodium salts ($NaCl$, Na_2SO_4) in solution. These soils are fairly friable and are not highly dispersed. The reason for this friability is found in the flocculating effects of the excess salts. The presence of sodium salts causes a repression of the "ionization" (thickness of the diffuse double layer) of the adsorbed sodium ions. This causes a decrease in potential, hydration, swelling, and dispersity. The colloidal material is flocculated.

Black alkali soils, however, present a different picture. Rainfall has just been sufficient to leach out the excess salts, and the soils are characterized by a Na-saturated exchange complex. The Na-saturated colloidal material hydrolyzes to form NaOH and Na_2CO_3. Such a system is highly dispersed, highly hydrated, and impervious to water. Its state of deflocculation is responsible for the poor physical properties. The dark color is due to the solubility of small amounts of organic matter because of the Na_2CO_3.

The relation of the adsorbed ions to alkali soil problems may be briefly summarized by the following schematic reactions:

White alkali soil:

$$\text{clay} \bigg\rangle \begin{array}{l} Na \\ Na \end{array} + \begin{array}{l} NaCl \\ Na_2SO_4 \end{array} \rightarrow \text{flocculation and friability}$$

Black alkali soil:

$$\text{clay} \bigg\rangle \begin{array}{l} Na \\ Na \end{array} + \begin{array}{l} NaOH \\ Na_2CO_3 \end{array} \rightarrow \begin{array}{l} \text{deflocculation and poor} \\ \text{physical properties} \end{array}$$

Reclamation of alkali soils:

$$\text{clay} \bigg\rangle \begin{array}{l} Na \\ Na \end{array} + CaSO_4 \rightarrow \text{clay) } Ca + Na_2SO_4 \text{ (leached)}$$

A different situation exists when the effects of calcium on the aggregation of acid soils are thoroughly analyzed. It has been generally accepted in the past that the flocculating effect of the calcium ion is the contributing factor for stable granulation. Experimental observations, however, indicate that the direct effect of the calcium ion on the aggregation of acid soils is not as important as was originally considered. Early experiments on the effect of exchangeable cations on the physical properties of soils indicated that H- and Ca-saturated soils were similar (Baver, 1928). In some cases, H soils were more flocculated than Ca soils; in other instances, the reverse was true. Later studies on the effect of calcium saturation on the properties of a colloidal clay distinctly showed that calcium had to be present in excess of the saturation capacity to bring about better flocculation than the hydrogen ion (Baver, 1929). Ca- and H-saturated clays do not differ much with respect to swelling and permeability (Lutz, 1934). In all instances, however, H clays are slightly more permeable than the corresponding Ca systems.

A statistical analysis of 77 different soils of the United States showed that there was no significant correlation between the amount of exchangeable calcium and granulation (Baver, 1935). Leaching the soils with either $0.01N$ HCl or $0.1N$ $CaCl_2$ and then with water, until they were free from chlorides, so as to give H- and Ca-saturated systems, did not cause dispersion where the calcium ions were replaced by hydrogen.

The fact that no direct effect of calcium upon the physical properties of the inorganic soil colloidal fraction could be obtained suggested that the influence of adsorbed calcium may be associated with the organic colloidal fraction. Experimental investigations, however, showed a close similarity between the properties of Ca and H humates (Baver and Hall, 1937). Organic colloids exhibited about the same differences between the Ca and H systems as clay. It was observed that dried Ca humus was more reversible than H humus, which suggested a greater stability of H humates in water. This fact was substantiated by Myers (1937), who reported that H humate formed more stable aggregates with clay than Ca-saturated humus.

Russell (1934) suggested that aggregate formation is dependent upon an interaction between exchangeable cations on the clay particle and the dispersion liquid. He visualized the clay particle as consisting of a central core with a negatively charged surface. This negative surface attracts water molecules which are oriented on the surface. The exchangeable cations also possess hulls of oriented water molecules. As the cations become more tightly adsorbed on the surface, the water molecules from both the surface and the cation are oriented in a joint field. This orientation is strong, since the negative end of the water dipole is attracted to the cation and the positive end to the surface of the particle. (The major facets of this concept have been discussed in Chapter 1.) This gives a linking system between particles consisting of particle-oriented water molecule–cation-oriented water molecule–particle. One should expect this type of linkage to be stronger with divalent cations such as calcium. This principle has been applied to explain the aggregation of soils by certain polyuronides (Peterson, 1947). The Ca ion becomes the link between the clay particle and the organic polymer. Stable aggregates are formed. Polyvalent cations can serve as bridges to form clay-organic complexes which act as the linking mechanism (Edwards and Bremner, 1967). Bonds in these complexes can be broken by sonic vibrations. The soils that are dispersed by these vibrations reaggregate upon drying. The bridging effect of calcium in conjunction with organic colloids will be discussed in the section on organic matter.

Experimental evidence indicates that the effect of calcium upon aggregate formation is indirect; that is, it affects the production and decomposition of organic matter as well as the mechanisms of binding action between organic colloids and clay particles.

Clay Particle Interactions

The soil colloidal material is responsible for the cementation of primary particles into stable aggregates. Stable aggregate formation cannot take place in sands or silts in the absence of colloids. The soil colloidal material may be divided into at least three distinct groups as far as its cementation effects are concerned. They are clay particles themselves, irreversible or slowly reversible inorganic colloids such as the oxides of iron and alumina, and organic colloids.

Cohesion between Clay Particles

It should be expected that the formation of secondary particles from primary separates would be related to the amount of finer particles in the soil that may serve as material to be aggregated. There is a high degree of correlation between the amount of $5\text{-}\mu$ clay in soils and the percentage of aggregates larger than 0.05 mm in diameter; this correlation for a large number of soils was found to be 0.566, with 0.21 being significant (Baver, 1935). The correlation of aggregation with $1\text{-}\mu$ clay (0.379) was not as significant as that with the $5\text{-}\mu$ fraction. Moreover, the correlation between the percentage of aggregates larger than 0.1 mm and clay was less than in the case of the smaller secondary particles. This correlation for the 5- and $1\text{-}\mu$ clay fractions was 0.323 and 0.23, respectively. Apparently the cementation effects of clay were more pronounced with the smaller aggregates. It was also observed that there was a higher correlation between clay content and aggregation as the amount of organic matter decreased. At the higher percentages of organic matter, the effect of clay in secondary particle formation becomes insignificant. The effect of organic matter is less significant in soils containing large amounts of clay. Polysaccharides produce aggregates of high stability in soils low in clay; the stabilizing effects are small when the clay content is high (Acton, Rennie, and Paul, 1963).

Not only the total quantity of aggregates in a soil but also the extent to which the clay is aggregated into larger secondary units are dependent upon the amount of clay. That is, the percentage of clay that is present in the form of aggregates varies with the amount of clay present. In the aforementioned group of soils there was a correlation of 0.428 (with 0.21 being significant) between the clay content and the percentage of mechanical separates larger than 0.05 mm that were aggregated into secondary particles larger than 0.05 mm.

Clay particles function as binding agents. The particle-oriented water-cation-oriented water–particle linkage proposed by Russell (1934) has already been discussed. He specified that a certain proportion of clay particles must be smaller than 1 μ; this reflects a high specific surface. The clay should also have a sizable base-exchange capacity; this denotes a high activity of the colloidal surfaces. The exchangeable cations must not be large since large organic cations inhibit aggregation. Consequently, as the soil is dehydrated, the number of interparticle links increases; they become stronger because they are shorter; and the cohesion between clay particles is greater. The molecular cohesive forces (primarily van der Waals) between the colloidal surface also exert a major influence in this cementation action. Clay is adsorbed on the surfaces of sand grains (Sideri, 1936); it is very slowly reversible after dehydration. The tenacity of the bonds between the clay and the sand increases with decreasing particle size of the clay (higher specific surface). The clay particles are oriented on the sand surfaces; there can be more than one layer of clay particles adsorbed. Capillary forces are primarily responsible for this orientation. The addition of colloidal clays to quartz sand and orthoclase, followed by dehydration, causes cementation of these primary particles into water-stable aggregates (Myers, 1937). The orientation of particles and desiccation or dehydration play significant roles in secondary particle formation (Hénin, 1938; Schofield, 1943).

The intercrystalline forces between the clay particles themselves may be sufficient to account for all the binding necessary in aggregate formation (Martin et al., 1955). It is significant that this concept concurs with the comments in Chapter 3 on soil cohesion which considered the van der Waals forces adequate to explain cohesion between soil particles. Martin and his associates visualized the contributions of organic matter on aggregate formation as modifying the surface properties of the clay.

Both quartz and clay are main constituents of soil crumbs (Emerson, 1959). Surface crumbs that are continually broken down by raindrops are reformed by interactions between the quartz grains and the dispersed clay. Groups of clay crystals form domains as a result of orientation and electrostatic attraction to each other. These domains function as a single unit. They are bonded to the surface of the quartz grains and to each other to form crumbs. Colloidal organic matter forms complexes between the surface of the quartz grains and the clay domains, thereby increasing the strength of the quartz-clay bonds. The various possibilities of bonding arrangements in crumb formation are illustrated in Figure 4-3.

Electrostatic attractive forces between the positive edges and the negative faces of clay minerals play a significant role in the aggregation of acid soils (Emerson and Dettman, 1960). Trivalent cations also appear to be important in greater clay-clay attractive forces under acid conditions.

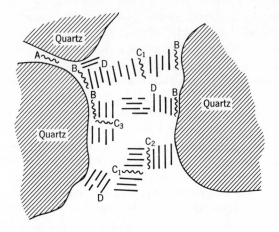

FIG. 4-3. Suggested arrangements of quartz particles, clay domains, and organic matters in a soil aggregate. (A) Quartz-organic colloid-quartz; (B) quartz-organic colloid-clay domain; (C) clay domain-organic colloid-clay domain; (C_1) face-face; (C_2) edge-face; (C_3) edge-edge; (D) clay domain edge-clay domain face. (After Emerson, 1959.)

Thus it is seen that cohesive forces between oriented clay particles are extremely important in aggregate formation. Dehydration is a basic requirement for the formation of secondary particles. Electrostatic and van der Waals forces play the dominant roles in clay-to-clay particle interaction.

Alternate Wetting and Drying

Variations in the moisture content of soils significantly affect both changes in aggregation and the development of different types of structural peds. Common experience has taught us that compact, tightly cemented clods slake down into smaller aggregates as a result of alternate wetting and drying. The farmer likes to have his "clay land" dry out thoroughly and then be rewetted slowly in order to produce a seedbed that has good tilth. What are the causes of this aggregate-forming effect of alternating wetting and drying?

Most experimental evidence shows that drying of soil colloids causes a shrinkage of the soil mass and a cementation of particles. The relationship of shrinkage to moisture content has been discussed in Chapter 3 and illustrated in Figure 3-6. It was observed that the decrease in volume of a soil equalled the volume of water lost down to the point where air entered the soil. Then the shrinkage became much less than the volume of water removed. Let us return to Figure 3-6 and compare the alternate drying and wetting curves, D, of the clay subsoil. It is seen that the volume of the rewetted soil is greater than that of the original. This increase in volume is permanent after complete dehydration and is undoubtedly due to the entrance of air into the soil mass; the air then becomes occluded in the pore spaces. These curves show that the drying process must extend below the point at which air enters the soil before rewetting occurs in order to obtain an increase in porosity.

The dehydration of a soil mass cannot be uniform, especially if the drying process is rapid. Consequently, unequal strains arise throughout the mass which will tend to form clods.

At least two processes operate to cause disruption of the clod into smaller units when the dried clod is wetted. The rapid intake of water causes unequal swelling throughout the clod, which produces fracturation and fragmentation along the cleavage planes. Moreover, the sorption of water into the capillaries results first in a compression of the occluded air and finally in a virtual explosion within the clod, as the pressure of the entrapped air exceeds the cohesion of the particles. The importance of entrapped air in the disruption of large aggregates and clods during the wetting process is supported by the facts that wetting in a vacuum or slow wetting by capillarity does not cause violent disruption of the clod (Yoder, 1936). When the clod is wetted slowly, the air is expelled from the larger pores as the smaller ones take up water by capillarity. If the clod is immersed in water, disintegration into smaller fragments takes place as the air is expelled. This disintegration of the clod is almost concurrent with the rise of air bubbles to the surface of the water.

Hénin (1938) attached much importance to the Yoder concept and suggested a mathematical expression for defining the stability and instability of aggregates and clods in water, in terms of capillary and cohesive forces. The condition for disruption is given by the expression $rC + C_1 < 2A$, where r is the diameter of the largest capillaries in the soil, C is the apparent cohesion of the soil, C_1 is the cohesion of water, and A is the affinity of the soil for water. Disruption is considered to take place only with the larger aggregates that possess capillaries in which the air can be compressed. Hénin considered both swelling and disruption as factors for the disintegration of clays on wetting. It is significant to note that disruption is decreased by the addition of alcohol to water. The amount of aggregates larger than 0.2 mm in diameter is correlated with the affinity of the soil for the liquid rather than with the surface tension of the different alcohol-water mixtures. This fact emphasizes the significance of unequal hydration and swelling in the slaking of desiccated clods.

Two major forces are involved in the water attack on dry cohesive systems (Winterkorn, 1942). First, there is the driving force for the entry of water, which is determined by the affinity of the internal surfaces for water. Second, there are the cohesive forces that hold the particles or aggregates together. As water moves into the system, and the affinity of the soil surfaces for water exceeds the cohesive forces, the soil aggregates lose their cohesion and the cementing bonds are destroyed. The rate of penetration is directly proportional to the affinity of the internal surfaces for water and to the fourth power of the effective radii of the soil pores. Prior to the entry of water, there is both free and adsorbed air in the pores. Water releases the adsorbed air, which is added to the free air in the pores. If there are no escape pores present, pressures build up within the aggregate due to the compression of the air. The size of the pores and the rate at which the cohesive bonds are destroyed in relation to the speed of water penetration determines to a great extent the type of slaking action that occurs. If the rate of bond destruction is equal to or greater than that of water penetration, there is an orderly, progressive slaking of the aggregates. If the rate is less than that of water entry and there is no buildup of air pressure because of the presence of large pores, there is no breakdown of the aggregates. If air is entrapped, however, there is an explosive effect and the aggregates shatter.

This analysis has been supported by several relevant investigations (Mazurak, 1950; Robinson and Page, 1950; Payne, 1954; Emerson and Grundy, 1954). There is no shattering of aggregates if CO_2 is substituted for air in the system because CO_2 is soluble in water and does not develop a gaseous pressure. Synthetic aggregates from kaolin, illite, and montmorillonite break down when wetted in air. When wetted under vacuum, however, there is no slaking of kaolin aggregates, only slight cracking of those from illite and slow but complete disintegration of the montmorillonite units. Apparently, air entrapment is solely responsible for the breakdown of kaolin and illite aggregates. Both air entrapment and differential swelling cause montmorillonite aggregates to disintegrate.

The breakdown of soil crumbs can be caused by (1) dispersion of the cementing material, (2) reduction in cohesion with increasing moisture content, (3) compression of entrapped air, and (4) stresses and strains set up by unequal swelling due to soil heterogeneity and nonuniform wetting. These factors are influenced by the initial moisture content of the soil and the method and speed of wetting. Water is pulled with so much force into the capillaries of soils that are dried at high moisture tensions as to cause an explosive effect that shatters the aggregates (Cernuda, Smith, and Vicente-Chandler, 1954). The initial moisture content of the aggregates is of great importance in determining the extent of the slaking. A linear relationship has been observed between the weakness of soil crumbs ($1/\text{cohesion}^{1/2}$) and the rate of wetting of arable and grassland aggregates in air (Emerson and Grundy, 1954). When these straight lines are extrapolated to zero rate of wetting, which represents no breakdown, the values obtained are the same as those procured with these crumbs when they are wetted under vacuum.

However, there is some evidence to suggest that even though the stability of soil crumbs is increased by wetting under vacuum, differential swelling may be the primary cause of aggregate breakdown (Panabokke and Quirk, 1957). Entrapped air may be just an aid to disintegration rather than the principal cause. Clays that exhibit aggregates and pore spaces that contain strongly oriented clay, either as thin coatings or as streaks bounding the pore spaces and cracks, have been observed to deteriorate even though they are wetted under vacuum (Brewer and Blackmore, 1956). This breakdown is attributed to the swelling of the oriented clay coatings or streaks. In the absence of oriented clay, disintegration is due to entrapped air and swelling of the whole aggregate. The differences in behavior of various clays may be associated with the nature of the fabric of the original aggregates.

Thus there appears to be considerable variation in the reaction of soil aggregates to rapid water entry. It seems that these variations are covered by the concepts of Winterkorn concerning the mechanism of water attack on dry, cohesive soils.

TYPES OF PEDS. Different types of peds in subsoil structures can develop as a result of soil shrinkage and swelling that arise from alternate wetting and drying of the soil material (White, 1966, 1967). As soils dry out, desiccation cracks develop and the interparticle spacings between sand, silt, and clay become shorter. These desiccation cracks are the origin of many ped faces. They form at moisture contents where water is available to plants. For example, as much as 50 percent of the total soil shrinkage can occur at moisture contents above the wilting point.

Wetting of the shrunken soil produces swelling which closes the cracks. Shearing forces develop in the boundary layer between the drier soil, which contains desiccation cracks, and the wetter soil where swelling is taking place. The density of the soil determines the frictional resistance encountered in the movement of soil particles over each other. This frictional resistance is small in soils with low bulk densities and large with high density of packing. In the case of high bulk densities, particles have to move up and over adjacent particles during shear. As discussed in Chapter 3, this causes an increase in volume, called dilatancy. Dilatancy requires less force in clays than in coarse-textured soils. This means that shear planes in clays should be closer together than in coarser material. When the swelling forces exceed those of internal friction, the shear planes transect the vertical faces that are formed during desiccation.

The density of packing increases with depth because of the pressures resulting from the weight of soil above the layer in question. This pressure fluctuates with the moisture content of the overlying soil. Consequently, shear planes should be less frequent and the structural peds should be larger in the deeper subsoil.

White has proposed the following mechanisms in the development of different types of peds in subsoils:

1. *Granular peds.* Swelling proceeds from the outside of the ped to the interior, starting with a concentric surface layer. There is plastic flow instead of shearing. The soil material has a high tensile strength and low swelling.

2. *Parallelepiped peds.* Development occurs in fine-textured soil materials with large total volume change potentials. Swelling exceeds that which is necessary to close desiccation cracks. Extensive shear planes are produced. These peds are generally found in all fine-textured subsoils.

3. *Blocky peds.* Formed under essentially the same conditions as parallelepiped peds except that localized shear planes resulting from swelling and shrinkage play the dominant role. These shear planes intersect during the swelling process. Blocky peds are not found at the deeper levels because overlying pressures prevent formation of localized shear planes.

4. *Prismlike peds.* Prisms are formed where the total volume-change potential is small. They develop from the cracking pattern due to desiccation which sets up horizontal tensions. These cracks form in the subsoil at moisture contents close to the wilting point while the upper horizons are subject to wetting and drying cycles. Thus they are found most frequently in arid and semiarid soils.

5. *Platy peds.* Platelike peds appear to develop under conditions where strong shear forces are absent and there is inadequate aggregation to produce granules.

White has used the total volume change (dry-clod bulk density minus wet-clod bulk density) and the volume increase per unit of water added freezing is beneficial for good soil tilth because of its granulating effects, to characterize swelling and serve as parameters in evaluating different types of peds. His concepts are based upon principles of soil mechanics and are an elaboration of the suggestions of Tiulin (1933) that swelling of soil colloids plays the major role in the formation of aggregates in subsurface horizons.

Alternate Freezing and Thawing

Farmers have long recognized the beneficial effects of freezing and thawing upon soil tilth when clay soils are plowed in the fall or late winter. Fall-plowed clays are friable and easily tilled in the spring. However, freezing and thawing may not always produce highly aggregated soils in the spring. The structure formed depends on the soil type and associated properties, the conditions of freezing and thawing, and the water content at the time of freezing.

Freezing may cause either aggregation or dispersion (Jung, 1931). The nature of crystallization of the ice is the determining factor. Crystallization is influenced by the rapidity of cooling. Large ice crystals form in the tension-free pore spaces with slow cooling. These crystals partially melt during thawing and then serve as nuclei for further freezing. Since there is about a 9 percent increase in volume when water changes to ice, the pores are enlarged, which promotes a loosening effect upon the soil. Water is drawn from around the clay particles to the ice crystals during this process, creating a dehydration effect. The combination of ice-crystal pressure and dehydration causes aggregation. On the other hand, large numbers of small crystals are formed if the cooling is rapid and the integrated effect of many expansions is the breakdown of soil aggregates.

Freezing of coarse-textured soils produces a homogeneous mass without microscopically visible different structures (Ceratzki, 1956). When fine-grained soils are frozen, ice crystallizes in layers within the soil material to give a layered or heterogeneous *Froststruktur* (frost structure). Water is drawn from the lower layers into the freezing zone to cause water saturation. This increased water movement to the freezing zone is stronger the larger the capillary conductivity. Therefore it is more pronounced in coarser-textured soils than in clays. Ceratzki observed a doubling of the water content in this layer in loess and loam soils with a sufficient water supply and the correct freezing conditions. This frost structure can remain in the frost-free soil if the aggregates are stable and can withstand the thawing of the overly wet soil. Excessive water can destroy not only the aggregates produced by frost action but also those formed by mechanical and biological processes. Even though the first action of one must consider the condition of the soil after thawing. The impact of thawing varies depending upon the amount of excess water in the frozen layer, the duration of the thawing, and the water stability of the aggregates formed.

Slater and Hopp (1949) challenged the concept of improving structure by alternate freezing and thawing. They found that repeated freezing and thawing decreased the water stability of aggregates, especially at the higher moisture contents. Although fall-plowed clay soils became more friable in the spring as a result of frost action on dense clods, they postulated that the bonds within the aggregates were destroyed and there was little water stability remaining to maintain good soil tilth during the growing season. The soil moisture content at the time of freezing, the effects of freezing and thawing at a given soil moisture, and the interaction between freezing and thawing cycles and moisture must be considered in evaluating freezing effect on soil structure (Logsdail and Webber, 1959). Freezing and thawing have been observed to decrease the mean-weight diameter of 2 to 3 mm natural soil aggregates (Sillanpää and Webber, 1961). Although fast freezing increases the mean-weight diameter of crushed aggregates <0.25 mm when they are dried at ⅓ and 15 atm, there is no difference between fast and slow freezing at saturation. The increase in aggregation of smaller aggregates by slow freezing apparently is the result of the formation of large ice crystals in the tension-free pore space. The crystal-growth pressure and dehydration produces stable aggregates.

Even though the concept may be correct in many cases that the two-way action of freezing and thawing on soil structure is of no great agricultural significance, one must not underrate the beneficial aspect of freezing (Ceratzki, 1956). The nature of the soil plays a significant role in the effects produced by freezing and thawing. The structure of a clay soil is not destroyed by thawing, although a loessial soil that is oversaturated is completely dispersed by freezing. A black earth exhibits characteristic horizontal "frost structure" in the spring along with remnants of crumb structure produced the previous fall, whereas a degraded brown earth shows no evidence of the fall structure because of overwetting during freezing. Compacted layers of clay loams are changed by freezing to good polyhedric structure which breaks up into polyhedric aggregates upon drying. The mellowing effect of freezing on free-lying clods of plowed soils originates with the formation of irregular aggregates in the surface because of ice bands (Ceratzki, 1956). They decompose in the spring to gritty aggregates as a result of freezing and thawing as well as drying on sunny days. Large, sharply angular, rhomboidal aggregates that are linked to each other are found immediately under the fine gritty material. Thus the clod is mellowed by freezing; further mechanical working must be done to form good tilth.

One must keep in mind that the aggregates produced by freezing and thawing are similar to those formed by alternate wetting and drying. They are temporary in nature and not too stable unless sufficient organic matter is present to stabilize them. The major beneficial effects of freezing and thawing are most evident during the spring season on soils that contain appreciable amounts of organic matter.

Iron and Aluminum Colloids

Experience in the chemistry of ferric hydroxide has shown that this hydrated colloid becomes almost completely irreversible upon dehydration. There is sufficient evidence to suggest that this irreversibility of colloidal iron hydroxide is the important factor in the production of stable aggregates in certain soils. This is especially true in lateritic soils, which are known both for their high degree of aggregation and for their large iron content. As much as 71 percent of the total particles in the lateritic Nipe clay from Puerto Rico consists of aggregates larger than 0.05 mm in diameter. As much as 95 percent of the total silt and clay is aggregated into units larger than silt.

The lateritic Davidson colloid is flocculated, irrespective of the nature of the cation on the exchange complex (Lutz, 1934). The colloid is only weakly hydrated. A very close relationship exists between the amount of "free" iron and the quantity of water-stable aggregates. This iron may serve a dual purpose in aggregation. That part in solution may act as a flocculating agent, and the other part, which is more gelatinous in nature, may exert a cementation action. Most of the iron is undoubtedly precipitated as a hydrated gel at the pH value of the soils that were investigated. Dehydration of these gels should form a good cement for binding the flocculated particles together.

The importance of sesquioxide-humus reactions in the formation of stable aggregates has been emphasized by many investigators (Tiulin and Korovkina, 1950; Beutelspacher, 1955; Filippovich, 1956; Alexsandrova and Nad', 1958). These effects will be discussed in the section on organic matter. Beutelspacher stated that hydrated colloidal ferric hydroxide adhered to clay minerals and cannot be removed mechanically. If ferrous salts are added to kaolin and then oxidized in air, the metal cation is adsorbed first to form crystallization nuclei of colloidal dimensions which are tightly held by the clay mineral. Humic acid then can be adsorbed.

Sesquioxide coatings on quartz grains increase the strength of clay-quartz bonds (Emerson, 1956). The aggregation of terra rosa and rendzina soils has been attributed to the cementation effects of precipitated and dehydrated, irreversible ferric oxide gels (McIntyre, 1956). This may be the result of the formation of organic mineral compounds by the interaction of humic acids and free sesquioxides. A high correlation has been observed between free iron oxides and the degree of aggregation (Chesters, Attoe, and Allen, 1957; Arca and Weed, 1966). Mixing Fe-oxyhydrate and humic acid to a silt-clay system has resulted in the formation of highly stable aggregates (Walter, 1965). Aging of the mixture not only increases the stability of the aggregates but also the quantity. Ca-, Mg-, and H-saturated silt-clay mixtures show positive microaggregate formation after joint adsorption of the Fe-humus complex. Divalent and trivalent cations increase the adsorption of polysaccharides (Saini and McLean, 1956). The amount of adsorption by Al- and Fe-kaolinites is about 3.3 times that for the Na-system; H-, Ca-, and Mg-kaolinites about 1.9 times as much. Aluminum bonding may contribute substantially to stable aggregate formation in acidic soils that contain 2:1 lattice-type clay minerals (Edwards and Bremner, 1967).

It is possible to change the properties of such a highly hydrated colloid as bentonite by treating with iron salts and then dehydrating the system. The bentonite assumes physical properties similar to those of lateritic colloids. The B-horizon of true podsols is usually aggregated to some extent as a result of the iron and humus compounds that have migrated from the upper layers. Cementation effects of these colloids are responsible for the binding together of the sand and silt grains generally predominant in these soils.

One of the most impressive effects of iron on soil aggregation takes place during the practice of flood fallowing in sugar cane production in Guyana. After the crop is harvested, the heavy marine clays are flooded to a depth of 3 to 12 in. for a period of 6 to 12 months. The anaerobic conditions thus produced reduce the iron and make it soluble. The water is then removed and the ferrous iron oxidizes to the ferric state. The surface soil is characterized by ferric iron streaks between the cleavage planes. An excellent granular structure is produced.

Organic Matter Effects

It has been recognized for a long time that organic matter serves as a granulating agent in soils. Baver (1935) observed a correlation of 0.559 (with 0.21 being significant) between the percentage of aggregates larger than 0.05 mm and the carbon content of a large number of different soils. This correlation for aggregates larger than 0.1 mm was 0.687, which indicates that organic matter is conducive to the formation of relatively large stable aggregates. If various soils are grouped according to their clay contents, the effect of organic matter is more pronounced in those soils containing the smaller amounts of clay. A very high correlation exists between organic matter and aggregation in soils containing less than 25 percent clay. For clay contents above 35 percent the correlation is significant but not nearly as high. The extent to which the finer mechanical separates are aggregated is also significantly correlated with the percentage of carbon.

Both clay and organic colloids are responsible for the major portion of soil aggregation. This suggests the possibility of interactions between the mineral and organic colloidal material to form clay-organic complexes. Greenland (1965) has reviewed the results of various investigators which showed that from 51.6 to 97.8 percent of the total soil carbon was in the form of these complexes.

It is significant to note that, even though desert soils are poorly aggregated, the amount of aggregates that is present is correlated with the small quantities of organic matter. The only group of soils in which a correlation has not been observed between organic matter and aggregation is oxisols, where dehydrated oxides of alumina and iron are responsible for stable aggregate formation.

Colloidal organic matter is more effective than clay in causing the formation of stable aggregates with sand (Demolon and Hénin, 1932) The addition of colloidal humus to quartz sand caused 71 and 94 percent of the sand to be aggregated for the Ca- and H-saturated systems, respectively, as compared with 28.5 and 33.5 percent, respectively, when colloidal clay was added (Myers, 1937). The mixing of 8 percent of colloidal humus to colloidal clay increased the aggregation of sand by about 25 percent over that produced by clay alone. Dehydration was essential to aggregate formation.

These beneficial effects on soil aggregation originate from the integrated activity of microorganisms, fauna, and vegetation.

Microbial Activity

The incorporation of organic matter in the soil brings into the picture the activities of soil microorganisms—fungi, actinomycetes, bacteria, and yeasts. Organic material itself without biological transformation has little if any effect on soil structure. Microorganisms without organic materials as sources of energy are ineffective in producing soil aggregation. Intense microbial activity takes place after the incorporation of organic materials to the soil. There is a large increase in the population of microorganisms. Fungi and actinomycetes produce mycelia. The metabolic processes of the microorganisms synthesize complex organic molecules. Decomposition products of the breakdown of organic materials are left in the soil. The result is the production of stable soil aggregates. Thus stability can result from (1) the mechanical binding action of the cells and filaments of the organisms, (2) the cementation effects of the products of microbial synthesis, or (3) the stabilizing action of the products of decomposition, acting individually or in combination.

DIRECT EFFECTS. Much research has been done to evaluate the aggregating effects of various soil microorganisms, studied both in pure and mixed cultures. Practically all of these investigations showed that the stabilization of aggregation depended upon the type of organism, the species within a given type, and the presence of both a source of energy and nitrogen for microbial metabolism. The results of several investigators are summarized in Table 4-4. Fungi are ranked first in their stabilizing

TABLE 4-4

The Ranking of Types of Microorganisms with Respect to the Stabilization of Soil Aggregates

| | Fungi | Actinomycetes | Bacteria | | Yeasts |
			Gum-producing	Other	
Peele and Beale (1940)	1	—	2	2	—
Martin (1945)	1	—	1	2	—
McCalla (1946)	1	2	3	5	4
Swaby (1949b)	1	2	2	3	3
Müller (1958)	1	—	2	2	—
Watson and Stojanovic (1965)	1	2	1	3	—

effects, although Martin (1945) found a strain of Bacillus subtilus that was equal to his best fungus, Cladosporium. Certain gum-producing bacteria are equal to the fungi; both are superior to actinomycetes (Watson and Stojanovic, 1965). The fungi and actinomycetes cause a mechanical binding of the aggregates by the mycelia that they produce. They also form products that have a stabilizing influence. The bacteria produce gums that have a cementation action. Martin estimated that about 50 percent of the stabilizing effect of fungi is due to mechanical binding and 50 percent to products of synthesis or decomposition. About 80 percent of the stabilizing influence of the bacteria is attributed to products of metabolism and only 20 percent to the cell bodies. The stabilizing effect of fungi on aggregation is directly related to the length of the mycelia (Swaby, 1949b). The impact of fungi and actinomycetes on soil aggregation is temporary and disappears as a result of bacterial decomposition of the mycelia. The stabilization of aggregates by a fungus apparently occurs only in the presence of sucrose as an energy source even though there is little difference in fungal numbers between sucrose-treated and nonsucrose soils (Harris et al., 1964). Fungi stabilize the larger aggregates, but this does not occur until there is mycelia growth of the fungus. At early stages of incubation bacteria are more important than fungi in the initial stabilization of small aggregates. Under anaerobic conditions, bacteria produce aggregates that are stable for a longer period of time than under an aerobic state. Fungi are not operative under anaerobic conditions and cannot stabilize the structure. Moreover, since the anaerobic bacteria cannot use the microbial products of sucrose metabolism as a source of energy, the cementation actions of these products are not subject to microbial attack. Aggregate stabilization seems to be a function of microbial synthesis of soil-cementing agents rather than the physical presence of bacteria or fungi in the soil. It appears therefore that the products of microbial metabolism are the major cause for stabilizing soil aggregates except for the mechanical binding effects of fungi mycelia in the early stages of organic matter decomposition.

PRODUCTS OF MICROBIAL SYNTHESIS. What is the nature of the products of microbial synthesis? Numerous investigators have attributed the major role of soil aggregate stabilization to the cementation action of polysaccharides (see Greenland, 1965, for a review). The polysaccharide theory is based upon several scientific observations:

1. When polysaccharides are added to the soil, there is an increase in the amount and stability of the aggregates.
2. There are many organisms in the soil that produce polysaccharides during the process of metabolizing energy sources.
3. Polysaccharides can be extracted from soils. Estimates have been made to indicate that 5 to 20 percent of the soil organic matter consists of plant or microbial polysaccharides.
4. There is a statistical correlation between aggregation and the polysaccharide content of the soil.

5. When polysaccharides are removed from soil by chemical treatment, there is a decrease in aggregation.

These polysaccharides are a mixture of large, linear, flexible polymers. They have large numbers of OH groups in the chain. They vary in their content of amino, carboxyl, phenolic, or other groups. They are subject to decomposition by bacterial action inasmuch as they can be used as a source of energy. This fact is well substantiated by the data of Harris and his associates (1963), who observed that the addition of nitrogen to soils treated with polysaccharides and then incubated reduced the stable aggregation after 20 weeks from 92 percent (no N added) to 35 percent. They can be considered as transitory products in the organic matter cycle, dependent upon the ecological conditions that favor active microbial activity. The impact of polysaccharides on the stability of soil aggregation can be evaluated by chemical treatment with sodium periodate.

Greenland, Lindstrom, and Quirk (1962) investigated the effect of destroying the polysaccharides in soils on the stability of aggregates. They used a modified technique of Dettman and Emerson (1959) in which the original permeability P_1 of Na-saturated aggregates to $0.05N$ NaCl was measured. The change in permeability of the control NaCl $+ Na_2B_4O_7$) and the periodate-treated soils ($NaIO_4 + Na_2B_4O_7$) at various times P_t was determined. The curves shown in Figure 4-4 present the results for 2-yr, 4-yr, and permanent pasture plots. It is seen that there was a slight decrease in stability of aggregation by treatment with sodium borate (curves 1A, 2A, 3A). This decrease was the highest with the 2-yr pasture. The large decreases in stability occurred with the periodate treatments, especially with the soils not under permanent pasture. In the rotation with only 2 yr of pasture, the stability of the aggregates was reduced by over 90 percent after 24–32 hr of leaching (curves 3A, 3B). The aggregates from soils under continuous wheat were completely destroyed by periodate treatment after 8 hr of leaching. On the other hand, the stability of the aggregates from the permanent pasture was reduced only about 30 percent by the periodate treatment. The data indicate that whereas polysaccharides appeared to be the major cementation agents in cultivated soils, other factors contributed to the stability of aggregates in old grasslands. Mehta and his co-workers (1960) could not destroy the aggregate stability of a brown earth forest soil with periodate, which indicated that polysaccharides were not the stabilizing medium. The stability was destroyed, however, by oxidation with chlorine dioxide (ClO_2), which does not oxidize polysaccharides but does oxidize lignin and humic compounds. Stability also deteriorated upon

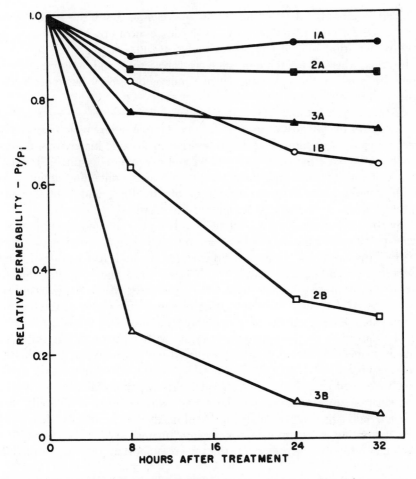

Fɪɢ. 4-4. Changes in relative permeability of soils with time. (A) After treatment with 0.02N NaCl + 0.2N Na$_2$B$_4$O$_7$; (B) after treatment with 0.02N NaIO$_4$ + 0.2N Na$_2$B$_4$O$_7$; (1) permanent pasture; (2) after 4 yr pasture in WWPPPP rotation; (3) after 2 yr pasture in WPPF rotation. (W = wheat; P = pasture; F = fallow.) (After Greenland, Lindstrom, and Quirk, 1962.)

soaking 7 days with sodium pyrophosphate, Na-EDTA, and NaOH followed by wet sieving in water or by treating with NH$_4$F, which caused partial destruction of the crystal. Clapp and Emerson (1965) also observed that periodate treatment of Minnesota grassland crumbs did not destroy the stability of the aggregates. They reasoned that the organic polymers were probably coordinated with polyvalent cations and the edges of the clay crystals. Pretreatment with sodium pyrophosphate before periodate oxidation caused a destruction of the aggregates. Thus the majority of the evidence confirms that polysaccharides contribute a major role in the stabilization of soil aggregates under a wide variety of conditions. However, there are situations such as forest soils and long-established grasslands where the stabilizing agent is not as transitory but more stable than polysaccharides.

PRODUCTS OF MICROBIAL DECOMPOSITION. Humic acids, which are fairly stable products of biological decomposition of organic materials, are generally extracted from soils with NaOH and precipitated with HCl. They are a mixture of high-molecular-weight polymers containing amino acid and phenolic groups. Electron micrographs by Flaig and Beutelspacher (1951) have shown them to be spherical in shape.

Swaby (1949a) separated α-humic acid from the soil and found it superior to all other humic fractions in soil aggregation. Fulvic acid, that part of the extracted humus not precipitated by acid, had a poor binding action. The addition of $CaCl_2$ or HCl improved the structure due to greater fixation of α-humate within the aggregates. Microscopic examination showed that the humus was either present as films around the macro-aggregates or as particles of precipitated humus. The humate film had to be spread over the soil before fixing with electrolytes in order to obtain cementing action. Swaby postulated the possibility of polar bonding between the NH_2, COOH, and OH groups of the humus with negatively charged clay. He treated the soil with different chemicals to obtain information on the nature of the groups on the organic colloids that are active in aggregate stabilization. Acid or alkaline hydrolysis destroyed the cementing substances in the organic materials; this suggested that colloidal proteins or polysaccharides were important in the formation of aggregates.

Ether extraction had no effect, thereby eliminating gums, fats, and waxes as active agents. Esterification had only a slight effect; this indicated that the COOH groups of proteins or polyuronides were of only small importance in polar bonding. Deamination reduced aggregation, which again suggested the binding action of proteins or aminopolysaccharides. Acetylation lowered the cementation effects, which indicated that phenolic or alcoholic OH groups were important. Oxidation decreased cementation, which suggested that ligninlike colloidal substances were important in aggregation. As a result of these chemical treatments, Swaby concluded that amino-polyuronides, polysaccharides, proteins, and ligninlike colloidal material all have cementation action. There is the possibility that fats, resins, and waxes may arrest slaking by making the aggregates waterproof and therefore increasing their stability. However, it is the cementation action of the foregoing complex organic compounds that are of major importance.

Mechanisms in Clay-Organic Complex Formation

In analyzing the mechanisms involved in the formation of clay-organic complexes, one must consider the nature of both the organic and clay mineral colloids. It has been shown that the organic matter phase consists of high-molecular-weight polymers of varying composition and shape. The polysaccharides, polyuronides, and so on, are mixtures of linear, flexible polymers. Humic acid is a mixture of spherical or globular polymers. There are two types of clay mineral of importance. One consists of clay minerals with mica-type lattices in which two silica sheets are held together by an alumina sheet. In this group are minerals such as montmorillonite which have expanding lattices; there also are the illites in which the lattices do not expand. Kaolinite is another type of clay mineral that has one silica and one alumina sheet. The crystal-lattice constitution and properties of these minerals have been thoroughly discussed in Chapter 1.

Polysaccharides are strongly adsorbed on the surfaces of the mica-type lattice clays. There is adsorption both on the external and internal surfaces of expanding-lattice montmorillonites. Those adsorbed on internal surfaces form interlamellar complexes. Polysaccharides with carboxyl groups do not form such complexes. They are adsorbed on the edges of the clay crystals. This is particularly true with kaolinite-type clay minerals. The energy with which these large, linear, and flexible molecules are adsorbed to the surface depends upon their flexibility, which determines the number of points of contact that can be made with the surfaces. The van der Waals forces play an important role in the adsorption process. They become more significant as dehydration brings the surfaces close together. There can be a hydrogen-bond linkage between the polymer and the oxygen atoms of the crystal. This means that the adsorbed polymer replaces water on the surface of the clay mineral. The surfaces are now covered with hydrocarbon chains and are most difficult to rewet. Polymers can be coordinated with exchangeable cations to give similar effects. Clapp and Emerson (1965) suggested that the stable aggregation of Minnesota soils was due to two kinds of polymers, one coordinated with exchangeable cations and the other H-bonded.

Humic acids are not adsorbed by clay minerals through processes of physical adsorption. They are negatively charged and should be repelled by negatively charged surfaces and attracted to positively charged sites. They do not form interlamellar complexes with expanding-lattice-type clays. There is evidence to suggest that they are adsorbed on positively charged sites. Beutelspacher (1955) observed in electron micrographs that they were adsorbed only on the edges of kaolinite, which are positive. They can be coordinated with exchangeable divalent and trivalent cations, in which these cations become bridges between the clay surfaces and the humic acid (see Greenland, 1965). Beutelspacher concluded that humic acid is not adsorbed on montmorillonite unless the surface has been activated by aluminum or iron oxides, which serve as bridges between the clay and the humus. Ca and Fe cause the humic acid to be precipitated on the surfaces of the clay (Flaig, 1958; Swaby, 1949a). Dehydration then causes the film to be more or less fixed on the surface. Tiulin and Korovkina (1950) divided aggregates into two groups that depended upon their association with divalent (Ca and Mg) or trivalent (Al and Fe) cations. The first group contains microaggregates in which the clay-organic matter complexes are saturated with Ca and Mg. The second group consists of aggregates in which the soluble oxides of Al and Fe are deposited on the surfaces of clay particles and form complexes with humus. Stability of soil aggregation is directly related to the proportion of the second group of aggregates. The accumulation of humus in soils may not improve structure unless colloidal, hydrated iron oxide is present (Filippovich, 1956). The beneficial effects of $CaCO_3$ on soil structure depends on the conversion of soluble iron salts into insoluble, hydrated, colloidal iron oxide. It is also possible that a reaction takes place between soluble humates and forms of Al and Fe hydroxides to produce complexes that are fixed upon clay surfaces by dehydration rather than by chemical linkage (Alexsandrova and Nad', 1958).

If the assumption is made that organic colloids can be linked to mineral surfaces by Al and Fe bonds, there appears to be more humus adsorption by soil colloids from igneous rocks than from basic rocks (Harradine and Jenny, 1958). This increase in humus substances in acid soils with rising rainfall is associated with a higher Al content, which accompanies the leaching of bases. Carboxylated polymers can be linked to the edges of montmorillonite crystals through Al ions (Emerson, 1963). Small quantities of complex Al ions on the basal surfaces of Na-montmorillonite also forms links with the polymers.

By way of summary, the mechanisms by which organic colloids stabilize soil structure can be attributed to the bonding of organic polymers to clay surfaces by (1) cation bridges, (2) hydrogen bonding, (3) van der Waals forces, and (4) sesquioxide-humus complexes. The major impact of these interactions is the changing of the properties of colloidal clay surfaces with respect to water. The organic colloids compete with water molecules for space on the surfaces, reduce wetting and swelling, and increase the strength of the aggregates through cementation effects.

Vegetation Effects

Vegetation is an important parameter in the genesis of soil structure. First, it produces residues that are the source of energy for microbial activity in the formation of humic compounds in the soil. Second, the root system not only contributes to the amount of residues produced but is a significant factor in the formation of stable aggregates. Third, the vegetative canopy has a protective effect on the stability of surface aggregates against the destructive action of raindrops.

ROOT INFLUENCES. Practical experience and scientific investigations have clearly demonstrated that sod crops promote stable granulation. Every student of soils should be familiar with the high state of granulation of chernozems and humid prairie soils. This was obvious in the curves in Figure 4-1. The entire soil mass of a bluegrass sod is penetrated by countless roots. Rounded crumbs are observed to be literally enmeshed by the root system. This condition approaches the ideal for stable granulation.

The exact mechanism of aggregate formation by plant root systems has not been established. The earliest explanation was based upon pressures exerted by growing roots which cause a separation of the soil particles adjacent to the root and a pressing together of these units into aggregates. In other words, each root hair that penetrates into large aggregates introduces a point of weakness. The penetration of sufficient root hairs produces granules. However, attempts to form aggregates by the application of artificial pressure have failed to duplicate the water stability of natural field aggregates (Rogowski and Kirkham, 1962). This suggests that mechanical pressures of roots or soil fauna may not be basic aggregating mechanisms. Another ingredient is necessary to stabilize aggregates that are temporarily formed under pressure.

Another possible aggregating factor is the dehydration of the soil in the vicinity of the root system as water is absorbed by the plant. This produces localized shrinkage and the formation of surfaces of fracture. The high state of aggregation of soils under grass sods has been attributed to the desiccation action of the extensive root system, which creates an alternate drying and wetting effect (Kolodny and Neal, 1940). Corn roots have been found to penetrate the claypan of Putnam soil and extend to depths of 42 to 48 in., provided that the amount of rainfall through the growing season is less than the amount of water which is drawn from the soil by evapotranspiration processes (Woodruff, 1968). Removal of water from the claypan causes shrinkage, which permits air to enter the clay and roots to follow. Optimum performance of corn on these soils is attained when the silt loam surface is maintained at low moisture tension by sprinkler irrigations in amounts that do not wet the soil beneath the surface horizon.

It is also possible that roots secrete substances that cement soil aggregates. This is an area practically devoid of experimental evidence. Egawa and Sekiya (1956) observed that there was more soluble humus in the root zone of well-aggregated grasslands. They suggested that the formation of large aggregates from smaller ones was due to cementing substances produced by roots and accompanying bacteria.

A fourth possibility is the binding action of products of microbiological conversion of plant-root secretions and the residues of the growing root itself. Gel'tser (1955) suggested that the formation of active humus and stable soil aggregates was primarily the result of the activity of bacteria in the rhizosphere of growing plants. Drying root hairs together with organic and inorganic root secretions supposedly provide the energy for this microbial activity. The plowing of sod crops and the subsequent decomposition of the dead roots does not cause further aggregation because the active humus is energetically decomposed by further microbial action. The extent of soil aggregation is dependent upon the length of time the plant growth is uninterrupted. This concept may have considerable merit if judged by the aggregate-stability curve for the prairie grass in Figure 4-6 and the soil structure profiles in Figure 5-6. Chernozems and humid prairie soils have excellent aggregation. Each year during their development a new root system was formed and the old one decomposed. The perennial grass had a long life on the soil. Root secretion per se, the decomposition of roots by bacteria in the rhizosphere, and alternate wetting and drying could all have played a part in this good structure.

It is significant to note that the soil particles within the volume occupied by the roots of rice plants are aggregated, even though the land is flooded. It seems obvious that localized dehydration cannot be important. It is doubtful if root pressure is significant. Root secretions might be operative. However, since the rice plant has a built-in mechanism to supply air to the submerged roots, it is possible that this aeration oxidizes the reduced iron which forms and stabilizes soil aggregates.

Earthworm Activities

Earthworms ingest and intermix soil and partially decomposed organic matter, excrete the ingested material as surface casts or subsurface deposits, and form a system of burrows that permeate the soil. The integrated effects of these activities promote good soil structure. Darwin (1881) published an interesting treatise on the habits and activities of earthworms. He estimated the annual worm-cast production on four different soils to range between 7.5 and 18.12 tons per acre. About 0.2 in. of soil was brought to the surface annually; this would cover the land to a depth of 1 to 1.5 in. in 10 years. Wollny (1890) first showed the beneficial effects of earthworm activity on soil structure. He obtained 5.9 to 31.2 percent increases in aeration porosity. The earthworm burrows raised the permeability of the soil to both air and water.

Dawson (1947) measured the stability of soil aggregates formed by earthworms by using the water-drop technique. Aggregates produced

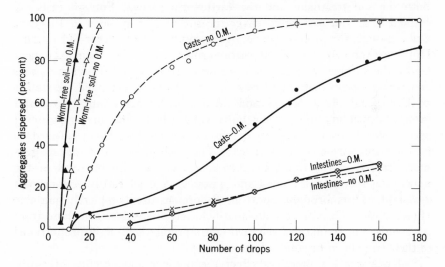

FIG. 4-5. Dispersibility of soil aggregates from earthworm intestines, excreted casts, and worm-free soil, with and without organic matter (O.M.). (Smooth curves after Dawson, 1947.)

within the intestines of the earthworms, in earthworm casts, and in worm-free soil were observed. All experiments were conducted with and without additions of organic matter (lespedeza hay). The results in Figure 4-5 indicate that maximum stability was exhibited by the aggregates from the intestines. No effect of organic matter on the stability of these aggregates was observed. Organic matter, however, had a significant influence on the stability of earthworm casts. There was a considerable increase in the number of bacteria in the casts as compared with the aggregates from the intestines. The aggregates in the worm-free soil were unstable, even in the presence of organic matter. (This passive effect of organic matter could have been due to the experimental technique since the soil-lespedeza hay mixture was only incubated 7 days at 6° C.) The data demonstrated clearly that earthworm activity had a pronounced effect upon the formation of stable aggregates.

The differences in the production of casts in fields of varying agricultural history depends upon the species of earthworm (Evans, 1948). Only 8 to 10 out of 25 different isolated species appear to be common in the fields (Satchell, 1958). Only three produce surface casts; the other field species excrete the ingested material into soil channels or cracks. There is an average annual production of 11 to 11.5 tons of casts per acre (Evans, 1948). This production varies considerably, depending upon both the soil treatment and the earthworm species. Enough casts are formed in an old pasture to cover the surface 2.4 in. deep in 10 years. Only enough for a 0.2-in. cover was produced in a 7-year-old pasture. However, the majority of the earthworms there voided their excrement below the surface. The aeration-porosity percentage was higher in soils containing species that produced casts. Guild (1955) extended the studies of Evans and his associates and divided earthworms into two major groups, depending upon their burrowing habits. The deep-burrowing species penetrated subsoils to depths of several feet. They proved to be the main surface-casting species. The shallow-working species generally stayed in the top 6 in. Aggregates taken from burrows of the noncasting species were small and unstable. Not all earthworm crumbs are stable. Casts brought up from B-horizons are completely unstable (Finck, 1952). Casts from cultivated fields are less stable than those from grasslands. The stability of the earthworm crumbs is directly proportional to their content in organic substances.

Soil management practices affect the number and activity of earthworms (Teotia, Duley, and McCalla, 1950). The data in Table 4-5 point out very clearly the impact of a surface mulch on the population of earthworms, the weight of worm casts formed per acre, and the quantity

TABLE 4-5

Effect of Tillage and Straw Mulch on Earthworm Activity (after Teotia, Duley, and McCalla, 1950)

Treatment	Earthworms (per acre, thousands)	Worm casts (per acre, tons)	Earthworm holes (no./ft²)
No straw, plowed	13	1.30	1
No straw, subtilled	56	11.92	6
Straw mulch, 2 tons	103	18.75	8
Straw mulch, 4 tons	169	29.12	18
Straw mulch, 8 tons	263	41.52	25

of worm channels. The stability of worm casts is much higher than that of natural soil aggregates. Alfalfa creates more stable aggregates than wheat straw. This observation suggests the presence of a nutritional factor in worm-cast production.

MECHANISMS IN WORM-CAST STABILITY. The stability of worm casts and artificial casts prepared from molded soil depends upon the treatment (Swaby, 1949c). There is a higher stability of the granules and worm casts from grassland as compared with those from arable land. Incubation increases the stability of grassland casts but not those from arable land. Artificial casts from a mixture of 0.5 percent grass roots and ground natural soil are not stable unless the casts are incubated. There is little difference in the stabilities of the incubated grassland and arable artificial casts. Fungal hyphae are the dominant stabilizing factors in incubated casts. Four possible mechanisms in the stabilization of worm casts have been suggested (Satchell, 1958). First, the casts could be mechanically bound together by filaments from vascular bundles and root fragments that are in the ingested plant material in the intestines of the earthworms. However, Swaby was unable to make stable artificial casts from root-soil mixtures. Second, soil particles could be cemented together into aggregates by secretions in the earthworm's intestines such as calcium carbonate or intestinal mucus. This possibility, however, cannot explain the difference between the stability of worm casts from grassland and arable land. Third, the casts could be stabilized by products of microbial synthesis within the intestines. Both Swaby and Finck stressed the importance of gum-producing bacteria in the intestines for increasing the stability of worm casts. Polysaccharides and related compounds were considered the causative agents. Swaby explained the increase in stability of incubated grassland casts on the basis of bacterial activity that results from a higher nutritive organic matter status. There were no fungi in the earthworm's intestines because of restricted aeration. Bacteria only produced stable aggregates when well supplied with nutritive organic matter. A fourth possibility could be the cementation action of Ca-humate that is synthesized in the intestinal tract from ingested decomposing organic matter and the Ca excreted by the calciferous glands of the earthworm. The layer of Ca-humate on the surface of aggregates is a strong stabilizing agent after it is dehydrated (Swaby, 1949).

Aggregate Stability-Time Relationships

Organic matter-soil aggregation relationships involve dynamic processes. Aggregate stability is continually changing as organic matter is added and decomposed. The cementing agents that are formed stabilize granules and are then decomposed to make the aggregates less stable. The changes in aggregate stability with time after incorporation of organic matter are summarized by the curves of Monnier (1965) in Figure 4-6. Stability has been based upon differences in aggregate breakdown in water, alcohol, and benzene. The peak of the curve represents the aggregation brought about by the microbial bodies in the soil. The major impact during this period of intense biological activity apparently is a mechanical binding action by the mycelia of fungi and actinomycetes and to some extent bacterial cells. This type of aggregate stability is only temporary since the mycelia and cells undergo bacterial decomposi-

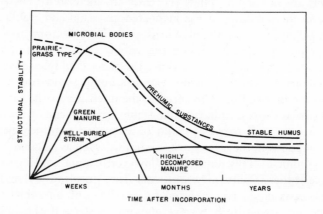

FIG. 4-6.   Stabilization of soil structure by organic matter. (After Monnier, 1965.)

tion as the intensity of the biological action declines. Stabilization of structure then enters the phase where the transitory products of microbial synthesis are responsible for aggregate stability. The specific activity of these organic substances in stabilizing structure is less but the aggregates formed are more resistant to further deterioration. These so-called prehumic substances are the polysaccharides and similar compounds which are subject to slow biological transformation. There is a decreased rate of decomposition which corresponds to a more durable effect on aggregate stability.

The final stage in the stabilization of structure is visualized as being dependent upon the stable humus that is formed. Stability is long lasting, even though the intensity of aggregate formation is not as high as in the other stages.

It should be noted that green manure, which is readily decomposable and has a low C/N, shows a sharp peak during the intense action at its fermentation but drops to zero in a relatively short time. Undoubtedly, fungal mycelia play a dominant role during the early stages of decomposition. The well-buried straw did not undergo intense biological fermentation, probably because of a higher C/N and a lower fungal activity. The highly decomposed manure had already been subjected to biological activity and consequently does not show an intensity peak. Stability of aggregation appears to be a function of the humic compounds produced.

The roots of the prairie-type grass produce a high stability of structure before the sod is incorporated in the soil. There is no flush of biological activity. Stabilization of the aggregates formed results from the transitory products of microbial metabolism and the final stable humus that is produced.

These curves emphasize the necessity for replenishing organic matter in soils to maintain stable granulation. They also point out the superiority of grass sods on the level of structure stability.

Synthetic Organic Polymers

The production of synthetic polyelectrolytes has been an outgrowth of basic research which showed that naturally occurring organic polymers, polysaccharides, and polyuronides are major contributing factors to soil aggregation. Polycations, polyanions, and nonionic polymers have been produced and tested for their ability to aggregate soils and stabilize soil structure. They have been referred to as *soil conditioners*.

Hedrick and Mowry (1952) observed that a hydrolyzed polyacrylonitrile, a polyanion, enhanced the workability of Miami silt loam by increasing the aggregate stability, the infiltration and permeability rates, and the plastic limit; stickiness, evaporation rate, crusting, and erosion were reduced. The same polyanion caused a shift in the particle-size distribution of Paulding clay from smaller to larger classes (Martin and associates, 1952). The untreated soil contained 55 percent particles < 0.25 mm. Soil treated with 0.05 percent polyelectrolyte had 53 percent of its particles in the 2- to 5-mm group. The treated soils were friable and loose. Runoff from Wooster silt loam was reduced from 58.8 to 7.4 percent; erosion from the treated soil was about $\frac{1}{100}$ that of the check. Hydrolyzed polyacrylonitrile approximately doubled the plastic and liquid limits of a silt loam, a clay, and an alluvial soil (Weeks and Colter, 1952). Soil erosion from the alluvial soil on a 34 percent slope over a 5-week period under about 19 in. of artificial rain amounted to about 88, 5.4, and 2.7 tons per acre for the control, straw-mulched, and polyanion-treated plots, respectively. One pound of soil conditioner per 100 ft² was approximately half as effective as 2 tons of straw per acre.

A surface treatment of the soil with the soil conditioner proved to be about twice as effective as mixing it into the surface with a rake. VAMA (vinyl acetate-maleic acid copolymer) decreased the breaking strength of soils at low bulk densities; the compressibility of soils with 1:1 type clay minerals was reduced (Jamison, 1954).

Crusting of high sodium soils has been prevented by intimately mixing VAMA or HPAN (hydrolyzed polyacrylonitrile) with the soil at the time of application (Allison and Moore, 1956). The rate of application for effective stabilization was directly related to both the specific surface and the clay content. Sodium removal from the top 6 in. of soil by irrigation water was twice as rapid on plots treated with soil conditioners. This was equivalent to a 2-ton application of gypsum.

MECHANISMS INVOLVED. Since soil conditioners are linear, flexible, organic polymers similar to soil polysaccharides, the mechanism of their adsorption by clay particles and their binding action on soil aggregates should be very similar to the naturally occurring polymers. The type of binding depends upon the nature of the clay mineral and of the polymer, that is, whether it is anionic, cationic, or nonionic.

Polycations are adsorbed by clays through cation exchange. Ca ions can act as bridges between the clay and the organic polymer (Greenland, 1965). Water-soluble polycations exhibit high flocculating power that follows the Schulze-Hardy valence rule (Ruehrwein and Ward, 1952).

Nonionic soil conditioners such as polyvinyl alcohol (PVA) form interlamellar complexes with common layer-lattice-type clays (Emerson, 1956, 1959). The polymer competes with water molecules for sites on the surface and swelling is suppressed. The strong, irreversible adsorption of PVA by montmorillonites has been characterized as a H bonding between the OH groups of the polymer and the O atoms of the clay surface (Greenland, 1963). The random coils of the polymer in solution collapse at the clay surface to form an adsorbed layer 10 Å thick. X-ray analyses of Na-montmorillonite have confirmed that interlamellar adsorption of the polymer increases the c-axis spacing 10 Å. There is limited interlamellar penetration of Ca-montmorillonite because the Ca ions restrict the accessibility to these inner surfaces. Suspensions of 1 percent show interparticle bonding that restricts the accessibility of polymer molecules to adjacent particles. If these particles are already aggregated, PVA is spread over adjacent surfaces like a coat of paint. The stability of soil aggregates is caused by lining of the pores with the polymer. The van der Waals forces are important in aggregate stabilization. These forces are additive at each point of contact between adsorbed polymer molecules and the clay surface (Greenland, 1965).

Polyanions do not form interlamellar complexes with clays. Their action appears to be confined to the formation of peripheral complexes that link clay crystals together in an edge-to-edge arrangement. Polyanions do not flocculate clays but stabilize clays that have been previously flocculated (Ruehrwein and Ward, 1952). Polyanions are adsorbed on the surface of individual dispersed particles. They bind particles together in clay aggregates because the molecules are adsorbed strongly enough and are sufficiently long to serve as bridges between particles. The saturation level of the polyanion approximates the anion-exchange capacity of the clay. These polyanions may be linked to edge atoms of the octahedral layers of clay crystals by a series of H bonds (Emerson, 1956). This bridging is lateral instead of vertical and not parallel to the c-axis. These polymers should be more efficient in acid soils since the edge O atoms and OH groups tend to coordinate better with H ions as the pH is lowered. Thus the particles may be linked together like a chain of beads (Greenland, 1965). The aggregation of montmorillonite by polyacrylic acid is almost equivalent to the estimated number of positively charged sites on the crystal edges (Warkentin and Miller, 1958). Addition of Na-metaphosphate to montmorillonite renders the positive charges on the edges electronegative and decreases the formation of PAA-clay complexes. The strength of the H bond linkage between soil and polymer appears to depend upon the nature of the active groups on the polymer (Vershinin, 1958). If the carboxyl group is taken as 1.0, amides are 1.6 and sulfonic groups 0.8. The adsorption of polyacrylonitrile is due to chemical bonds between the polymer and the atoms in the clay surface.

The activity of soil conditioners has been related to their functionality (Montgomery and Hibbard, 1955). This is the ratio of the number of active groups per repeating polymer segment to the total weight of the segment. This ratio increases to an optimal value and then decreases. There are not enough active groups to bind clay particles together at low functionality values. At high values the groups become too water-soluble and are leached. The molecular weight of the polymer is important in aggregation since it determines the number of secondary valences or H bonds involved. The polymer chain length seems to be related to the effectiveness of PAA as an aggregating agent (Warkentin and Miller, 1958). Even though substitutions within cellulose derivatives are important in making them water-soluble, the degree of polymerization has more influence on aggregate-stabilizing activities of the polymers than either the type or extent of the substitutions.

Summarizing, large organic molecules with different specific groups in the molecule interact with sites on the surface of the clay; there can be a number of interactions occurring. The extent to which the adsorbed molecules can collapse on the surface determines the number of contacts (Greenland, 1965). These organic polymers can be ionic or nonionic. Irrespective, they compete with water molecules on the surface. Nonionic polymers form interlamellar complexes with expanding-lattice type clays. Cationic polymers enter into base-exchange reactions with clay particles and form clay-polymer bridges to stabilize aggregation. Polyanions form edge-to-edge bridging to increase the stability of the aggregates.

References

Acton, C. J., D. A. Rennie, and E. A. Paul (1963). The relationship of polysaccharides to soil aggregation. *Can. J. Soil Sci.*, **43**:201–209.

Alexsandrova, L. N., and M. Nad' (1958). The nature of organo-mineral colloids and methods of their study. *Pochvovedenie*, **10**:21–27.

Allison, L. E., and D. C. Moore (1956). Effect of VAMA and HPAN soil conditoners on aggregation, surface crusting, and moisture retention in alkali soils. *Soil Sci. Soc. Am. Proc.*, **20**:143–146.

Arca, M. N., and S. B. Weed (1966). Soil aggregation and porosity in relation to contents of free iron oxide and clay. *Soil Sci.*, **101**:164–170.

Baver, L. D. (1928). The relation of exchangeable cations to the physical properties of soils. *J. Am. Soc. Agron.*, **20**:921–941.

Baver, L. D. (1929). The effect of the amount and nature of exchangeable cations on the structure of a colloidal clay. *Missouri Agr. Exp. Sta. Research Bull.* 129.

Baver, L. D. (1934). Classification of soil structure and its relation to the main soil groups. *Am. Soil Survey Assoc. Bull.* XV, pp. 55–56.

Baver, L. D. (1935). Factors contributing to the genesis of soil microstructure. *Am. Soil Survey Assoc. Bull.* XVI, pp. 55–56.

Baver, L. D., and N. S. Hall (1937). Colloidal properties of soil organic matter. *Missouri Agr. Exp. Sta. Research Bull.* 267.

Beutelspacher, H. (1955). Wechselwirkung zwischen anorganischen und organischen Kolloiden des Bodens. *Z. Pflanzenernähr., Düngung Bodenk*, **69**:108–115.

Blackmore, A. V., and R. D. Miller (1961). Tactoid size and osmotic swelling in calcium montmorillonite. *Soil Sci. Soc. Am. Proc.*, **25**:169–173.

Bradfield, Richard (1936). The value and limitations of calcium in soil structure. *Am. Soil Assoc. Bull.* XVII, pp. 31–32.

Brewer, R. (1964). *Fabric and Mineral Analysis of Soils*, John Wiley and Sons, New York.

Brewer, R., and A. V. Blackmore (1956). The effects of entrapped air and optically oriented clay on aggregate breakdown and soil consistence. *Aust. J. Appl. Sci.*, **7**:59–68.

Brewer, R., and J. R. Sleeman (1960). Soil structure and fabric. *J. Soil Sci.*, 11:172–185.

Ceratzki, W. (1956). Zur Wirkung des Frostes auf die Struktur des Bodens. *Z. Pflanzenernähr, Düngung Bodenk.*, 72:15–32.

Cernuda, C. F., R. M. Smith, and Jose Vicente-Chandler (1954). Influence of initial soil moisture condition on resistance of microaggregates to slaking and to water-drop impact. *Soil Sci.*, 77:19–27.

Chesters, G., O. J. Attoe, and O. N. Allen (1957). Soil aggregation in relation to various soil constituents. *Soil Sci. Soc. Am. Proc.*, 21:272–277.

Clapp, C. E., and W. W. Emerson (1965). The effect of periodate oxidation on the strength of soil crumbs. *Soil Sci. Soc. Am. Proc.*, 29:127–130.

Darwin, C. (1881). *The Formation of Vegetable Mould, through Action of Worms, with Observations on Their Habits.* Murray, London.

Dawson, Roy C. (1947). Earthworm microbiology and the formation of water-stable aggregates. *Soil Sci. Soc. Am. Proc.*, 12:512–515.

DeBoodt, M., and L. DeLeenheer (1958). Proposition pour l'evaluation de la stabilite des aggregates sur le terrain. *Int. Symposium on Soil Structure Proc.*, Ghent, pp. 234–241.

Demolon, A., and S. Hénin (1932). Recherches sur la structure des limons et la synthese des aggregates. *Soil Research*, 3:1–9.

Dettman, Margaret G., and W. W. Emerson (1959). A modified permeability test for measuring the cohesion of soil crumbs. *J. Soil Sci.*, 10:215–226.

Edwards, A. P., and J. M. Bremner (1967). Microaggregates in soils. *J. Soil Sci.*, 18:64–73.

Egawa, Tomuji, and Kozo Sekiya (1956). Studies on humus and aggregate formation. *Soil and Plant Food*, 2:75–82.

Emerson, W. W. (1956). Synthetic soil conditioners. *J. Agr. Sci.*, 47:117–121.

Emerson, W. W. (1959). The structure of soil crumbs. *J. Soil Sci.*, 10:235–244.

Emerson, W. W. (1963). The effect of polymers on the swelling of montmorillonite. *J. Soil Sci.*, 14:52–63.

Emerson, W. W., and M. G. Dettman (1960). The effect of pH on the wet strength of soil crumbs. *J. Soil Sci.*, 11:149–158.

Emerson, W. W., and G. M. F. Grundy (1954). The effect of rate of wetting on water uptake and cohesion of soil crumbs. *J. Agr. Sci.*, 44:249–253.

Evans, A. C. (1948). Studies on the relationships between earthworms and soil fertility. II. Some effects of earthworms on soil structure. *Ann. Appl. Biology*, 35:1–13.

Filippovich, Z. S. (1956). Absorption of colloids by soils and the formation of structure. *Pochvovedenie* 2:16–26.

Finck, A. (1952). Ecological and pedological studies on the effect of earthworms on soil fertility (G). *Z. Pflanzenernähr. Düngung Bodenk.* 58:120–145.

Flaig, W. (1958). Zur Chemie der anorganischen und organischen Komponenten der Krümelbildung. Probleme der Krümelstabilitätmessung und der Krümelbildung. *Tagungsberlichte Nr. 13, Inst. für Acker, u. Pflanzenbau,* Münchenberg, pp. 225–244.

Flaig, W., and H. Beutelspacher (1951). Zur Kenntnis der Huminsäuren II. Elektromikroskopiche Untersuchungen an natürlichen und synthetischen Huminsäuren. Z. Pflanzenernähr, Düngung Bodenk., 52:1–21.

Gel'tser, F. Yu (1955). The importance of annual and perennial herbaceous plants in the production of soil fertility. Pochvovedenie, 5:44–53; Soils and Fert., 18:497.

Greenland, D. J. (1963). Adsorption of polyvinyl alcohols by montmorillonite. J. Colloid Sci., 18:647–664.

Greenland, D. J. (1965). Interaction between clays and organic compounds in soils. Part I. Mechanisms of interaction between clays and defined organic compounds. Soils and Fert., 28:415–425.

Greenland, D. J., G. R. Lindstrom, and J. P. Quirk (1962). Organic materials which stabilize natural soil aggregates. Soil Sci. Soc. Am. Proc., 26:366–371.

Guild, W. J. Mc L. (1955). Earthworms and soil structure. Soil Zoology, Nottingham School Agr. Sci., Academic Press, New York, pp. 83–98.

Harradine, Frank, and Hans Jenny (1958). Influence of parent material and climate on texture and nitrogen and carbon contents of virgin California soils. I. Texture and nitrogen content of soils. Soil Sci., 85:235–243.

Harris, R. F., O. N. Allen, G. Chesters, and O. J. Attoe (1963). Evaluation of microbial activity in soil aggregate stabilization and degradation by the use of artificial aggregates. Soil Sci. Soc. Am. Proc., 27:542–545.

Harris, R. F., G. Chesters, O. N. Allen, and O. J. Attoe (1964). Mechanisms involved in soil aggregate stabilization by fungi and bacteria. Soil Sci. Soc. Am. Proc., 28:529–532.

Hedrick, R. M., and D. T. Mowry (1952). Effect of synthetic polyelectrolytes on aggregation, aeration, and water relationships of soils. Soil Sci., 73:427–441.

Hénin, S. (1938). Etude physico-chimique de la stabilite structurale des terres. Monograph National Center of Agronomic Research, Paris, pp. 52–54.

Jamison, Vernon C. (1954). The effect of some soil conditioners on friability and compactibility of soils. Soil Sci. Soc. Am. Proc., 18:391–394.

Jenny, Hans (1930). A study of the influence of climate upon the nitrogen and organic matter content of soil. Missouri Agr. Exp. Sta., Research Bull. 152.

Jenny, Hans, and C. D. Leonard (1934). Functional relationship between soil properties and rainfall. Soil Sci., 38:363–381.

Jung, E. (1931). Untersuchungen über die Einwirkung des Frostes auf den Erdboden. Kolloidchem. Beihefte, 32: 320–373.

Kolodny, L., and O. R. Neal (1940). The use of micro-aggregation or dispersion measurements for following changes in soil structure. Soil Sci. Soc. Am. Proc., 6:91–95.

Kononova, M. M. (1961). Soil Organic Matter. Pergamon Press, New York, p. 176.

Kubiena, W. L. (1938). Micropedology. Collegiate Press, Ames, Iowa.

Logsdail, D. E., and L. R. Webber (1959). Effect of frost action on structure of Haldimand clay. Can. J. Soil Sci., 39:103–106.

Lutz, J. F. (1934). The physico-chemical properties of soil affecting soil erosion. *Missouri Agr. Exp. Sta. Research Bull.* 212.

McCalla, T. M. (1946). Influence of some microbial groups on stabilizing soil structure against falling water crops. *Soil Sci. Soc. Am. Proc.*, 11:257–263.

McIntyre, D. S. (1956). The effect of free ferric oxide on the structure of some terra rosa and rendzina soils. *J. Soil Sci.*, 7:302–306.

Marshall, T. J. (1962). The nature, development, and significance of soil structure. *Trans. Joint Meeting Com. IV and V, Intern. Soc. Soil Sci.*, New Zealand, pp. 243–257.

Martin, James P. (1945). Microorganisms and soil aggregation: I. Origin and nature of some aggregating substances. *Soil Sci.*, 59:163–174.

Martin, James P., William P. Martin, J. B. Page, W. A. Raney, and J. D. DeMent (1955). Soil aggregation. *Advances in Agronomy*, 7:1–37, Academic Press, New York.

Martin, W. P., G. S. Taylor, J. C. Engibous, and E. Burnett (1952). Soil and crop responses from field applications of soil conditioners. *Soil Sci.*, 73:455–471.

Mazurak, Andrew P. (1950). Effect of gaseous phase on water-stable synthetic aggregates. *Soil Sci.*, 69:135–148.

Mehta, N. C., H. Streuli, M. Müller, and H. Deuel (1960). Role of polysaccharides in soil aggregation. *J. Sci. Food Agr.*, 11:40–47.

Monnier, G. (1965). Action des materies organiques sur la stabilite structurale des sols. *Ann. Agron.*, 16:327–400, 471–534.

Montgomery, P. S., and B. B. Hibbard (1955). Theoretical aspects of the soil-conditioning activity of polymers. *Soil Sci.*, 79:283–292.

Müller, G. (1958). Beziehungen zwischen Biologie und Struktur des Bodens. *Tagungsberichte Nr. 13, Inst. für Acker. Pflanzenbau*, Münchenberg, pp. 167–192.

Myers, H. E. (1937). Physico-chemical reactions between organic and inorganic soil colloids as related to aggregate formation. *Soil Sci.*, 44:331–359.

Nikiforoff, C. C. (1941). Morphological classification of soil structure. *Soil Sci.*, 52:193–211.

Panabokke, C. R., and J. P. Quirk (1957). Effect of initial water content on stability of soil aggregates in water. *Soil Sci.*, 83:185–195.

Payne, D. (1954). Some factors affecting the breakdown of soil crumbs on rapid wetting. *Trans. 5th Int. Cong. Soil Sci.*, Leopoldville, 2:52–58.

Peele, T. C., and O. W. Beale (1940). Influence of microbial activity upon aggregation and erodibility of lateritic soils. *Soil Sci. Soc. Am. Proc.*, 5:33–35.

Peterson, J. B. (1947). Calcium linkage, a mechanism of soil granulation. *Soil Sci. Soc. Am. Proc.*, 12:29–34.

Robinson, D. O., and J. B. Page (1950). Soil aggregate stability. *Soil Sci. Soc. Am. Proc.*, 15:25–29.

Rogowski, A. S., and Don Kirkham (1962). Moisture, pressure and formation of water-stable soil aggregates. *Soil Sci. Soc. Am. Proc.*, 26:213–216.

Ruehrwein, R. A., and D. W. Ward (1952). Mechanism of clay aggregation by polyeletrolytes. *Soil Sci.*, 73:485–491.

Russell, E. W. (1934). The interaction of clay with water and organic liquids as measured by specific volume changes and its relation to the phenomena of crumb formation in soils. *Phil. Trans. Roy. Soc. London*, 233A:361–389.

Saini, G. R., and A. A. MacLean (1966). Adsorption-flocculation reactions of soil polysaccharides with kaolinite. *Soil Sci. Soc. Am. Proc.*, 30:697–699.

Satchell, J. E. (1958). Earthworm biology and soil fertility. *Soils Fert.*, 21:209–219.

Schofield, R. K. (1943). The role of soil moisture in soil mechanics. *Chem. Ind.*, 62:339–341.

Schofield, R. K., and H. R. Samson (1954). Flocculation of kaolinite due to the attraction of oppositely charged crystal faces. *Disc. Faraday Soc.*, 18:135–145.

Sideri, D. I. (1936). On the formation of structure in soil: II. Synthesis of aggregates; on the bonds uniting clay with sand and clay with humus. *Soil Sci.*, 42:461–481.

Sillanpää, Mikko, and L. R. Webber (1961). The effect of freezing-thawing and wetting-drying cycles on soil aggregation. *Can. J. Soil Sci.*, 41:182–187.

Slater, Clarence S., and Henry Hopp (1949). The action of frost on the water stability of soils. *J. Agr. Res.*, 78:341–346.

Soil Survey Staff (1951). *Soil Survey Manual, U.S.D.A. Handbook* No. 18.

Swaby, R. J. (1949a). The relationship between microorganisms and soil aggregation. *J. Gen. Microbiology*, 3:236–254.

Swaby, R. J. (1949b). The influence of humus on soil aggregation. *J. Soil Sci.*, 1:182–193.

Swaby, R. J. (1949c). The influence of earthworms on soil structure. *J. Soil Sci.*, 1:194–197.

Teotia, S. P., F. L. Duley, and T. M. McCalla (1950). Effect of stubble mulching on number and activity of earthworms. *Nebraska Agr. Exp. Sta. Research Bull.* 165.

Tiulin, A. F. (1933). Considerations on the genesis of soil structure and on methods for its determination. *Trans. 1st Com. Int. Soc. Soil Sci.*, Moscow, Vol. A., pp. 111–132.

Tiulin, A. F., and A. V. Korovkina (1950). The different quality of water-stable aggregates in relation to the group composition of secondary particles smaller than 0.01 mm. *Pochvovedenie*, pp. 142–150.

Vershinin, P. V. (1958). Synthetic soil conditioners. *Pochvovedenie*, 10:28–37.

Walter, B. (1965). Über Bildung und Bindung von Mikroaggregaten in Böden. II. Mitteilung. *Z. Pflanzenernähr., Düngung Bodenk.*, 110:43–49; III. Mitteilung. 110:49–63.

Warkentin, B. P., and R. D. Miller (1958). Conditions affecting formation of the montmorillonite-polyacrylic acid bond. *Soil Sci.*, 85:14–18.

Watson, J. H., and B. J. Stojanovic (1965). Synthesis and bonding of soil aggregates as affected by microflora and its metabolic products. *Soil Sci.*, 100:57–62.

Weeks, Lloyd E., and William G. Colter (1952). Effect of synthetic soil conditioners on erosion control. *Soil Sci.*, 73:473–484.

White, E. M. (1966). Subsoil structure genesis: Theoretical consideration. *Soil Sci.*, **101**:135–141.

White, E. M. (1967). Soil age and texture in subsoil structure genesis. *Soil Sci.*, **103**:288–298.

Winterkorn, Hans F. (1942). Mechanism of water attack on dry cohesive soil systems. *Soil Sci.*, **54**:259–273.

Wollny, E. (1890). Untersuchungen über die Beeinflussing der Fruchtbarkeit der Ackerkrume dürch die Thätigkeit der Regenwürmer. *Forschungen a.d. Gebeite der Agriculturphysik*, **13**:381–395.

Woodruff, C. M. (1968). Personal communication.

Yoder, R. E. (1936). A direct method of aggregate analysis of soils and a study of the physical nature of erosion losses. *J. Am. Soc. Agron.*, **28**: 337–351.

Zakharov, S. A. (1927). Achievements of Russian science in morphology of soils. *Russ. Pedolog. Investigations*, LL. Acad. Sci., USSR.

CHAPTER 5

Soil Structure—
Evaluation and
Agricultural
Significance

EVALUATION OF SOIL STRUCTURE

Persons who are interested in structure and structural stability as related to erosion, infiltration, root penetration, aeration, or engineering applications concerned with water penetration and load- bearing evaluate soil structure by methods that correlate best with those factors associated with the intended use. The agriculturist's interest is associated with "tilth," which is a general term that signifies the ability of the granules or aggregates to withstand destruction by the impact of implements, raindrops, or running water so that water penetration, aeration, and the penetration to the roots are maintained at favorable levels. Tilth also involves water retention and the workability of soils, both of which depend to a large extent upon the basic textural nature of the soil.

Soil structure can be evaluated by determining the extent of aggregation, the stability of the aggregates, and the nature of the pore space. These characteristics change with tillage practices and cropping systems. They play a significant role in affecting soil-plant relationships.

Aggregation

In evaluating the aggregation of soils, one is interested in the size distribution, quantity, and stability of the aggregates. These parameters of aggregation are important in determining both the amount and distribution of the pore spaces associated with the aggregates and the susceptibility of the aggregates to water and wind erosion.

Aggregate Analysis

TECHNIQUES EMPLOYED. An aggregate analysis aims to measure the percentage of water-stable secondary particles in the soil and the extent to which the finer mechanical separates are aggregated into coarser fractions. In general, three techniques are employed to accomplish such an analysis. They are wet and dry sieving, elutriation, and sedimentation.

Direct dry sieving of soils as they occur in the field has been used to evaluate the distribution of clods and aggregates (Keen, 1933; Cole, 1939). Air-dry sifting is considered to give a better picture of aggregation of arid California soils than wet sieving, since the aggregates are so weakly held together in the moist condition that the mechanical action of sieving is sufficient to destroy them. Clogging of the flat sieves and breaking up of weak aggregates by the mechanical action required in the sieving operation are major problems with this technique. These difficulties have been overcome by the use of a rotary sieve (Chepil, 1962). Dry sieving of aggregates gives an important index for characterizing the susceptibility of soils to wind erosion. No special preparation of the sample is required for consistent results.

The wet-sieving technique of Tiulin (1928) is the best known of the earlier endeavors to find a measure of soil aggregation. According to his method, the soil is slowly wetted by capillarity for 30 minutes and is then transferred onto a nest of sieves immersed in water. The sieves are slowly raised and lowered in the water 30 times. The weight of soil on each sieve is then determined. The bottom sieve in this nest has an opening of 0.25 mm. Mechanical methods have been devised to raise and lower the sieves (Yoder, 1936; Kemper, 1965). The nest of sieves is raised and lowered through a distance of 1¼ in. at a rate of 30 oscillations per minute for 30 minutes. Wet sieving is well adapted to the separation of large aggregates. It can be used to screen out aggregates as small as 0.1 mm, although 0.25 mm is more satisfactory as the lower limit of size.

The greatest problem in wet sieving is the method of wetting the sample for analysis. Air drying decreases the percentage of large aggregates in favor of smaller ones (Yoder, 1936; Russell, 1938). This effect is greater, the more intense the drying process. The more rapidly the soils are wetted, the greater is the breaking up of the larger aggregates. Immersing the soil in water causes more destruction of the larger aggregates than wetting by capillarity. Spraying water onto the aggregates with an atomizer produces the least destruction of any of these three methods. The mechanics of the shattering effect of air that is displaced during rapid wetting was discussed earlier. It was shown that wetting under vacuum usually solved this problem. This disruptive effect can be obviated by pretreatment of aggregates with ethyl alcohol to displace the air before wet sieving (Hénin, Robichet, and Jongerius, 1955). (See Kemper and Chepil, 1965, for detailed directions for the wet sieving of aggregates.)

Elutriation may be used for separating aggregates with diameters between 1 and 0.02 mm. It is particularly useful for making separations below the limit where wet sieving cannot be employed (Baver and Rhoades, 1932; Demolon and Hénin, 1932).

Sedimentation methods have been used to determine the aggregate distribution in the finer fractions that cannot be separated by sieving. They are limited to aggregate sizes <1 mm. Both the pipette and hydrometer methods have been used for the separations. Two important difficulties exist in the use of sedimentation. There is a varying density of the aggregates, especially with the larger secondary particles, and there is the possibility of flocculation during sedimentation because of the downward motion of the larger aggregates.

EXPRESSION OF RESULTS. There are several ways to express the aggregation of soils. The author suggested the term "state of aggregation" to designate the percentage of aggregates larger than a certain specified size in a given weight of soil. It is obvious that a sandy soil cannot contain as many aggregates as a well-granulated silt loam, because it possesses such a high content of coarse mechanical separates. Nevertheless, it is possible for all the silt and clay in coarse-textured soils to be present in the form of aggregates. Therefore, in order to have a measure of the percentage aggregation of the fine mechanical separates, a value is used which is obtained by dividing the percentage of aggregates larger than a given size (0.05 to 0.1 mm) in the soil by the percentage of mechanical elements smaller than this size. This gives the "degree of aggregation" of the fine particles. In other words, it represents the percentage of particles smaller than a given size that are aggregated into stable units larger than this size.

Examples of the structure capacities of several widely different soils are given in Table 5-1. These data point out that chernozems and lateritic soils possess a high state of aggregation. Similarly, a high percentage of the small particles is aggregated. Podsols, however, are poorly aggregated. This is especially true in the A_2-horizon. The effects of organic matter and iron are noticeable in the A_1- and B-horizons, respectively. The importance of recognizing the extent to which the smaller mechanical separates are aggregated is shown by comparing the surface layers of the two chernozemlike soils. One is a fine sandy loam and the other a silt loam. The state of aggregation of both is about the same in spite of the differences in texture. This is due to the fact that 75.5 percent of the particles smaller than 0.05 mm in the fine sandy loam are aggregated, whereas only 50 percent are aggregated in the silt loam.

TABLE 5-1

The Structure Capacity or Extent of Aggregation of Various Soil Profiles

Soil type	Depth (in.)	Mechanical separates larger than 0.05 mm (percent)	State of aggregation Aggregates larger than 0.05 mm (percent)	Degree of aggregation Mechanical separates smaller than 0.05 mm aggregated into units larger than 0.05 mm (percent)
Chernozem	0–3	41.1	44.2	75.5
(fine sandy	3–9	55.3	30.4	68.0
loam)	9–15	55.2	32.5	72.1
	18–28 (lime)	37.1	33.6	53.4
Chernozemlike	0–6 (A₁)	13.3	43.4	50.0
silt loam	7–12 (A₂)	7.6	70.9	76.5
Lateritic clay	0–5	29.1	67.1	94.6
	5–30	14.1	70.6	82.1
	30–60	28.7	37.7	52.9
Podsol	0–4 (A₁)	72.8	8.3	24.5
(fine sand)	4–10 (A₂)	64.8	3.9	11.1
	10–16 (B₁)	77.5	10.9	51.2
	16–28 (B₂)	85.7	7.7	53.6
	28–36 (B₃)	85.6	8.1	35.8
	36+ (C)	74.5	0.7	26.8

van Bavel (1949) introduced the mean weight-diameter (MWD) of soil aggregates as an index of aggregation. The proportion by weight w_i of a given size fraction of aggregates is multiplied by the mean (average) diameter $\bar{x}_i$ of that fraction. The sum of these products for all size fractions is called the mean weight-diameter (MWD).

$$\text{MWD} = \sum_{i=1}^{n} \bar{x}_i w_i \qquad (5\text{-}1)$$

The MWD also may be determined graphically from accumulation-frequency curves. The upper limit of separation in a given class is plotted against the accumulated frequency. The MWD index for virgin land, corn in rotation, and continuous corn on Webster silty clay loam was 1.604, 0.432, and 0.288, respectively. The geometric mean diameter (GMD) has been proposed as an index of aggregation (Mazurak, 1950). The weight of the aggregates in a given size fraction is multiplied by the logarithm of the mean diameter of that fraction. The sum of these products for all size fractions is divided by the total weight of the sample to give the GMD.

Schaller and Stockinger (1953) compared five methods for expressing aggregation data and obtained correlations of 0.958 and 0.913 between the MWD and the percentage of aggregates >2 mm and >1 mm, respectively. The corresponding correlations with GMD were 0.866 and 0.882. In other words, there was a highly significant correlation between MWD or GMD and the state of aggregation which indicates that either expression gives a good index of aggregation.

Stability of Structure

GENERAL PRINCIPLES. The stability of structure refers to the resistance that the soil aggregates offer to the disintegrating influences of water and mechanical manipulation. The stability of the aggregate is of utmost importance in forming and preserving good structural relationships in soils. It is recognized that "aggregate stability" is not necessarily synonymous with "structural stability" because soil aggregates may be altered by a variety of destructive forces. Moreover, it is often not obvious which of these forces are involved in the use of "aggregate stability." It is known that water content is often a crucial factor in structural stability. It is almost always a factor in determining the degree to which particular mechanical forces will cause structural breakdown. For example, rather compact and coherent aggregates may be found in the dry state, but if these secondary particles disintegrate in water, the aggregation is not very stable. Water may cause the deterioration of aggregation in two ways. First, there is the hydration effect of water, which causes a disruption of the aggregate through the processes of swelling and the exploding of entrapped air. The mechanisms of these effects have been discussed earlier. The second manner in which water destroys aggregation and deteriorates soil structure is by falling rain. The impact of falling drops of water on exposed soil exerts a significant dispersive action on the aggregates. The dispersed particles are then carried into the soil pores, causing increased compaction and decreased porosity. Intense rains destroy the granulation and open structure of the top inch or more of soil to form a dense, impervious surface known as a *crust*. This type of structure degradation is least common with those aggregates that are stabilized with humus or iron compounds. Insufficient emphasis has been placed upon the deteriorating action of falling raindrops; in many instances, raindrops are the major cause of the dispersion of soil aggregates. Their immediate influence is confined to a shallow layer in the surface, but the structure of this layer may be broken down to limit the air and moisture relations of the entire profile.

Cultivation and other tillage operations generally cause a continued decrease in the stability of aggregates unless the organic matter level of the soil is kept relatively high and mechanical manipulation of the soil is performed at optimum moisture contents.

METHODS OF EVALUATION. Several methods have been proposed to characterize the stability of soil aggregates. They may be divided into four categories:

1. Stability against disruption during wet sieving.
2. Stability against the impact of falling drops of water.
3. Stability against disintegration during leaching with dilute NaCl solutions.
4. Stability against slaking when pretreated with alcohol or other organic liquids.

Wet sieving has been used extensively to determine the size distribution and the stability of aggregates. Stability can be evaluated in several ways using this technique. Russell and Feng (1947) oscillated aggregates in sieves vertically in water for different periods of time and found that the relation between aggregate stability and length of oscillation time was exponential according to the equation

$$\log W = a - b \log T \qquad (5\text{-}2)$$

where W is the weight of water-stable aggregates, T is the oscillation time, a is $\log W$ at zero oscillation time, which was called the "initial stability," and b is the slope of the regression equation, which was termed "rate of disintegration." Soils with high initial stabilities and low rates of disintegration had stable aggregates. This concept is a refinement of the elutriation technique to measure aggregate stability (Baver and Rhoades, 1932). The mean diameter of the aggregates at the 50 percent level on the accumulation curve has also served as an index of stability (Robinson and Page, 1950). The change in mean weight-diameter from dry sieving to wet sieving also can characterize aggregate stability (DeLeenheer and DeBoodt, 1954). Stability has also been determined by wetting aggregates 1 to 2 mm in size under vacuum, wet sieving them through a 0.25-mm screen for 5 min, and then subtracting the amount of sand contained in the stable aggregates that remain on the screen (Kemper, 1965).

The action of raindrops on aggregate disintegration has been simulated in the water-drop method (McCalla, 1944). Stability is measured by the number of drops of water 4.7 mm in diameter, falling from a height of 30 cm on a lump or clod of soil, that are required to completely destroy the clod.

Emerson (1954) measured the stability of soil crumbs by determining the concentration of NaCl that caused the aggregates to disperse and render them impermeable. Soil crumbs were wetted under a suction of 15 cm of $0.5N$ NaCl and then leached with the same solution to replace any exchangeable cations present. They were then percolated with increasing dilutions of NaCl and changes in permeability were noted. That concentration at which the crumbs were completely dispersed and their permeability was zero was called the *critical concentration*, which was an index of stability. The lower the concentration, the higher was the stability. For example, the critical concentrations for soils from plots receiving no fertilizer or manure was $0.034N$; that for the manured and fertilized plot was $0.005N$; and that for permanent grass was $0.0003N$. The test was later modified to require only a single concentration of NaCl (Dettman and Emerson, 1959). A $0.05N$ solution was just high enough not to promote flocculation but low enough to produce decreases in permeability. The index of cohesion, or aggregate stability, is equal to the ratio K_2/K_1, where K_1 is the initial permeability of the crumbs before leaching 3 liters of $0.05N$ NaCl and K_2 is the final permeability after leaching. A soil from a field at Jealots Hill, which had been in grass 100 years, had 68 percent crumbs >2 mm and an aggregate stability index of 90; these values for an area under continuous cultivation were 33 and 35, respectively. The data in Figure 4-4 were obtained with a modification of this technique.

Hénin, Robichet, and Jongerius (1955) approached stability from the point of view of the destructive forces involved in the disintegration of the crumbs, particularly those associated with rapid wetting. They determined the percentage of aggregates >0.2 mm by wet sieving in water with no pretreatment, pretreatment with ethyl alcohol, and pretreatment with benzene. The alcohol pretreatment replaced the air in the pores to eliminate explosive effects. Values thus obtained were considered as representing the initial state of stability. The benzene decreased the wettability of the aggregates as it was fixed by any organic matter present. It had a much greater effect on soils containing organic matter than on those rich in clay. The alcohol pretreatment exhibited the highest percentage of aggregates. For example, an untreated plot showed 2.42, 6.88, and 1.36 percent aggregates >0.2 mm for no pretreatment, alcohol, and benzene, respectively. These values for the manured plot were 5.32, 8.11, and 5.69 percent. The index of stability is calculated according to the equation

$$S = \frac{A + L}{\left(\dfrac{A_{al} + A_{air} + A_b}{3}\right) - 0.9(SG)} \qquad (5\text{-}3)$$

where $A + L$ is the maximum amount of particles <0.002 mm that come from the sieving of aggregates >0.2 mm, A_{al}, A_{air}, A_b are the aggregates from pretreatment with alcohol, no pretreatment, and pretreatment with benzene, respectively, and SG is the coarse sand in the sample. They obtanied a negative linear correlation between log S and log K (permeability of the aggregates).

Soil Porosity

Basic Principles

The porosity of the soil can be calculated from the real and apparent specific gravity. The real specific gravity of the solid particle is determined by means of a pycnometer or specific-gravity bottle. It represents the weight of 1 cm³ of solid particles. The apparent specific gravity or volume weight is measured by weighing a given volume of soil in its natural structure. It represents the weight of 1 cm³ of soil and pore space. Naturally, since only the solid particles in a dried soil contribute to this weight, it is possible to make an easy calculation of the total pore space. This is done by the well-known formula

$$\text{percent pore space} = 1 - \frac{\text{bulk specific gravity}}{\text{real specific gravity}} \times 100 \qquad (5\text{-}4)$$

This expression gives the volume percentage of the pore space, but does not characterize the size of the pores. The total porosity of an average soil varies in the neighborhood of 50 percent. Sands usually have less than this amount; clays and organic soils have higher porosities. Porosity varies with the size of particles and state of aggregation. One is interested not so much in the total porosity as in the size distribution of the pores, especially the larger ones that contribute to the aeration porosity. These large pores increase with soil aggregation and with the size of the aggregates. Aeration porosities below 10 percent are restrictive to root proliferation.

The ideal soil should have the pore space about equally divided between large and small pores. Such a soil would have sufficient aeration, permeability, and water-holding properties. A clear picture of the porosity of soil profiles is shown in Figure 5-1. There are several significant points in this figure that should be emphasized. First, let us compare the Marshall and Shelby profiles. It is seen that the Marshall silt loam profile has a uniform total and aeration porosity. This is not true of the Shelby loam since the maximum porosity occurs in the upper 12 in. The Shelby soil has a higher soil volume than the Marshall, which is indicative of its greater apparent (bulk) density. The most impermeable layer in the Shelby occurs at about 22 to 26 in., where the areation porosity is only about 5 percent of the total volume. The minimum aeration porosity

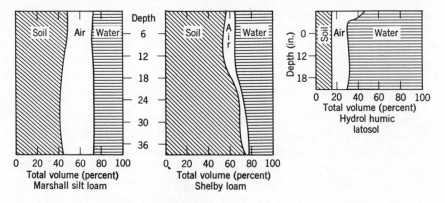

Fig. 5-1. Profile pore-space relationships.

of the Marshall is about 25 percent. This makes the ratio of the aeration porosity of the Marshall to that of the Shelby equal to about 5.

Approximately 50 percent of the total pores in the Marshall silt loam represent aeration porosity. The Shelby, on the other hand, has a total porosity of about 25 percent in the subsoil, only 11 percent of which is aeration pores. Its water capacity is too high for its air capacity to ensure adequate aeration and drainage.

The structure profile of the tropical hydrol humic latosol is quite different from the other two. The total porosity of the first 20 in. is about 85 percent. The aeration porosity of the first 3 in. is approximately 30 percent; that of the next 17 in. is about 15 percent. The soil volume is only about 15 percent since the bulk density is only around 0.5. Although this soil is rather permeable, its high water-holding capacity makes it susceptible to easy puddling and destruction of the aggregates.

It should be kept in mind that these schematic presentations of pore-space relationships do not take into consideration the size distribution, shape, continuity, or tortuosity of the pores. Aeration porosity is separated from the water pores on the basis of the pore size that will hold water under ⅓ atm tension. This type of presentation does give a relative differentiation between structure under various soil situations.

Methods of Evaluation

Most soil-porosity determinations have been based upon determinations of the apparent specific gravity or bulk density at some arbitrary moisture content. Total porosity has been calculated from the real and apparent specific gravity. More refined techniques have made it possible to obtain relatively accurate characterizations of the soil pore space.

SIZE DISTRIBUTION OF PORES. The size distribution of pores is based upon the moisture-tension curve that relates the amount of water in the soil in equilibrium with the tension forces applied. It suffices at this time to call attention of the capillary-rise equation upon which pore-size calculations are based:

$$h = \frac{2\gamma \cos \Theta}{\rho g r} \tag{5-5}$$

where h is the heighth of rise in a capillary tube with radius r, γ is the surface tension of water with density ρ, g is the acceleration due to gravity, and Θ is the contact angle between water and soil pore (assumed to be zero). This equation simplifies to the usable expression $d = 3h$, where d is the diameter of the pore in millimeters and h is the amount of tension in centimeters of water with which the water in the pore is in equilibrium. The volume of water removed from a given volume of soil at a specified tension represents the volume of pores of the size indicated by that tension (Vomocil, 1965).

AIR PYCNOMETER. The volume of air-filled pores can be measured directly by using an air pycnometer. Both water and volume of soil as well as the volume occupied by the soil air can be determined in 2 minutes by a relatively simple laboratory technique that is based on Boyles law (Kummer and Cooper, 1945) (see Figure 5-2). Since the product of the pressure and the volume of a gas is constant (Boyle's law), all one needs is an instrument in which a quantity of gas at a known pressure and volume can expand into a larger volume. The new volume can be calculated from the resulting pressure change. Vessel

A in Figure 5-2 may be compared with a compression chamber, which includes an air tube that leads to valve *E*. Vessel *B* may be compared with the air-volume chamber which includes a manometer and mercury well (or sensitive gauge). With valve *E* closed, the pressure-volume rela-

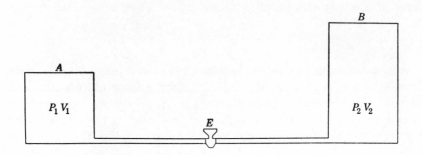

FIG. 5-2. Schematic diagram of pycnometer principle (After Kummer and Cooper, 1945.)

tionship in *A* is P_1V_1 and in *B* is P_2V_2. When valve *E* is opened, the pressures in *A* and *B* equalize. The resulting conditions for the two chambers would be as follows:

$$\text{In } A = P_3V_1$$
$$\text{In } B = P_3V_2$$

But

$$P_1V_1 + P_2V_2 = P_3V_1 + P_3V_2$$

and

$$P_3 = \frac{P_1V_1 + P_2V_2}{V_1 + V_2} \tag{5-6}$$

Since P_1, P_2, and V_2 are kept constant, the value of V_1 can be determined if P_3 is known. P_3 is measured either by a mercury manometer or by a sensitive gage attached to chamber *B*. If there is a decrease in the volume of V_1 due to the placing of the sample in the chamber, there will be an increase in P_3.

A pressure pycnometer has been designed that is portable and can be used for measurements in the field (Page, 1947). Such determinations can be made quite rapidly. It requires more time to take the samples. After the field measurements are made, the samples may be taken to the laboratory and the aeration porosity determined by the blotter method. The direct determination of the soil pore space by the pycnometer may give more reliable data than the blotter technique if one is interested in infiltration rates, gaseous diffusion, the effect of tillage on structure, and water and air movement under irrigation. Porosity measurements in the field with the pycnometer are generally higher than those obtained with the blotter technique. The difference is attributed to the lack of complete saturation when the blotter is used, since soil swelling takes place during saturation.

The air-pycnometer technique (see Vomocil, 1965, for details) avoids many of the criticisms of the usual method for measuring soil porosity. When tension techniques are employed, there is always the question of the swelling effect of water on porosity. However, soils in the air-dried state adsorb oxygen and the air pycnometer gives erroneous results on soils near or below air dryness unless the proper calibrations have been made (Jamison, 1953).

BULK DENSITY DETERMINATIONS. Bulk density is the ratio of soil mass to its volume. It is usually expressed in terms of grams per cubic centimeter (g/cm³). Soil engineers express bulk density as pounds per cubic foot. Measurements of bulk density generally require soil cores or clods in their natural structure. Engineers, however, use a technique in which they weigh the soil from an excavation of measured volume. The discovery of gamma radiation has made it possible to measure bulk density *in situ*.

The Core Technique. The obtaining of core samples is relatively simple if no roots or stones occur in the soil. The core sampler is pushed or driven into the soil to the desired depth and then removed. Since the volume of the sample is known, it is easy to obtain the bulk density and the air and water capacity of the core at any desired moisture content. Many samplers are provided with a brass or celluloid casing which holds the core and permits easy removal and handling of the sample during wetting and weighing.

There are several factors that should be considered in obtaining core samples by this technique. The forcing of the sampler into the soil may shatter or compress the soil, depending upon the moisture content and the method of sampling. Hammering tends to cause shattering, and rapid pushing produces compression. The size of the sampler affects the amount of compression. Narrow sampling tubes favor compression; a 3- to 4-in. diameter is a satisfactory and convenient size for most work. Very slight effects on the natural soil structure are produced by hammering the instrument into the soil. A wooden plank on top of the heavy steel ring aids in absorbing the shock of the hammer. Many soil cores may be extracted by having a large number of brass inner rings. After the sample is removed from the soil, the core is placed in a paraffined pint-sized ice cream container and sealed. In this way the cores can be stored until complete porosity measurements can be made, without serious moisture losses. A simple form of such a sampler has been designed that takes core samples that are very handy to use in field and laboratory studies (Lutz, 1947).

A modification of the usual King type of core sampler has also been perfected (Veihmeyer, 1929). The diameter of the cutting edge is slightly less than that of the tube so that the soil slides through the tube easily. There is also a small enlargement on the outside of the tool above the cutting edge to make the tool more easily extracted from the soil, especially when deep core samples are taken. This sampler has the distinct advantage of making deep sampling possible. The bulk density is equal to the weight of the dried soil in its natural state divided by the volume of the core.

The Clod Technique. The bulk densities of natural blocks of soil that were covered with paraffin were measured by Shaw in 1908 (Shaw, 1917). This basic principle was used by Sekera (1931) to determine the porosity and water capacity of soils. After the clod is allowed to saturate by capillarity with water under a tension of about 7 cm, its volume is determined by measuring the displacement of the wet clod in mineral oil. If the wet clod is then removed from the soil, washed with xylene until free from oil, dried, and weighed, it is possible to determine the aeration porosity. This method is well adapted to the study of the porosity of the larger units into which the soil crumbles. It may give an erroneous concept of the porosity of the profile as a whole if there are considerable pores between these aggregates or fragments. In order to prevent slaking, it is essential not to dry the clods before saturating them by capillarity. The Soil Survey Laboratory of the Soil Conservation Service, United States Department of Agriculture, in 1959 began to use a plastic coating of natural clods to measure bulk density (Brasher et al., 1966). The plastic coating is permeable to water vapor but practically impermeable to liquid water. It is simple to use, is durable and flexible, and does not melt like paraffin. It is well adapted to field usage. The plastic coating can be removed from the flat surface of the clod and the clod placed upon a tension table to determine the aeration porosity at different tensions. The clod technique permits measuring the bulk density in soils where rocks or roots would interfere with taking core samples. Its flexibility makes it a reliable tool in evaluating the porosity parameters of soil structure.

Gamma Radiation Transmission. The transmission of gamma rays through a material is a function of its density. Vomocil (1954) applied this principle to soils and obtained excellent correlations with bulk densities determined by the core method. The method involves preparing two holes in a soil at a specified distance apart. A probe containing a Geiger counter is lowered into one hole; this detects the radiation transmitted through the soil from a gamma-ray source that has been lowered in the second hole. The transmission of gamma rays follows Beer's law:

$$\frac{I}{I_0} = e^{-ux} \tag{5-7}$$

where I_0 is the initial radiation intensity at the source of the gamma rays, I is the transmitted intensity at distance x from the source, and u is the factor that is proportional to the soil bulk density. Since the transmission intensity is also a function of the soil moisture content, determinations of soil water at the time of measurement must be made. This can be done in the same holes with the neutron-scattering method (van Bavel, 1958). This technique makes possible obtaining bulk density values without appreciable soil disturbance. Changes in bulk density with time can be easily measured. Once the instrument is calibrated in a given soil, it can serve as a measure of soil moisture.

Microscopic Evaluations

Microscopic methods that utilize the principle of filling the soil pore space with fixing liquids are making many contributions to a visual analysis of soil structure.

Pigulevsky (1930) devised a method of soil fixation whereby thin sections can be examined under the microscope. A mixture of three parts of paraffin and one of naphthalene was forced into the soil-pore space. After cooling, the mass was cut and polished for study under the microscope. Harper and Volk (1936) modified Pigulevsky's technique by substituting lacquer or Bakelite varnish as the cementing material to fill the pores. Thin sections were prepared, and photographs were taken to give a magnification of 4.5 diameters. The aeration porosities of two distinctly different soils are shown in Figure 5-3. It is seen that the Marshall silt loam profile exhibits a rather uniform aeration porosity throughout all horizons. This is in accordance with the porosity data in Figure 5-1. The Parsons very fine sandy loam has a high percentage of aeration porosity in the surface layer but very little in the subsoil horizons. These layers give an appearance similar to that in Figure 4-2b. The particles are so tightly packed that little granulation or aeration porosity is perceptible.

The classical investigations of Kubiena (1938), Altemüller (1956a,

MARSHALL SILT LOAM   PARSONS VERY FINE SANDY LOAM
Depth

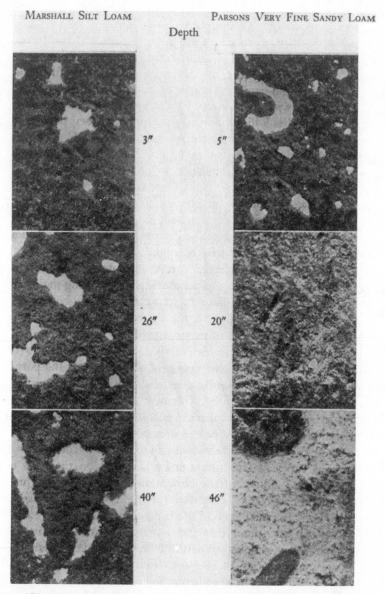

FIG. 5-3. Photomicrographs of aeration porosity in soil profiles (magnification 22 times). (Courtesy of H. J. Harper.)

1956b), and Brewer (1964) have resulted in improved techniques, better interpretations, and broader applications of the thin-section approach to the study of soil structure and related problems. New improved resins have been available to prepare the thin sections. The pore-size distribution of Marshall silt loam as determined by the micrometric technique for visually evaluating the amount and size of the pores compares very closely with the distribution calculated from moisture-tension curves (Swanson and Peterson, 1942). The data in Figure 5-4 confirm that the percentage of pores >50 μ, as measured micrometrically, is nearly the same as the aeration porosity at 60 cm moisture tension.

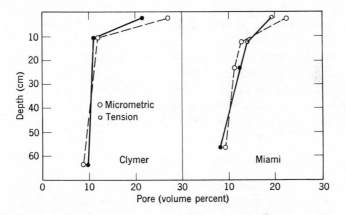

Fig. 5-4. Comparison of pore volume percentages by micrometric ($>50\mu$) and moisture-tension (60 cm water) techniques. (After Blevins, 1967.)

AGRICULTURAL SIGNIFICANCE

Cultural Changes Affecting Soil Structure

Improving Soil Tilth

Tilth is generally defined as the physical condition of the soil in its relation to plant growth. These soil physical conditions include ready infiltration of rainfall, sufficient moisture, adequate aeration, and favorable soil temperatures. This means that tilth is related to the size distribution of the soil aggregates (Russell, 1938). Good soil tilth is manifested in the friable range of soil consistency. It is a dynamic soil condition and tends to deteriorate under the usual cropping and tillage operations.

Tillage can produce good tilth by loosening and granulating the soil. By breaking up the soil mass, infiltration of rainfall and aeration are increased and soil strength is decreased. It is important to point out that the range of friable consistency in which tilth reaches an optimum is the same range in which the soil can be tilled with the least output of power and with its best effects upon granulation.

Various methods have been proposed to measure soil tilth. The size distribution of clods, obtained through dry sieving (Keen, 1930; Chepil, 1962), gives a picture of the relative fineness or coarseness of the tilled layer. The data can be expressed as mean weight-diameter, as discussed in Chapter 4. Yoder (1937) attempted to translate the impact of clod size in terms of cotton yields and observed that there was a relationship between the production of cotton and the aeration porosity of the different clod seedbeds.

The penetrometer, as discussed in Chapter 3, has been used as an index of tilth (Hénin, 1936; Shaw, Haise, and Farnsworth, 1942). Scott-Blair (1938) compressed the soil to determine tilth. The more the tilled soil could be compressed, the better the tilth; the amount of compression denoted the looseness of the soil. Both the penetrometer and compression techniques measure consistency and structure. In other words, compressibility and penetration vary with consistency at a given structure.

Tillage Operations to Promote Tilth

PLOWING. The mechanical functions of the plow, listed in the order in which they occur during the passage of the plow through the soil, consist of the cutting loose, granulation, and inversion of the furrow slice, and the turning under of residues, weeds, and green manures (Nichols, 1929, 1934). Thus, from a purely mechanical point of view, the plow has been perfected to turn down surface material and to cause a loosening of the surface soil as a result of the breaking loose and inversion of the furrow slice. In the absence of surface material the primary function of the plow is to sever and granulate a given volume of soil, which is usually referred to as the furrow slice. The cutting loose of the furrow slice takes place at the edge and shin of the plowshare. Granulation occurs throughout the entire moldboard, although a considerable part of the granulating action may take place at the front portion of the plow surface. Lifting and inversion of the furrow slice take place through the length of the moldboard. The completeness of coverage of materials depends upon the success with which the furrow slice is lifted and inverted.

DEEP PLOWING. The depth of plowing generally varies with the type of plow and the power available. Animal plowing usually means shallow plowing. This is especially true in the developing countries where the depth of plowing may not exceed 3 to 4 in. Although tractor power makes possible deeper plowing, normal depths average about 8 in. Deeper plowing has been advocated on many soils to extend the depth of the rootbed, both for increasing root elongation and proliferation as well as soil-moisture storage. There have been large variations in the success of deep-plowing operations. One must consider the type of soil, the nature of the crop, the climate, and the economics of the operation in evaluating the success or failure of deeper plowing.

Soils may be categorized into at least three groups from the standpoint of deep plowing. There are the more or less homogenous soil profiles with little distinct horizonation and possessing about the same texture in the surface and the subsoil. The low humic latosols (oxisols) of Hawaii and the grumusols of Texas fall in this category. Observations have shown that the rooting depth of sugarcane in latosols is generally restricted to the depth of tillage. Deep plowing on these soils therefore is essential for optimum water storage, root penetration, and cane growth. Increased cotton and sorghum yields have been obtained by physically modifying the profile of the grumusol Houston black clay (Burnett and Tackett, 1968). Root development to lower depths was increased with greater utilization of subsoil water.

There are also soils with heterogenous profiles; those with compacted traffic soles in the subsoil and those with naturally occurring dense clay-pans. Traffic soles can be broken up by deeper plowing. Increased yields of corn and cotton have resulted from deep tillage and placement of fertilizers on soils where root penetration was restricted by a traffic pan (Patrick et al., 1959). There was increased utilization of subsoil moisture during periods between rains by the deeper root system.

Numerous experiments on claypan soils have produced data that question the economic advisability of deep plowing or subsoiling along with deeper placement of fertilizers. There has been evidence that deep placement of lime increases the rooting of alfalfa and sweet clover. Even though subsoil applications of lime and fertilizers increased the yields of alfalfa in Wisconsin, there were no increases in the yields of corn and oats (Englebert and Truog, 1956). Subsoil tillage of fragipans in the cornbelt of the United States has rarely produced increased yields of corn. Deeper placement of fertilizers has not been as effective as plowing them under (Larson et al., 1960).

Robertson and Volk (1968) failed to obtain increased root penetration or higher corn yields by deep profile mixing of a Florida spodosol, Leon fine sand. Lutz (1968) reported decreases in the yields of corn, cotton, and tobacco from deep plowing a Coastal Plain soil where the turned up subsoil caused serious surface crusting. He called attention to the fact that about three-fourths of the root systems of common field crops is in the upper 15 cm of soil and that the majority of summer rains consist of light showers. Both factors militate against deep plowing having significant effects upon root development except where a physical barrier may occur in the surface 20 to 30 cm.

Smith, Woodruff, and Whitt (1947) plowed Putnam silt loam, a heavy claypan soil, to a depth of 18 in. by having two plows follow each other with the second plowing the bottom of the furrow of the first. This resulted in a shattering of the subsoil without causing much of it to come to the surface. Liming and fertilizing the subsoil produced deep rooting of sweet clover. Yields of corn were increased by 31 percent by this deep plowing.

SUBSOILING. The use of subsoilers to loosen dense subsoils is a highly power-consuming operation. The major factor in this power requirement is the horizontal pressure of the standard against the soil which increases the force required for fragmentation or shear. Nichols and Reaves (1958) obtained decreases in draft from 7 to 22 percent by using curved rather than vertical standards. Power requirements depend upon the moisture content of the soil. Dry, consolidated soils have high draft requirements because of large cohesive forces. On the other hand, subsoiling soils in the moist, plastic state accomplishes few beneficial effects. There is no shattering of the subsoil. In fact, there is a puddling action that tends to seal the grooves made by the shoe of the implement and the standard. Water runs into the subsoiler paths under rainy conditions and accumulates there. It is standard field practice in Trinidad and Jamaica to subsoil heavy plastic clays to improve subsurface tilth. This subsoiling operation is carried out during the dry season so that there is shattering to a depth of 18 to 24 in. A considerable amount of power is required to pull the subsoiler through a clay soil dry enough to shatter. Cambered beds are used to remove excess rainfall so that the beneficial tilth created by the subsoiling operation does not deteriorate rapidly.

Subsoiling has been a widely used operation in the Hawaiian sugar industry. Subsoilers have been employed in many instances as a prelude to plowing in order to obtain better penetration and granulating action of disk plows. The performance of these subsoilers produced some interesting facts (Trouse and Humbert, 1959). First, there was incomplete shattering of the soil between the tines of the implement even when the standard shoe was replaced by duckfoot or sweep attachments. There were islands of unshattered soil between subsoiler paths where the field was subsoiled in two directions. Second, shear-plane development and shattering of soils at moisture contents above the permanent wilting percentage usually was confined to the upper 6 in. and plastic-flow shear extended to the remaining depth of subsoiling. This highly power-consuming operation should be considered in terms of specific objectives sought, the type and moisture content of the soil, and the costs involved.

Seedbed-Rootbed Preparation

The seedbed is usually considered the plowed layer of soil which has been so prepared that planted seeds will readily germinate and the young plants will have a satisfactory conditon for root development and growth. Unfortunately, the term seedbed has emphasized the importance of preparing a few surface inches so that the seeds will germinate and emerge. Insufficient emphasis has been placed upon the structural qualities of the plowed layer that will permit and promote effective growth after emergence. This becomes particularly obvious when one considers that the so-called seedbed functions in this capacity for only a week or 10 days. Therefore "rootbed" would be a much more appropriate term since it would refer to a favorable structure not only for germination and emergence but also for the complete development of the plant.

The Seedbed-Rootbed Profile

Let us first analyze the major requirements of a good seedbed-rootbed and then visualize the type of profile necessary to satisfy them. Advances in tillage techniques have not changed significantly since the basic requirements suggested by Slipher (1932). These are:

1. *Permit the rapid infiltration and satisfactory retention of usable rainfall.* With the exception of the sandy types, most soils have a sufficient water-holding capacity. However, the presence of a thin compacted layer at the immediate surface reduces the amount of infiltration, or intake of rainfall, and enhances surface runoff. This runoff, which may reach as much as 50 percent of the rainfall for certain torrential summer storms in the cornbelt, restricts the amount of water that may be available to the plant. It is significant to note that a compacted layer ¼ in. thick at the surface of the soil may determine the rate of water movement throughout the entire profile.

2. *Afford an adequate air capacity and a ready exchange of soil air with the atmosphere.* Adequate aeration is essential for the seed, as well as the growing plant, to carry on normal respiration activities. Moreover, the aerobic microbiological processes depend upon a sufficient supply

of oxygen. Thus root development and normal growth are definitely related to aeration. Poorly aerated soils, however fertile from the standpoint of nutrients, are not productive. Fortunately, the same physical qualities of granulation that permit the rapid intake of rainfall also favor adequate aeration.

3. *Offer little resistance to root penetration.* Although factors other than compaction within the seedbed may be more responsible for the development of extensive root systems, nevertheless, any physical resistance to penetration tends to hinder the normal growth of roots. A good granular structure provides a more suitable medium for root development than a fine-grained compacted seedbed.

4. *Resist erosion.* Granular soils resist the dispersing action of raindrops. Consequently, the structure of the surface horizon is maintained, and runoff is reduced. Smaller runoffs cut down soil losses due to washing.

The lower part of an ideal seedbed-rootbed should contain the finest granules and possess the firmest degree of settling. The coarseness of the granules, in most cases, should increase as one approaches the surface. The immediate surface should consist of distinctly coarse granules to absorb the shock of the impact of raindrops and thereby preserve an open structure. It is true that the granules should not be so coarse in the vicinity of the seed that sufficient moisture cannot be had for the purpose of germination. On the other hand, the granules should not be so fine that inadequate aeration will prove to be the limiting factor. But what are the limits of coarseness or fineness? This question cannot be answered satisfactorily in light of existing data. Yoder (1937) found that the emergence of cotton plants from seedbeds with different degrees of granulation was most rapid when about half the granules were $\frac{1}{8}$ to $\frac{1}{4}$ in. in diameter and the other half were smaller. A mixture of granules of varying sizes showed the best effects on stand.

This ideal type of seedbed aims at keeping the soil receptive to rainfall and favorable to air exchange. Its firmness underneath establishes sufficient contact with the subsoil to permit easy root development downward and an optimum movement of water within the soil profile. It must be realized, however, that various crops require different types of seedbed. Small seeds, such as vegetables and grasses, require a fine, fairly compact seedbed so that the seed and seedling will be supplied with the necessary water and plant nutrients. Crops with large seeds do not need such a firm seedbed; they often respond better to a coarser than to a very fine seedbed.

Any surface compaction in excess of 0.5 psi suppresses seedling emergence because of poor aeration and the inability of the seedling to break through the compacted layer (Stout, Buchele, and Snyder, 1961). On the other hand, compaction at the seed level at 5 to 10 psi improves emergence as a result of better soil moisture conditions around the seed. Planting machines should pack the soil at the seed level, press the seeds into this firmed area, and then cover the seeds with loose soil. Johnson and Henry (1964) also observed that even though surface compaction decreased the size of the granules and reduced overall drying rates, it increased soil strength and decreased emergence. They suggested that the soil should be firmed 1 in. above the seed. A diffusion barrier is thereby created, the overall drying rate is reduced, and seedling emergence is improved. They found compaction above the seed to be superior to that at the seed level.

Tillage Practices for Seedbed-Rootbeds

Building a favorable seedbed-rootbed for the growth of crops begins before plowing and continues long after planting. In other words, the stable granular structural condition of the soil, which is a primary prerequisite for obtaining good tilth in the seedbed-rootbed, must be produced before plowing. The plowing operation simply permits a falling apart of the furrow slice into granules that should already be present. Agronomic practices should be designed to take advantage of the structure-forming qualities of well-developed root systems and organic matter.

NORMAL TILLAGE. Plowing is generally the first operation in the preparation of a seedbed-rootbed. It can contribute materially in obtaining good tilth. If the plowing operation produces a fairly satisfactory seedbed that will require only a small amount of surface tillage, then good plowing will aid materially in the preservation of good tilth. If the plowing operation produces a poor type of seedbed that requires a considerable amount of surface preparation, then poor plowing will result in the gradual deterioration of soil structure.

The success of the plowing operations in obtaining good granulation of the furrow slice depends primarily upon the granular nature of the soil and the moisture content at which the soil is plowed. There is an optimum range of soil moisture for each soil that permits the most effective results in plowing. This is the moisture range of the friable consistency. If plowed when it is not friable, a soil will usually be left in a cloddy condition. The breaking down of such clods requires many surface tillage operations, such as disking, rolling, and harrowing, for a quick preparation of the seedbed. Fall or early spring plowing is often practiced on heavy soils that are difficult to find at the right moisture content. Nature then causes a breakdown of the clods by alternate freezing and thawing or wetting and drying.

Nichols and Reed (1934) observed that different physical conditions of the soil at the time of plowing materially affect the action of the plow upon soil granulation. They described different soil conditions and the type of reaction to the plow under each. Three of these conditions are the following:

1. Hard Cemented Soils. This is a condition which is frequently experienced when soils are plowed too dry. The normal packing of the soil is magnified to a considerable extent by the shrinkage due to the severe drying. The soil has a harsh consistency, and the cementation effects are large. Such soils break into large, hard blocks; there is no definite arrangement of the planes of fracture. Plowing soils in this condition makes the preparation of the desired type of seedbed most difficult. Even after the slaking action of rains, considerable surface fitting is required. The seedbed is generally a mixture of clods and dust.

2. Heavy Sods. Well-developed root systems of grasses form a dense mat in the surface of the soil, which tends to hold the furrow slice together. Even though it appears that the normal shear planes do not develop under these conditions, Nichols and Reed found that careful examination showed the presence of these planes below the zone of greatest root concentration. Good granulation seems to be obtained, even though the furrow slice is bound together by the roots. Decomposition of the roots, even to a small extent, results in an ideal granular seedbed. If sods are plowed too wet, the amount of this granulation is decreased; sods usually are easy to prepare into a good seedbed if the plowing operation is performed far enough in advance of planting. This is more true of grass sods than of legumes. Stiff sods require gently sloping moldboards in order to obtain successful inversion of the furrow slice.

3. Compacted Surfaces. A compacted surface of crust on soil that is somewhat loose underneath does not break regularly during plowing. Blocks of the surface layer are lifted up by the shear planes and shoved to one side. Granulation of the surface crust is difficult to achieve during the plowing operation; however, such a condition is uncommon.

MINIMUM TILLAGE. Minimum tillage refers to the planting of row crops in plowed land with no seedbed preparation except in the row where the seed is placed. This means that the remainder of the field serves as a rootbed for the growing plant in about the same structural state that existed after plowing. There are three variations of minimum tillage: strip processing, wheel-track planting, and plow plant.

Strip processing is the tilling of only a strip of soil in which the seed is planted. It is observed in Table 5-2 that the hand tilling of small areas in a plowed sod in which to place corn seeds produced the highest yields of corn. Strip processing seeks to achieve this same effect mechanically. Strips 6 to 10 in. wide are prepared by using sections of disks, rotary hoes, or harrows to break down the large clods and fill in the

TABLE 5-2

The Effects of Different Methods of Seedbed Preparation on Soil Properties and on Yield of Corn (Page, Willard, and McCuen, 1946)

| | | | Physical properties of soil | |
Plot number	Method of seedbed preparation	Eight-yr average yield of corn (bu/A)	Non-capillary porosity (percent)	Relative compact-ness
1	Plowed, followed by discing, standard treatment	46.7	25.9	100
2	Plowed, complete inversion of furrow slice, corn planted directly in surface	47.2	24.9	101
3	Rototilled	42.1	24.2	132
4	Subsurface tillage, corn planted in trash	39.1	19.2	183
5	Surface treatment only—shallow roto-tilling or discing only	34.5	14.2	180
6	Standard treatment plus straw mulch after first cultivation	50.5	26.9	91

open spaces between the furrow slices. Good seed-soil contact is thereby provided. The rotary hoe is an especially good tool since it leaves on the surface large aggregates, which are more resistant to the impact of raindrops than smaller ones. Crusting is reduced. The rotary hoe functions over a wide range of soil conditions ranging from fall-plowed to freshly plowed land.

Wheel-track planting depends upon the pressure of a loaded wheel to crush large aggregates, prepare a smooth seedbed in the row, and provide the necessary seed-soil contact at the seed level for favorable moisture conditions. This procedure should follow plowing as closely as possible for maximum results in order to take advantage of favorable soil moisture. The tractor wheels precede the row-planting implement. This type of planting in narrow wheel marks should permit seeding of small grains with minimum tillage.

In the plow-plant method, both operations are accomplished simultaneously. Planters are mounted either on the plow or on the side of the tractor. There is generally no strip processing to accommodate the seed. Consequently, the plow-plant technique may not be satisfactory on heavy soils that are not well aggregated. Promising results from plow planting corn have been reported (Bolten and Aylesworth, 1957; Meyer and Mannering, 1961).

TABLE 5-3
Effect of Tillage Practices on Corn Yields (after Schaller, 1951)

Tillage practice	Marshall soil (alfalfa-brome sod) (bu/A)	Marshall, Ida, Monona soils (corn-stubble land) (bu/A)
Plowed	96.1	73.0*
Hard-ground listed	72.3	77.7*
Subsurface tilled	70.9	70.3*
Loose-ground listed	96.2	—
Plowed	—	84.1†
Loose-ground listed	—	87.3†

* 32 fields.
† 18 fields.

Listing is another type of minimum tillage that is used considerably on the medium-textured, well-drained soils of western Iowa and adjacent areas. The listing machine has from two to four furrow openers that look like double-moldboard plows. Corn is planted in the bottom of each furrow. Contour listing conserves moisture and reduces soil runoff and erosion. If listing is done on unplowed land, it is known as hard-ground listing. In this case, listing and planting can be done in one operation. Loose-ground listing is often done on plowed land by attaching a disk-type furrow opener to a corn planter. Listing of corn stubble land generally has given larger yields than conventional plowing followed by the usual seedbed operations (Schaller; Table 5-3). Fall plowing of sod land followed by spring listing has produced similar yields to conventional plowing. Under the preceding conditions, maximum corn yields have been obtained with a minimum of power and labor and runoff and erosion have been controlled.

ZERO TILLAGE. The advent of chemical herbicides has made it possible to kill plants in sods without having to turn them under with a plow. This has stimulated interest in planting corn with zero tillage. Special grassland corn planters have been developed (Triplett, Johnson, and Van Doren, 1963). Triplett et al. (1964) planted corn in a nonplowed alfalfa-timothy sod that had been killed with herbicides. They found no significant differences between the zero-tillage and conventional corn tillage systems. There were 23 replicated experiments involving 8 soil types. Average corn yields for complete tillage were 98.0 bu/A; those for non-tillage were 99.2 bu/A. Later experiments with corn-stover mulches (Triplett et al., 1968) on Wooster silt loam showed that mulch protection was necessary to maintain no-tillage corn grain yields. These beneficial effects were attributed to increased soil moisture under the mulch. It seems apparent that corn can be grown on many soils with zero tillage, which will not only reduce power and labor costs but will also protect the soil from runoff and erosion. When corn is planted in sods or residues by direct planting of the seeds in the untilled soil, it can be visualized as a form of mulch tillage.

MULCH TILLAGE. Mulch tillage, or stubble mulching, attempts to achieve the objectives of plowing without turning under the surface residues. It is a system of management of crop residues that uses subsurface tillage to loosen the soil and at the same time leave the residues on the surface for erosion control. This subsurface loosening is usually accomplished by sweeps. It is a system that was developed in the arid and semiarid wheat belt of the United States. It received considerable emphasis following the dust storms of the middle 1930s. Although it is primarily an arid to semiarid agricultural practice, it has been extended to the more humid areas as an erosion control measure.

McCalla and Army (1961) summarized the available data on the various impacts of stubble-mulch tillage on soil properties and crop yields. The more important effects are outlined here.

1. Soil Temperature. Mulches reduce soil temperatures in the spring and early summer as a result of the reflection of solar energy, their insulating influences, and the greater heat capacity of the soil beneath them due to increased moisture.

2. Soil Structure. There is little difference between aggregation and porosity of stubble-mulched and plowed lands. The percentage of water-stable aggregates has been observed to increase under stubble mulching; the amount of dry aggregates has been shown to decrease. Crusting of soils can be controlled by stubble mulching, which increases the infiltration rate as well as the emergence rate of seedlings.

3. Erosion Control. The control of both wind and water erosion is perhaps the most important effect of mulch tillage. Water erosion is controlled by protecting the soil surface against the dispersive action of falling raindrops and increasing the infiltration rate. Wind erosion is controlled because the force of the wind at the surface of the soil may be reduced 5 to 99 percent by anchored crop residues.

4. Crop Yields. The impact of mulch tillage on crop yields depends considerably on the climate. There has been a tendency toward increased yields in the semiarid to arid climates. Plowing usually produces higher yields in subhumid to humid climates. This is illustrated by the data in Table 5-2, where corn yields on the subsurface-tilled plots are lower than those on the plowed ones. These results indicate that mulch tillage for corn in humid areas should be confined to well-drained soils where there is an erosion hazard.

SEQUENCES OF OPERATIONS. Surface tillage operations in the preparation of seedbeds for row crops usually consist of disking or harrowing, leveling, and compacting. The number and type of these operations vary with the supposed condition of the soil, the nature of the crop to be planted, and the opinion and prejudices of the individual farmer. It cannot be overemphasized that the larger the number of surface manipulations, the greater will be the eventual breakdown of soil granulation. This is true irrespective of the types of implements used. The disk and the harrow will loosen a compacted surface so that it is friable. These same implements will recompact a loose, plowed soil. Tilth produced through surface manipulations is only temporary and superficial, however essential. Surface tillage operations should aim at producing the desired tilth with the least agitation of the structure left by the plow.

There were 26 sugar plantations in Hawaii in 1964. A survey of tillage sequences (Trouse and Baver, 1965) showed that three plantations on hydrol humic latosols prepared seedbed-rootbeds by plowing with a moldboard plow and then fitting the soil with a rotary hoe. Two plantations on humic latosols disk-harrowed the soil and then plowed with a gyrotiller. Each of the remaining 21 plantations followed its own sequence. These ranged from a total of eight different operations on a heavy clay soil to a combined subsoiler-disk harrow operation on a friable reddish prairie soil.

SUMMARY. The major principles in good seedbed-rootbed preparation for corn and other row crops are indicated by the data in Table 5-2. There are several significant points in this pioneering experiment. First, corn yields on Plot 2, where no seedbed preparation other than plowing took place, were equal to those on Plot 1, which was farmed according to the standard procedures in the cornbelt. Both plowed plots were the most porous and mellow. Second, the yields of corn on those plots where organic residues were left on the immediate surface, Plots 4 and 5, were much lower than those on the plowed plots. The subsurface-tilled and disked plots were compact and the most poorly aerated. Subsurface tillage did not adequately kill the sod plants, which presented a problem. Corn on the disked plots exhibited severe potash deficiencies. Third, the rototilled Plot 3 did not produce as high yields as the plowed plots, although they were always higher than those that were subsurface tilled. The rototilled plot was left in such a loose and fluffy condition that the first intense rain left the surface in an extremely tight and compact state. Fourth, Plot 6, which was mulched with straw after the first cultivation, gave the highest yields. This was due to better soil structure and moisture conditions under the mulch. The mulch preserved the good structure that was formed by plowing and cultivating.

Cultivation

The maintenance of good tilth through cultivation during the growing season is of major importance, especially since it regulates the water and air regimes of the soil. Rapid infiltration of rainfall and adequate aeration are the two most significant reasons for cultivating crops.

Increasing the Infiltration of Water

Previous discussions have shown that certain soils crust under the impact of falling raindrops. These crusts reduce water infiltration and impede aeration under some circumstances. Traffic from seedbed-preparation and postplanting operations cause surface and subsurface compaction that also restrict water infiltration and adequate aeration. Cultivation increases the infiltration rates of crusted and compacted soils.

The investigations of Musgrave and Free (1936) and Neal (1937) provided basic data on the relationships of the condition of the soil surface to water infiltration that are still valid. Infiltration rates of water into a field soil of Marshall silt loam (1) in its natural packing, (2) cultivated 4 in. deep, and (3) cultivated 6 in. deep, were 0.85 in., 1.77 in., and 1.87 in. per 15 min, respectively. The first hour rates for these same conditions were 1.49 in., 2.31 in., and 2.93 in., respectively. Cultivation only increased the initial rates of infiltration; after 1 hr the curves were practically parallel. This means that loosening the surface soil and maintaining a high porosity at the surface cut down runoff by permitting the rapid intake of water at the beginning of the storm. Once this cultivated layer was saturated, further infiltration was limited by the rate of percolation through the profile.

TABLE 5-4
Effect of the Condition of the Soil Surface on Runoff and Soil Losses (Neal, 1937)

Surface treatment	Cumulative soil and water losses at the end of the minutes indicated							
	10		20		30		40	
	Soil (lb)	Runoff (per-cent)	Soil (lb)	Runoff (per-cent)	Soil (lb)	Runoff (per-cent)	Soil (lb)	Runoff (per-cent)
Normal	0.20	3.6	1.09	30.3	2.42	45.3	7.61	65.3
Dry pulverized	0	0	0.12	9.1	0.48	20.9	2.84	42.6
Dry hard baked	1.86	41.2	5.50	63.2	8.71	72.7	16.68	83.6
Rough spaded	0	0	0	0	0.19	4.0	3.74	39.3

Neal (1937) demonstrated that the condition of the soil surface at the beginning of a rain is an important factor in determining the amount of runoff and erosion. His experimental results, which were obtained by artificial applications of rain on a small plot, are given in Table 5-4. The normal surface treatment consisted of cultivating to a depth of 4 in. and raking with an ordinary garden rake; the moisture content varied between one-fourth and one-half of the capillary capacity in the surface inch and between two-thirds and three-fourths of the capillary capacity throughout the rest of the cultivated layer. The very dry pulverized surface was worked and dried throughout the upper 4 in. The rough-spaded surface was prepared by spading to a depth of 4 to 5 in. after the soil was dry and allowing it to dry out without further treatment. The dry, hard-baked surface was prepared by allowing the soil to dry out after a heavy application of water. This type of surface simulated field conditions after an intense rain. Water was applied at the rate of 2 in./hr on an 8 percent slope.

It may be seen that the hard surface lost the most water as well as the most soil. About 84 percent of the total rainfall was lost. On the

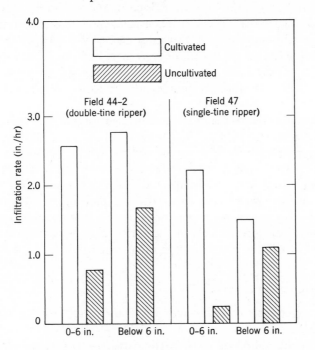

Fɪɢ. 5-5. Effect of cultivation on infiltration rate of soils. (After Trouse and Baver, 1965b.)

other hand, the rough-spaded surfaces, which simulated the condition of the soil in the field after cultivation, lost only about 39 percent. Runoff on the hard-baked surface occurred 4 min after the rain started; that on the rough-spaced surface took place 26 min after the rain began.

Trouse and Baver (1965b) observed that the cultivation of ratoon fields of sugarcane was essential to restore good tilth to soils that were compacted during the harvesting operation. Cultivation reduced the erosion hazard and improved water storage. The data in Figure 5-5 indicate that the infiltration capacity of the surface 6 in. was increased threefold to sevenfold. There was a sizable increase in the permeability of the soil beneath 6 in. It should be emphasized that cultivation of compacted soils must penetrate below the depth of compaction into the more permeable subsoil to be effective in runoff and erosion control.

Enhancing Aeration

Although little direct information exists concerning the extent to which cultivation increases the normal aeration of the root zone, there is reason to believe that the air supply to plant roots as well as to soil microorganisms is greatly augmented. Cultivation decreases the bulk density of the soil, which means that total porosity is increased. Moreover, the percentage of large pores (aeration porosity) is increased to a greater extent than the total pore space. Consequently, the total volume of air within the seedbed is raised.

Cultivation for better aeration is important on flat as well as on rolling lands. As a matter of fact, it is usually more important on flat lands, since the standing of water on the soil tends to facilitate the formation of a compacted layer at the surface. Common experience has demonstrated that cultivation of compacted soils gives marked crop responses that do not seem to be associated with soil moisture. This is particularly true of such high air-requiring crops as sugar beets and potatoes. The loosening of compacted soils in the early spring speeds up bacterial decomposition and the mineralization of nitrogen.

Degradation of Structure

The aggregation of cultivated forest and prarie soils is much lower than soils from adjacent virgin areas. Not only are there fewer aggregates present in the cultivated soils, particularly the larger granules, but also there is a lower percentage of the finer particles in the aggregated state. Degradation of aggregates is highly correlated with decreases in organic matter. Cultivation not only exposes the accumulated organic matter in virgin soils to greater microbial decomposition but also removes the source of annual accretions to the organic matter cycle in the soil and the structure-forming activities of earthworms that are always abundant in grasslands and under forest litter.

Sixty years of cultivation of Putnam silt loam in Missouri led to a 38 percent decrease in organic matter, a 33 percent reduction in available bases, and a corresponding diminution of aggregation (Jenny, 1933). The porosity of a clay soil in northwestern Ohio decreased from 16 to 18 percent below that of relatively virgin forest during 40 years of farming to intertilled crops. A porosity of 60.3 percent in the surface foot of a soil, producing its first corn crop after the sod had been plowed, gave a corn yield of 80 bu/A as compared with 20 bushels in the cultivated field across the fence, where the soil had a total porosity of 50.5 percent. This difference in yield was not primarily due to fertility, since fertilization gave little response on this soil unless its physical properties were first improved. A virgin sod in Australia exhibited 80 to 100 percent aggregation throughout the upper 6 in. of soil (Figure 5-6). Continuous

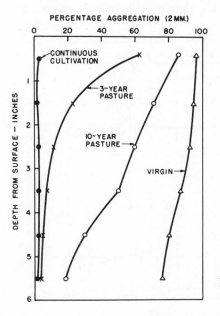

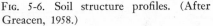

Fig. 5-6. Soil structure profiles. (After Greacen, 1958.)

cultivation reduced the number of aggregates to about 2:5 percent at the same depths. These typical examples point out quite clearly the impact of normal agricultural practices on the deterioration of soil structure.

Restoration of Structure

Cropping Systems

The restoration of stable granulation is a relatively slow process and many years are usually necessary to bring about the good stable granulation that approaches that of virgin conditions. Such restoration should rely considerably upon grass crops because of the strong aggregating effects of their extensive root systems. Although many results have been reported to show the effect of different cropping systems on soil aggregation, only a few will be cited to illustrate the major principles.

The data in Figure 5-7 illustrate the effect of rotations on the structure of Putnam silt loam from experimental plots on Sanborn Field (Woodruff, 1939). They were obtained by wet sieving 1- to 2-mm air-dry aggregates in water for different lengths of time. The rotation of corn and wheat, either with sweet clover as a catch crop or with manure, was much more highly aggregated than continuous corn. After 200 min of sieving, the rotation plots contained more stable aggregates than were in the corn

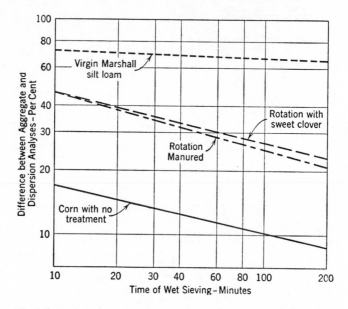

Fig. 5-7. The effect of rotations on soil aggregation. (Woodruff, 1939.)

plot after only 10 min. The slopes of the curves indicate that the aggregates from the different plots which were water-stable had about equal resistances to abrasion. It required 100 years of grass for the percentage of soil particles <2 mm that were aggregated into granules >2 mm to increase from 12 percent in a cultivated soil in Scotland to 73 percent under grass (Low, 1955). This improvement under grass was linked with earthworm activities as well as with root effects. The data in Figure 5-6 indicated that 3 years of pasture increased aggregation considerably in the upper 2 in. of soil; the major effect was in the first inch. This is in the area of maximum concentration of roots. The soil under 10 years of pasture showed improved aggregation at all depths with the first inch closely approaching the stability of structure of the virgin area. These curves point out that the root effects of grasses become deeper with the age of the sod as the root systems proliferate.

Cropping systems play a significant role in soil aggregation primarily through the impact of the granulating effects of root systems, the protective influences of vegetative canopies, and the production of organic materials that promote biological activity. Generally speaking, intertilled annual crops do not have the potential of contributing positively to this role as perennial crops. Rotations have been most effective in promoting and maintaining good soil structure. However, present cultural practices of growing corn continuously under conditions of large stalk populations, heavy fertilization, the control of weeds with herbicides, and the return of residues to the soil indicate that good soil tilth can be maintained under continuous culture with an annual crop.

Effects of Lime, Manure, and Fertilizers

Lime and manure have long been looked upon by farmers and scientists as practical solutions of soil-structure problems. Attention has previously been called to the fact that field and laboratory data do not confirm any direct effect of lime on soil structure. It is recognized, however, that liming promotes greater development of vegetation and production of organic matter, which usually causes a regeneration of structure. Experimental evidence in Belgium observed that plow-sole compaction and aggregate-stability deterioration was twice as great on soils low in lime as compared with those with visible $CaCO_3$ present (DeLeenheer, 1964). The effect of lime was minimized on fields where the cattle density was higher than 1.3 animals per hectare; this was due to the beneficial influences of the manure. The recommendation has been made to add 50 to 60 tons of $CaCO_3$ per hectare on all mechanized farms experiencing soil-structure deterioration.

Applications of farmyard manure have produced varying results on soil structure. Many investigators have obtained no beneficial physical effects of manure on soil properties and have attributed crop responses to the plant nutrients in the manure. Others have noted significant effects of manure on soil structure. The data in Figure 5-7 point out that manure was equivalent to a sweet-clover catch crop in the rotation in promoting stable granulation. About 46 percent of the soil particles from the untreated continuous-corn plot passed through a 2-mm screen as compared with only 29 percent from the manured continuous-corn plot. The method of application of manure affects the aeration porosity of the surface layers of clay (Baver and Farnsworth, 1940). The results in Table

5-5 indicate that manure significantly increased the percentage of larger pores in the 3-in. layer in which it was incorporated. The influence of the manure almost disappeared by the end of the growing season, with the exception of the surface application. Apparently, the mulching effect stabilized the surface aggregates against the destructive impacts of raindrops. The manure curve in Figure 4-6 showed that aggregate stability was increased slightly several months after incorporation of the manure. It made a slight contribution to the stable humus in the soil.

TABLE 5-5
The Effect of Manure on the Aeration Porosity of Brookston Clay
(Baver and Farnsworth, 1940)

Plot treatment	Depth (in.)	Aeration porosity (percent by volume)	
		June 11, 1940	Sept. 25, 1940
Fall-plowed-check	0–3	8.7	5.9
	3–6	6.2	5.6
Fall-plowed-20 tons manure under	0–3	10.4	8.7
	3–6	17.5	8.7
Fall-plowed-20 tons manure disked in surface in spring	0–3	16.8	13.9
	3–6	6.4	7.2

In general, it appears that manure exerts a favorable influence on granulation and aeration in the soil but that this effect is not permanent. With the possible exception of heavy applications, the fertility factor usually outweighs the physical. Applications of manures to sods increase the growth of these structure-improving crops and benefit structure in this way.

Little is known concerning the effects of fertilizer on soil structure. Theoretically, large applications of sodium salts should cause a dispersion of the soil aggregates. Such a deterioration of structure will probably not occur under normal agricultural practices. Increased foliage and root production as a result of fertilizer applications undoubtedly have a great influence on the preservation and partial restoration of structure. This should be especially true of sod crops.

Lutz and associates (1966) obtained significant effects of phosphate fertilizers on soil tilth. Soil strength was decreased and the water-holding capacity increased. They attributed this action to the replacement of

OH ions by phosphate ions on the edges of the clay particles which caused aggregation of the particles.

Changing the Structure Profile of Deep Sands

The "single-grained" structure of sands causes them to drain rapidly and to have a very low water-retention capacity. Erickson, Hansen, and Smucker (1968) placed impermeable asphalt barriers in deep sands at depths of 55 to 60 cm in order to change the structure profile. These barriers not only decreased the depth of penetration of water but also increased the water retention in the plant root zone. Apparently, water drains down to the barrier or to the temporary perched water table above the barrier. It then moves off the edges of the barrier. An appreciable amount of water that is held at low tensions remains above the barrier, which accounts for the increased water-retention capacity of the sand.

Soil Structure and Root Development

There are several factors that influence root penetration and proliferation in soils. The availability of plant nutrients plays an important role in the proliferation of the root system, even though roots can penetrate zones with a low nutrient status. Water availability also affects root elongation. Roots do not penetrate dry soils and the rate of elongation decreases with increasing tension of the soil water. If toxic substances, such as excessive amounts of soluble aluminum, are present, root development can be restricted. The two main factors affecting the development of a good root system in compacted soil are aeration and mechanical impedance. In other words, the ability of plant roots to penetrate the soil is a function of its porosity and compressibility.

Aeration Effects

It is difficult to separate the effects of a lack of aeration and mechanical impedance since both are affected by increased bulk densities that result from soil compaction. Diminution of root elongation is highly correlated with aeration (Trouse and Baver, 1962). There is substantial evidence to support the concept that aeration is a prime factor in root development. Lawton (1945) obtained a 65 percent reduction in root growth of corn plants when Clarion loam was compacted to lower the aeration porosity from 37 to 1 percent. Bertrand and Kohnke (1957) established a close correlation between the oxygen-diffusion rate and the growth of corn plants in a compacted soil. They concluded that restriction of root development was due primarily to aeration. Wiersum (1960) found that no roots elongated when the oxygen-diffusion rate was below a certain minimum. There was a good correlation between rooting depth and oxygen diffusion. Gardner and Danielson (1964) obtained a correlation of 0.998 between root penetration of cotton roots and percentage of aeration porosity.

Mechanical Impedance

Observations in the field have indicated that the density of the soil itself must play a significant role in root penetration. Roots of many species of plants have been traced down to a "pan" and then observed to change direction and grow parallel to the friable soil-pan interface. Veihmeyer and Hendrickson (1948) studied the relationship between soil density of different types of soil and the penetration of sunflower roots. They found that the "threshold bulk density" of sands was about 1.75 g/cm^3; that of clays varied between 1.46 and 1.63 g/cm^3. Since roots penetrated saturated noncompacted soils from which most of the air had been expelled, Veihmeyer and Hendrickson suggested that the failure of penetration in compacted soils may be due to the size of pores rather than to a lack of oxygen.

Wiersum (1957) investigated the ability of roots of young oat plants to penetrate pores of various diameters and rigidity. The data showed conclusively that such roots can penetrate only pores with diameters exceeding those of very young roots. The plasticity of the young root tip of grasses does not permit it to undergo constriction and then pass through a narrow pore. The question arises how to explain the presence of roots in soil layers that have pore sizes below the limit of root diameters. The penetrating root has sufficient pressure to push aside obstructing soil particles if the plasticity and the density of the soil permit (Gill and Miller, 1956). When restriction due to pore rigidity occurs, roots have wrinkled tip surfaces and there is profuse branching behind the root tips. The experiments of Gill and Miller proved that root growth was practically stopped by mechanical impedance when the oxygen concentration was low; 15 psi (1 kg/cm^2) decreased root growth by 71 percent at an oxygen value of 1 percent; growth almost ceased at 30 psi. For a given pressure, root growth increased with oxygen content up to an apparent optimum level. This appeared to be above about 10 percent oxygen. Mechanical impedance seemed to be the main factor affecting root development at pressures greater than 20 psi.

The results of Barley in Figure 5-8 point out several significant facts concerning root elongation. First, at low pressures where mechanical impedance is a minor factor, the concentration of oxygen determines the extent of root growth. Second, roots do not elongate at high levels of mechanical impedance, irrespective of the oxygen levels. Third, at a given oxygen concentration, root elongation decreases logarithmically with increased pressure (mechanical impedance). A correlation of −0.951 was obtained between the amount of pressure on the diaphragm and log length of the seminal roots. These data are very similar to those obtained by Gill and Miller, who used a similar technique.

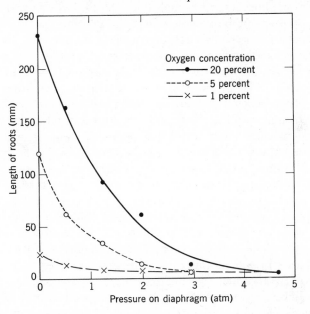

FIG. 5-8. The effects of pressure on the diaphragm and oxygen concentration on the growth of corn roots. (After Barley, 1962.)

The penetration of a root into the soil has been likened to the cylindrical compression of a sharp probe (Barley and Greacen, 1967; Greacen, Farrell, and Cockroft, 1968) (see Chapter 3). This distribution of bulk density with distance from the center of the root channel follows that predicted by the model for cylindrical compression. Cylindrical probes, however, provide only a relative measure of resistance to penetration. Growing roots (1) have flexibility, (2) have a different shape, which depends on the resistance of the soil, (3) possess different adhesion and friction properties, and (4) increase soil strength during growth by absorbing water, which changes the pore-water pressure. A pressure of about 7 bars is required to expand a root cavity in a soil just dense enough to halt root penetration. This is similar to the maximum pressure a root can exert. Apparently a plant root penetrates the soil by cylindrical compression but possesses some mechanism for overcoming soil-root friction.

Elongation of corn-seedling roots in soils in which bulk densities ranged from 0.94 to 1.30 at soil-moisture tensions of 10 and 100 cm of water decreased with increasing bulk density (Phillips and Kirkham, 1962). Elongation was correlated with bulk density at the 100-cm tension, since aeration was nonlimiting at this moisture content. Growth was correlated with both aeration and bulk density at 10-cm tension.

Experiments on the penetration of cotton roots into compressed cores where aeration was nonlimiting demonstrated that root elongation at a given bulk density depended upon the soil-moisture tension (Taylor and Gardner, 1963). The higher the moisture tension, the lower was the percentage root penetration into the cores. As would be expected, soil strength increased with increasing bulk densities and soil-moisture tension. A highly significant linear correlation of —0.96 was obtained between root penetration and soil strength; the correlation with bulk density was only —0.59. A high bulk density per se apparently was not the controlling factor in restricted root development; it was the augmented soil strength that results from this increase in density.

The data in Figure 5-9 show that the penetration of seedling cotton roots into soil cores compacted to different bulk densities is affected by varying levels of oxygen percentages. Bulk densities above 1.5 caused sharp decreases in root penetration. The curve relating root growth at a given oxygen percentage to bulk density was sigmoidal in shape. At the lower bulk densities, root elongation increased rapidly with rising oxygen concentration to a maximum at about 10 percent oxygen. Aeration was the dominant factor at these densities. Mechanical impedance played the major role at the higher densities. These results are comparable to those of Barley previously mentioned.

Interactions between Aeration Mechanical Impedance, and Soil-Moisture Stress

The suitability of the soil as an environment for the growth and functioning of plant root systems depends on (1) the availability of nutrients, oxygen, and water, (2) the degree of mechanical impedance to root proliferation, and (3) temperature. There is considerable interaction between these factors as they relate to root proliferation. For example, soil water directly affects the growth and functioning of the root system; it indirectly affects aeration, mechanical impedance, and soil temperature.

The aeration-mechanical impedance interaction has been apparent from previous discussions of the effect of aeration as well as mechanical impedance on root development. The aeration-mechanical impedance-soil moisture interactions have been analyzed by Eavis and Payne (1968) and Eavis (1970). Summaries of the important features of these investigations are illustrated in Figures 5-10 and 5-11. The curves in Figure 5-10 depict

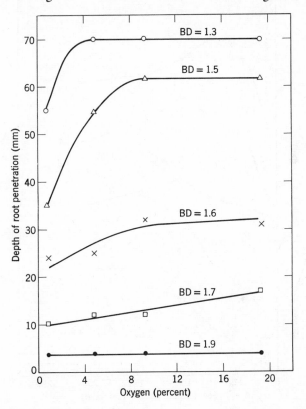

FIG. 5-9. The effect of oxygen concentrations and bulk density (BD) on depth of root penetration. (After Tackett and Pearson, 1964.)

the effect of increasing soil moisture tension on the resistance of soil to penetration (A) and to root elongation (B) as related to bulk density. These results point out that resistance to penetration (mechanical impedance) increases with bulk density and increasing moisture tension; this follows expectations. Root penetration is approximately inversely proportional to the level of mechanical impedance. Root elongation (curve B) increases with moisture tension to a maximum and then decreases as the moisture tension increases. The initial increase to the maximum point is associated with improving the aeration as the water is removed from the pores. The subsequent decrease is related to the increase in soil strength.

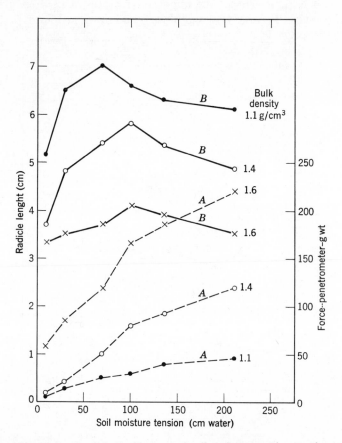

Fig. 5-10. The effect of soil bulk density and soil moisture tension on the resistance of soil to penetration (*A*) and root elongation (*B*) (Eavis, 1970.)

Figure 5-11 points out that only aeration and soil moisture stress restrict root growth in a low density soil (optimum physical condition, BD = 1.0) where mechanical impedance is not a factor. The mechanical impedance factor becomes more important as the bulk density (BD) becomes greater (curves BD = 1.4 and BD = 1.6). Lack of aeration effects are more strongly expressed at matric potentials below about 100 cm of water. Mechanical impedance is the major factor that restricts root elongation

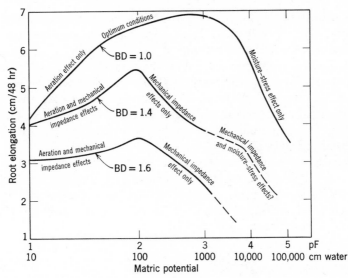

Fɪɢ. 5-11. Contributions of mechanical impedance, aeration, and moisture stress to root restriction. (Eavis, 1970.)

above this matric potential. Moisture stress effects may be linked with mechanical impedance at the higher tension values.

Soil Crusting

Mechanisms Involved

Soil crusts usually are formed as a result of compaction at the immediate surface due to an externally applied force. This force is supplied primarily by the impact of raindrops as the soil is wetted and the radiant energy of the sun as the soil dries. When the raindrops fall on dry soils, there is almost instantaneous slaking of the soil aggregates, followed by dispersion and orientation of the finer particles, and the clogging of the pores as these particles are carried into the soil. A compacted zone of higher bulk density is formed at the surface. The soil crust formed by raindrop impact consists of two distinct parts (McIntyre, 1958). First, there is a thin skin approximately 0.1 mm thick formed on the surface due to the compaction that results from the impact. Second, the dispersed particles that arise from impact are washed into the soil with the infiltrating water and clog the pores immediately beneath the surface to give a layer of decreased porosity. The water permeability of the "washed-in" zone is reduced about 200 times below that of the undisturbed soil beneath; that of the surface skin is reduced 2000 times. Aggregates from virgin soils with rather stable structure slake but do not disperse under raindrop impact. Consequently there is no washed-in zone and the water permeability of the crust is a function of the permeability of the thin surface skin. The zone immediately below the thin surface skin consists of oriented clay particles with very few isolated air pores (Evans and Buol, 1968; Wilding and Schmidt, 1968). It seems therefore that dispersion of aggregates rather than slaking is responsible for soil crusting. Upon drying, surface tension forces cause particle interaction and orientation as shrinkage takes place.

Particle rearrangement at the immediate surface can also be brought about by intense slaking and dispersion of aggregates when the soil is wetted to saturation, as is often the case in surface irrigation. A compacted zone is formed during infiltration of the turbid water, which becomes a hard crust upon drying.

Evaluating Crust Strength

Evaluation of the strength of soil crusts can be made in at least three ways. The most popular appears to be the modulus of rupture test to simulate crusting (Richards, 1953). The resistance to penetration of a soil probe can also be used (Parker and Taylor, 1965). These two procedures have been discussed in Chapter 3. The determination of crust strength can be approached from the point of view of the emerging seedling (Morton and Buchele, 1960; Arndt, 1965a, 1965b). Morton and Buchele developed a penetrometer to simulate a mechanical seedling and measured the force required to push it upward through a 3-in. layer of soil compacted to different bulk densities. Using various sizes of probes to represent different seedling diameters, they found that the emergence energy increased directly with seedling diameter, degree of compaction, initial soil-moisture content, and depth of planting.

Arndt designed an instrument that records the force required to bring about emergence of a mechanical probe that is buried prior to the formation of a surface crust. This type of probe does not penetrate the crust as it is forced upward. In the case of a wet crust, there is a conical rupture as a cone-shaped mass of soil is forced out of the crust ahead of the probe. In a dry soil, there is first a cracking of the crust, followed by the formation of a dome-shaped structure of tilted, broken pieces of crust. The modulus of rupture values, which reflect tensile strength, were only about 20 percent of the force required for emergence of the *in situ* probe through the crust. This probe produces both compression and shear stresses.

Crust strength, as measured by the modulus of rupture technique, has been observed to increase as the rate of drying decreases (Lemos and Lutz, 1957; Hillel, 1960). These results are not in agreement with the observations of Arndt, who found that larger impedance values were obtained with his method when the evaporation rates were high. Crust strength is greater when the rain falls on air-dry rather than wet soil (Hanks, 1960). There is a direct relationship between clay content and crust strength when the soil does not contain organic matter.

The mechanical composition of the surface soil plays an important role in crust formation, determining the strength of the crust as well as the frequency and width of cracking on drying (Arndt, 1965a). One must also consider the tremendous effect of organic matter on increasing the resistance of soil aggregates to the destructive impact of raindrops.

Infiltration and Soil Crusting

The soil structural relationships of the immediate surface have a major effect on the infiltration capacity of a soil. The data in Table 5-4 emphasize the importance of a compact soil surface on infiltration rates and water runoff. The impact of raindrops forms a surface crust that rapidly reduces infiltration (McIntyre, 1958). Wischmeier and Smith (1958) introduced the expression "accumulated rainfall since the last tillage operation" in their erosion-loss equation to provide a measure of the influence of soil compaction and crusting of the surface on runoff (see Chapter 13).

Seedling Emergence and Crust Characteristics

Restriction of seedling emergence can take place in two ways. First and foremost is the effect of mechanical impedance due to crust strength. Resistance at the surface may be so great that the seedlings buckle, grow in a horizontal direction and fail to emerge.

The emergence of grain sorghum seedlings decreased if the soil strength, as measured with a penetrometer, exceeded 3 bars and ceased when the strength reached 13 to 18 bars; this depended on the type of soil (Parker and Taylor, 1965). A crust strength of 273 mbars (modulus of rupture technique) prevented the emergence of bean seedlings in Pachappa fine sandy loam (Richards, 1953). On the other hand, critical modulus of rupture values ranged from 1200 to 2500 mbars for the emergence of sweet corn on Pachappa loam that contained exchangeable sodium (Allison, 1956). It is necessary to specify the moisture content in establishing a critical crust strength for restricting seedling emergence (Hanks, 1960). For example, modulus of rupture values for a silt loam soil vary for both crop and moisture content of the crust (Table 5-6). These data emphasize the difficulties of establishing critical soil strengths because of the variations encountered due to the nature of the plant, the moisture content of the crust at time of emergence, and the experimental technique itself.

Arndt made a study of the morphology of soil crusts in relation to cracking and seedling emergence. He observed that the natural pattern of cracking and the size of the seedlings may be more important factors in emergence than the strength of the soil crust. The location of the

TABLE 5-6
The Effect of Crop and Moisture Content
on Critical Crust Strength for Seedling Emergence
(after Hanks, 1960)

	Critical modulus of rupture range (mbars)	
Crop	25 percent moisture	14.5 percent moisture
Wheat	3,210–6,410	800–1,600
Grain sorghum	1,610–3,210	0–800
Soybeans	6,410	0–800

seedling in relation to the natural cracks in the crust is of considerable significance. In the case of thin seedlings, the cracks must be frequent enough to enable the seedlings to emerge freely. For coarse seedlings, both the size and frequency of the cracks are important for emergence. The cracks must be wide enough so that there is no jamming of the plates between cracks as the coarse seedlings lift plates of soil during emergence.

Strong crusts with high bulk densities can impede aeration under moist conditions if the pores in the crusted layer contain sufficient water to prevent effective diffusion of oxygen into the soil. Diffusion is not restricted significantly in the case of dry crusts. If lack of aeration does become a problem, germination of the seeds would be impaired, which would also account for decreases in emergence.

Crusting of soils can be controlled by surface mulches, which protect the soil from the impact of raindrops. As previously stated, organic matter promotes the formation of stable aggregates that resist dispersion. Certain artificial soil conditioners also reduce soil crusting by producing stable aggregation.

Soil Structure and Plant Growth

The soil as a medium for plant growth provides the environment for the germination and emergence of seedlings and for the development and functioning of the plant's root system. Not only must the soil furnish the nutrients required for the metabolic processes of plant growth but it must also provide favorable air-water regimes for proper functioning of the plant. In too many instances, the nutrient uptake from the soil is limited by a lack or excess of water, a deficiency of oxygen, or inadequate root proliferation resulting from high bulk densities that impede the growth of roots. The combination of increased density and decreased

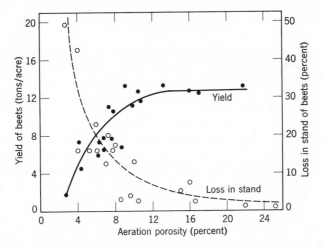

Fig. 5-12.   The relation of aeration porosity to the growth of sugar beets.

aeration not only restricts the proliferation of roots and their normal absorption of nutrients and water but also impedes microbiological activity. Unfavorable soil structure therefore becomes the limiting factor in crop production and should be considered a parameter of soil fertility. The integrated effects of poor soil structure on crop production are well illustrated in Figure 5-12, which shows the relation of sugar beet yields to soil aeration on a heavy clay soil that had received different organic matter treatments (Baver and Farnsworth, 1940). These curves point out two significant impacts of aeration on plant production. First, adequate aeration markedly decreased the loss in stand of beets due to a blackroot rot fungus. The deleterious effects of these organisms practically disappeared at aeration porosities greater than 10 percent. Second, the yields of beets mounted rapidly as aeration increased. Maximum yields were reached at about 10 percent aeration porosity. Not only the total yield of beets but also the percentage of sugar in the beets were raised. Short, stubby beets were produced on soils of low aeration and long, tapering ones on the aerated soil. Apparently this was due to a combination of increased soil strength and decreased aeration in those plots with poor soil structure. The well-aerated plots exhibited serious nitrogen deficiencies. Thus, on the same field, there were plots that were showing no response to fertilization because of a lack of aeration, while other plots were suffering from nutritional deficiencies after adequate aeration had been provided.

References

Allison, L. E. (1956). Soil and plant responses to VAMA and HPAN soil conditioners in the presence of high exchangeable sodium. *Soil Sci. Soc. Am. Proc.*, **20**:147–151.

Altemüller, H. J. (1956a). Neue Möglichkeiten zur Herstellung von Bodendünn-schliffen. *Z. Pflanzenernähr. Düngung, Bodenk.*, **72**:56–62.

Altemüller, H. J. (1956b). Mikroskopische Untersuchung einiger Löss-Bodentypen mit Hilfe von Dünnschliffen. *Z. Pflanzenernähr., Düngung, Bodenk.*, **72**:152–167.

Arndt, W. (1965a). The nature of the mechanical impedance to seedlings by soil surface seals. *Aust. J. Soil Res.*, **3**:45–54.

Arndt, W. (1965b). The impedance of soil seals and the forces of emerging seedlings. *Aust. J. Soil Res.*, **3**:56–58.

Barley, K. P. (1962). The effects of mechanical stress on the growth of roots. *J. Exp. Botany*, **13**:95–110.

Barley, K. P., and E. L. Greacen (1967). Mechanical resistance as a soil factor influencing the growth of roots and underground shoots. *Advances in Agronomy*, **19**:1–43, Academic Press, New York.

Baver, L. D., and R. B. Farnsworth (1940). Soil structure effects in the growth of sugar beets. *Soil Sci. Soc. Am. Proc.*, **5**:45–48.

Baver, L. D., and H. F. Rhoades (1932). Aggregate analysis as an aid in the study of soil structure relationships. *J. Am. Soc. Agron.*, **24**:920–930.

Bertrand, A. R., and H. Kohnke (1957). Subsoil conditions and their effects on oxygen supply and the growth of corn roots. *Soil Sci. Soc. Am. Proc.*, **21**:135–140.

Blevins, R. L. (1967). Micromorphology of soil fabric at tree root-soil interfaces. Ph.D. Thesis. Ohio State University.

Bolton, E. F., and J. W. Aylesworth (1957). A comparison of some tillage methods for corn on Brookston clay soil. *Can. J. Soil Sci.*, **37**:113–119.

Brasher, B. R., D. P. Franzmeier, V. Valassis, and S. E. Davidson (1966). Use of Saran Resin to coat natural soil clods for bulk-density and water-penetration measurements. *Soil Sci.*, **101**:108.

Brewer, R. (1964). *Fabric and Mineral Analysis of Soils*, John Wiley and Sons, New York.

Burnett, Earl, and J. L. Tackett (1968). Effect of soil profile modification on plant root development. *Trans. 9th Int. Cong. Soil Sci.*, 3:329–337.

Dawson, Roy C. (1947). Earthworm microbiology and the formation of water-stable aggregates. *Soil Sci. Soc. Am. Proc.*, 12:512–515.

Dettman, Margaret G., and W. W. Emerson (1959). A modified permeability test for measuring the cohesion of soil crumbs. *J. Soil Sci.*, 10:215–226.

Chepil, W. S. (1962). A compact rotary sieve and the importance of dry sieving on physical soil analysis. *Soil Sci. Soc. Am. Proc.*, 76:4–6.

Cole, R. C. (1939). Soil microstructure as affected by cultural treatments. *Hilgardia*, 12:429–472.

DeLeenheer, L. (1964). Preservation of soil structure on mechanized farms in Belgium. *Trans. 8th Int. Cong. Soil Sci.*, Bucharest, pp. 561–570.

DeLeenheer, L., and M. DeBoodt (1954). Discussion on the aggregate analysis of soils by wet sieving. *Trans. 5th Int. Cong. Soil Sci.*, Leopoldville, 2:111–117.

Eavis, B. W. (1970). Soil physical conditions affecting seedling root growth. I. Separation of the roles of mechanical impedance, aeration and moisture stress. In press.

Eavis, B. W., and D. Payne (1968). Soil physical conditions and root growth. *Proc. 15th Easter School Agr. Sci.*, University of Nottingham.

Emerson, W. W. (1954). The determination of the stability of soil crumbs. *J. Soil Sci.*, 5:233–250.

Englebert, L. E., and E. Truog (1956). Crop response to deep tillage with lime and fertilizer. *Soil Sci. Soc. Am. Proc.*, 20:50–54.

Evans, D. D., and S. W. Buol (1968). Micromorphological study of soil crusts. *Soil Sci. Soc. Am. Proc.*, 32:19–22.

Gardner, H. R., and R. E. Danielson (1964). Penetration of wax layers by cotton roots as affected by some soil physical conditions. *Soil Sci. Soc. Am. Proc.*, 28:457–460.

Gill, W. R., and R. D. Miller (1956). A method for study of the influence of mechanical impedance and aeration on the growth of seedling roots. *Soil Sci. Soc. Am. Proc.*, 20:154–157.

Greacen, E. L. (1958). The soil structure profile under pastures. *Aust. J. Agr. Res.*, 9:129–137.

Greacen, E. L., D. A. Farrell, and B. Cockroft (1968). Soil resistance to metal probes and plant roots. *Trans. 9th Int. Cong. Soil Sci.*, 1:769–779.

Greenland, D. J., G. R. Lindstrom, and J. P. Quirk (1962). Organic materials which stabilize natural soil aggregates. *Soil Sci. Soc. Am. Proc.*, 26:366–371.

Hanks, R. J., and F. C. Thorp (1956). Seedling emergence of wheat as related to soil moisture content, bulk density, oxygen diffusion rate, and crust strength. *Soil Sci. Soc. Am. Proc.*, 20:307–310.

Hanks, R. J. (1960). Soil crusting and seedling emergence. *Trans. 7th Int. Cong. Soil Sci.*, 1:340–346.

Harper, H. J., and G. W. Volk (1936). A method for the microscopic examination of the natural structure and pore space in soils. *Soil Sci. Soc. Am. Proc.*, 1:39–42.

Hénin, S. (1936). Quelques résultats obtenus dans l'etude des sols a l'aide de la sonde dynamometrique de Demolon-Hénin. *Soil Research*, Vol. 5, No. 1.

Hénin, S., O. Robichet, and A. Jongerius (1955). Principes pour l'evaluation de la stabilite de la structure du sol. *Ann. Agron.*, 6:537–557.

Hillel, D. (1960). Crust formation in loessial soils. *Trans. 7th Int. Cong. Soil Sci.*, 1:330–339.

Jamison, Vernon C. (1953). The significance of air adsorption by soil colloids in picnometric measurements. *Soil Sci. Soc. Am. Proc.*, 17:17–19.

Jenny, Hans (1933). Soil fertility losses under Missouri conditions. *Missouri Agr. Exp. Sta. Bull.* 324.

Johnson, W. H., and J. E. Henry (1964). Influence of simulated row compaction on seedling emergence and soil drying rates. *Trans. Am. Soc. Agr. Engr.*, 7:252–255.

Keen, B. A. (1931). *The Physical Properties of Soils.* Longmans, Green and Co., London.

Keen, B. A. (1933). Experimental methods for the study of soil cultivation. *Empire J. Exp. Agr.*, 1:97–102.

Kemper, W. D. (1965). Aggregate stability. *Methods of Soil Analysis, Agronomy Monograph*, No. 9, Part 1. Academic Press, New York, pp. 511–519.

Kemper, W. D., and W. S. Chepil (1965). Size distribution of aggregates. *Methods of Soil Analysis, Agronomy Monograph*, No. 9, Part 1. Academic Press, New York, pp. 499–510.

Kubiena, W. L. (1938). *Micropedology.* Collegiate Press, Ames, Iowa, pp. 134, 140, 156.

Kummer, F. A., and A. W. Cooper (1945). The dynamic properties of soils. IX. Soil porosity determinations with the air pressure pycnometer as compared with the tension method. *Agr. Eng.*, 26:21–23.

Larson, W. E., W. G. Lovely, J. T. Pesek, and R. E. Burwell (1960). Effect of subsoiling and deep fertilizer placement on yields of corn in Iowa and Illinois. *Agron. J.*, 52:185–189.

Lawton, K. (1945). The influence of soil aeration on the growth and absorption of nutrients by corn plants. *Soil Sci. Soc. Am. Proc.*, 10:263–268.

Lemos, Petezval, and J. F. Lutz (1957). Soil crusting and some factors affecting it. *Soil Sci. Soc. Am. Proc.*, 21:485–491.

Low, A. J. (1955). Improvements in the structural state of soils under ley. *J. Soil Sci.*, 6:179–199.

Lutz, J. F. (1947). Apparatus for collecting undisturbed soil samples. *Soil Sci.*, 64:399–401.

Lutz, J. F. (1968). Water movement-root relations in heterogeneous soils. *Trans. 9th Int. Cong. Soil Sci.*, 3:339–346.

Lutz, J. F., Rafael A. Pinto, Ricardo Garcia-Lagos, and H. Gill Hilton (1966). Effect of phosphorus on some physical properties of soils: II. Water retention. *Soil Sci. Soc. Am. Proc.*, 30:433–437.

McCalla, T. M. (1944). Water-drop method of determining stability of soil structure. *Soil Sci.*, **58**:117–121.

McCalla, T. M., and T. J. Army (1961). Stubble mulch farming. *Advances in Agronomy.* **13**:125–196, Academic Press, New York.

McIntyre, D. S. (1958). Permeability measurements of soil crusts formed by raindrop impact. *Soil Sci.*, **85**:185–189.

Mazurak, Andrew P. (1950). Effect of gaseous phase on water-stable synthetic aggregates. *Soil Sci.*, **69**:135–148.

Meyer, L. D., and J. V. Mannering (1961). Minimum tillage for corn: Its effect on infiltration and erosion. *Agr. Eng.*, **42**:72–76.

Morton, C. T., and W. F. Buchele (1960). Emergence energy of plant seedlings. *Agr. Eng.*, **41**:428–431.

Musgrave, G. W., and G. R. Free (1936). Some factors which modify the rate and total amount of infiltration of field soils. *J. Am. Soc. Agron.*, **28**:727–739.

Neal, J. H. (1937). The effect of degree of slope and rainfall characteristics on runoff and soil erosion. *Missouri Agr. Exp. Sta. Research Bull.* 280.

Nichols, M. L. (1929). Methods of research in soil dynamics as applied to implement design. *Alabama Agr. Exp. Sta. Bull.* 229. Also (1932), *Agr. Eng.*, **13**:279–285.

Nichols, M. L., and C. A. Reaves (1958). Soil reaction: To subsoiling equipment. *Agr. Eng.*, **39**:340–343.

Nichols, M. L., and I. F. Reed (1934). Soil dynamics. VI. Physical reactions of soils to moldboard surfaces. *Agr. Eng.*, **15**:187–190.

Page, J. B. (1947). Advantages of the pressure pycnometer for measuring the pore space in soils. *Soil Sci. Soc. Am. Proc.*, **12**:81–84.

Page, J. B., C. J. Willard, and G. W. McCuen (1946). Progress report on tillage methods in preparing land for corn. *Ohio Agr. Exp. Sta. Agron. Mimeo*, No. 102, April.

Parker, J. J., Jr., and H. M. Taylor (1965). Soil strength and seedling emergence relations. I. Soil type, moisture tension, temperature and planting depth effects. *Agron. J.*, **57**:289–291.

Patrick, W. H., Jr., L. W. Sloane, and S. A. Phillips (1959). Response of cotton and corn to deep placement of fertilizer and deep tillage. *Soil Sci. Soc. Am. Proc.*, **23**:307–310.

Phillips, R. E., and Don Kirkham (1962). Mechanical impedance and corn seedling root growth. *Soil Sci. Soc. Am. Proc.*, **26**:319–322.

Pigulevsky, M. C. (1930). Soil as an object for work in agriculture. *Proc. 2d Int. Cong., Soil Sci.*, Leningrad, **1**:82–93.

Richards, L. A. (1953). Modulus of rupture as an index of crusting of soils. *Soil Sci. Soc. Am. Proc.*, **17**:321–323.

Robertson, W. K., and G. M. Volk (1968). Effect of deep profile mixing and amendment additions on soil characteristics and crop production of a Spodosol. *Trans. 9th Int. Cong. Soil Sci.*, **3**:357–366.

Robinson, D. O., and J. B. Page (1950). Soil aggregate stability. *Soil Sci. Soc. Am. Proc.*, **15**:25–29.

Russell, E. W. (1938). Soil structure. *Imp. Bur. Soil Sci. Tech. Commun.* 37.

Russell, M. B., and C. L. Feng (1947). Characterization of the stability of soil aggregates. *Soil Sci.,* 63:299–304.

Schaller, Frank W. (1951). Plow, list or disk? *Iowa Farm Sci.,* 5:175–177.

Schaller, F. W., and K. R. Stockinger (1953). A comparison of five methods for expressing aggregation data. *Soil Sci. Soc. Am. Proc.,* 17:310–313.

Scott-Blair, G. W. (1938). A new laboratory method for measuring the effects of land amelloration processes. *J. Agr. Sci.,* 28:367–378.

Sekera, F. (1931). Die nutzbare Wasserkapazität und die Wasserbeweglichkelt im Boden. *Z. Pflanzenernähr. Düngung Bodenk.,* 22A:87–111.

Shaw, B. T., H. R. Haise, and R. B. Farnsworth (1942). Four years' experience with a soil penetrometer. *Soil Sci. Soc. Am. Proc.,* 7:48–55.

Shaw, Charles F. (1917). A method for determining the volume weight of soils in field conditions. *J. Am. Soc. Agron.,* 9:38–41.

Slipher, J. A. (1932). The mechanical manipulation of soil as it affects structure. *Agr. Eng.,* 13:7–10.

Smith, D. D., C. M. Woodruff, and D. M. Whitt (1947). Building a soil deeper. *Agr. Eng.,* 28:347–348.

Stout, B. A., W. F. Buchele, and F. W. Snyder (1961). Effect of soil compaction on seedling emergence under simulated field conditions. *Agr. Eng.,* 42:68–71.

Swanson, C. L. W., and J. B. Peterson (1942). The use of micrometric and other methods for the evaluation of soil structure. *Soil Sci.,* 53:173–185.

Tackett, J. L., and R. W. Pearson (1964). Oxygen requirements of cotton seedling roots for penetration of compacted soil cores. *Soil Sci. Soc. Am. Proc.,* 28:600–605.

Taylor, Howard M., and Herbert R. Gardner (1963). Penetration of cotton seedling taproots as influenced by bulk density, moisture content, and strength of soil. *Soil Sci.,* 96:153–156.

Tiulin, A. F. (1928). Questions on soil structure. II. Aggregate analysis as a method for determining soil structure. *Perm. Agr. Exp. Sta. Div. Agr. Chem.,* Report 2, pp. 77–122.

Triplett, G. B., Jr., W. H. Johnson, and D. M. Van Doren, Jr. (1963). Performance of two experimental planters for no-tillage corn culture. *Agron. J.,* 55:408–409.

Triplett, G. B., Jr., D. M. Van Doren, Jr., and W. H. Johnson (1964). Non-plowed, strip-tilled corn culture. *Trans. Am. Soc. Agr. Engr.,* 7:105–107.

Triplett, G. B., Jr., D. M. Van Doren, Jr., and B. L. Schmidt (1968). Effect of corn (Zea Mays L.) stover mulch on no-tillage corn yield and water infiltration. *Agron. J.,* 60:236–239.

Trouse, A. C., Jr., and L. D. Baver (1962). The effect of soil compaction on root development. *Trans. Joint Meeting Com. IV and V, Intern. Soc. Soil Sci.,* New Zealand, pp. 258–263.

Trouse, A. C., Jr., and L. D. Baver (1965). Tillage problems in the Hawaiian sugar industry. IV. Seedbed preparation and cultivation. *Exp. Sta. Hawaiian Sugar Planters' Assoc., Tech. Suppl. to Soils Report* No. 12.

Trouse, A. C., Jr., and R. P. Humbert (1959). Deep tillage in Hawaii. I. Subsoiling. *Soil Sci.*, **88**:150–158.

van Bavel, C. H. M. (1949). Mean weight diameter of soil aggregates as a statistical index of aggregation. *Soil Sci. Soc. Am. Proc.*, **14**:20–23.

van Bavel, C. H. M. (1958). Soil densitometry by gamma transmission. *Soil Sci.*, **87**:50–58.

Veihmeyer, F. J., and A. H. Hendrickson (1948). Soil density and root penetration. *Soil Sci.*, **65**:487–493.

Vomocil, James A. (1954). In situ measurement of soil bulk density. *Agr. Engr.*, **35**:651–654.

Vomocil, James A. (1965). Porosity. *Methods of Soil Analysis, Agronomy Monograph*, No. 9 Part I. Academic Press, New York, pp. 299–314.

Wiersum, L. K. (1957). The relationship of the size and structural rigidity of pores to their penetration by roots. *Plant and Soil*, **9**:75–85.

Wiersum, L. K. (1960). Some experiences in soil aeration measurements and relationships to depth of rooting. *Neth. J. Agr. Sci.*, **8**:245–256.

Wilding, L. P., and B. L. Schmidt (1968). Private communication.

Wischmeier, Walter H., and Dwight D. Smith (1958). Rainfall energy and its relationship to soil loss. *Trans. Am. Geophys. Union*, **39**:285–291.

Woodruff, C. M. (1939). Variations in the state and stability of aggregation as a result of different methods of cropping. *Soil Sci. Soc. Am. Proc.*, **4**:13–18.

Yoder, R. E. (1937). The significance of soil structure in relation to the tilth problem. *Soil Sci. Soc. Am. Proc.*, **2**:21–33.

CHAPTER 6

Soil Aeration

Soil aeration refers to the exchange of carbon dioxide and oxygen gases between the soil pore space and the aerial atmosphere. CO_2 is produced and O_2 is consumed in the soil through the respiratory processes of plant roots and of microbial activity. Soil aeration therefore consists of exchanging O_2 from the atmosphere for CO_2 in the gaseous phase of the soil. In the preceding chapter on soil structure, attention was called to the fact that root development and plant growth were related to the aeration porosity of the soil.

Pore Types Involved in Aeration

In order to evaluate the significance of porosity in the gaseous interchange between soil and air, it is essential to understand the nature of the pore spaces involved. There are two types of pores in a soil with

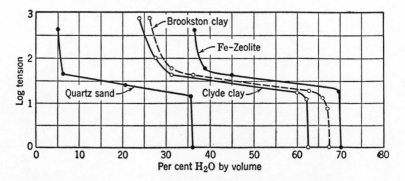

FIG. 6-1. Log-tension curves of 40- to 60-mesh systems of various materials.

a well-developed structure, that is, a soil that contains aggregates or crumbs. There are the pores between the crumbs, or *intercrumb pores*, and those within the crumbs, or *crumb pores*. The presence of these two types of pore is clearly illustrated by the curves in Figure 6-1 (Nelson and Baver, 1940). The percentage of water by volume of 40- to 60-mesh crumbs of Brookston and Clyde clays, 40- to 60-mesh quartz sand, and 40- to 80-mesh Fe-zeolite is plotted against the logarithm of the tension used to drain the pores. It is obvious from these curves that all four separates drain at approximately the same tension. The quartz sand represents the only system in which the separates are nonporous. The aeration porosity of 30 percent constitutes the space between the sand grains. If 30 percent is subtracted from the saturation percentages of the other systems, the resulting values fall approximately at the flex point for all the curves. A summary of the pertinent data is given in Table 6-1. The water withdrawn from the aggregated systems below

TABLE 6-1

Intercrumb and Crumb Porosities of 40- to 60-mesh
Separates (after Nelson and Baver, 1940)

Separate	Total porosity at saturation (percent by volume)	Intercrumb porosity (percent by volume)	Crumb porosity (percent by volume)
Quartz sand*	37.0	30.0	7.0
Clyde clay	63.5	31.5	32.0
Brookston clay	68.0	31.0	37.0
Fe-zeolite†	70.0	30.0	40.0

* The 7.0 percent water held at tensions above the flex point undoubtedly represents film water around the sand grains. The intercrumb porosity is interparticle porosity.

† 40- to 80-mesh separates.

the flex point may be considered to be that coming from the pores between the aggregates or crumbs. The water removed at the flex point may include some water at the contact points of the particles (especially true of the quartz-sand separates). This means that the individual crumbs act like solid particles as far as interparticle porosity is concerned. Desaturation by applying tension is the same as that occurring in a single pore-type system (Currie, 1961b).

Composition of Soil Air

The composition of the soil air depends upon the respiration of microorganisms and plant roots, the solubility of CO_2 and O_2 in water, and the rate of gaseous exchange with the atmosphere. Russell and Appleyard (1915) studied the composition of the soil air under various conditions of cropping and fertilization. They drove a hollow cylindrical tube in the soil and extracted the air from the 6-in. depth by means of a simple mercury pump arrangement. They gave the following percentages by volume as representing the mean composition of the soil air: $N_2 = 79.2$, $O_2 = 20.6$, and $CO_2 = 0.25$; for atmospheric air these percentages were $N_2 = 79.0$, $O_2 = 20.97$, and $CO_2 = 0.03$. The oxygen content of the soil air was only slightly less than that of the atmosphere. However, the CO_2 content of the soil air was six to seven times greater than that of the atmosphere. These values change with the season, soil, crop, tillage, and biological activity.

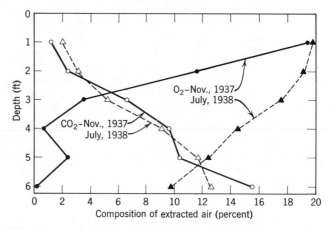

Fig. 6-2. The effect of depth on the composition of soil air. (Data of Boynton and Reuther, 1938.)

The CO_2 and O_2 percentages of the soil air vary with soil depths at different times of the year (Boynton and Reuther, 1938) (Figure 6-2). The CO_2 percentage increases with depth during all seasons of the year. At the 1-ft level, the CO_2 concentration varied from 0.15 to 3.0 percent from spring to fall; at the 3-ft depth, the values ranged from 15.5 percent in November, 1937, to 10.6 percent in September, 1938. The two curves in the figure are typical of the changes of CO_2 with depth in that the variations with seasons follow the same pattern.

The changes in the O_2 percentages with depth are more pronounced than with CO_2 and are affected more by the season, especially at depths below 1 ft. The O_2 concentration varied from 20.15 to 15.3 percent from March to September at the 1-ft level; at the 3-ft depth, the values ranged from 0.3 to 9.95 percent over the same time interval; at 6 ft, the O_2 percentages varied from 0.2 percent in November, 1937, to 9.0 percent in September, 1938. These decreased O_2 contents were associated with increased moisture in the subsoil during the wet months. The data showed that the percentages of CO_2 and O_2 at the deeper levels were about equal during the growing season.

CO_2 Production

The CO_2 production in soils reaches a maximum in the late spring and early fall and a minimum in the summer and winter months (Figure 6-3). These variations are related to fluctuations in the rate of biochemical

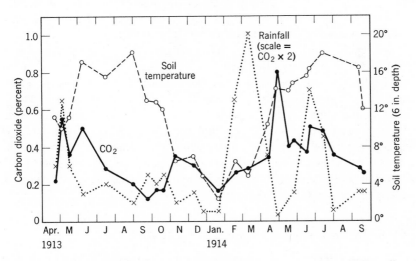

FIG. 6-3. Seasonal variation in the CO_2 content of soil air. (Data of Russell and Appleyard, 1915.)

changes in the soil. In the fall, the bacteria increase first and then the CO_2 concentration rises. The CO_2 production from November to May follows variations in soil temperature. Rainfall is the dominant factor from May to November. Rainfall brings dissolved oxygen into the soil with the infiltrating water. This facilitates biochemical changes.

There is a considerable microbiological impact upon the CO_2 content of the soil (Romell, 1922). Romell estimated that the CO_2 concentration at a depth of 20 cm due to bacteria would double in $1\frac{1}{2}$ hr and increase 10 times in 14 hr if gaseous exchange between the soil air and the atmosphere were prevented. Therefore normal aeration would require a complete renewal of soil air every hour to a depth of 20 cm to maintain the usual average composition of the soil air. About 90 percent of the CO_2 production in a Muscatine soil has been attributed to microbial activity (Papendick and Runkles, 1966). CO_2 evolution varies with the time after dry soils are wetted; it is highest after wetting and then declines with time to a constant value.

Although the data of Russell and Appleyard showed that the CO_2 content of the soil air from the Hoos Field wheat plots was always considerably higher than the air from the fallow plots, they attributed these differences not to the direct effect of the crop but to the physical differences of the soil in the two plots. The fallow plot was left rough, which allowed the ready escape of CO_2. The wheat plot became compact after seeding, which was less favorable to gaseous diffusion. They concluded that there was no evidence from their studies that the growing crop markedly increased the amount of CO_2 in the soil air.

The depth of tillage affects CO_2 production (Tamm and Krzysch, 1964). Shallow plowing was observed to increase CO_2 formation in the topsoil (1 to 15 cm) but to decrease the biological activities in the lower layers. Deep plowing caused a favorable microbial activity in the deeper layers. If the concentration of CO_2 in the topsoil was considered equal to 100, that of the 15- to 20-cm and the 28- to 40-cm depths was 49 and 24, respectively, for shallow plowing; the values for deep plowing at these lower depths were 92 and 44, respectively. The earlier work of Russell and Appleyard was confirmed; manuring increases the CO_2 content of the soil air in the topsoil.

O_2 Requirements and Consumption

CO_2 production and O_2 consumption are synonymous in the respiration process. It was observed in Figures 6-2 and 6-3 that increases in the CO_2 concentration of the soil air were always accompanied by decreases in the O_2 content. The location of organic matter in or on the soil determines the O_2 consumption, as measured by the decreased O_2 percentage in the soil air (Epstein and Kohnke, 1957). When the different organic matter placements are listed in order of the decreasing O_2 concentration of the soil atmosphere, they follow the series check > surface mixed > placed in a layer 2-in. deep > placed on the surface. Mixing organic matter with the surface soil gives the best O_2 concentration of the soil air.

Brown, Fountaine, and Holden (1965) measured the mean daily O_2 requirements of mature potatoes, kale, and tobacco and found them to be 2.8, 5.6 and 3.0 liters/m², respectively. The mean daily consumption of an undisturbed sandy clay loam was 2.2 liters/m²; that of a peat top soil was 10.8 liters/m². Oxygen consumption increased with the amount of organic matter in the soil and with cultivation.

These data on the composition of the soil air point out the necessity of good aeration to maintain a satisfactory medium for plant growth.

GASEOUS EXCHANGE—THE RENEWAL OF SOIL AIR

Most of the present data on the composition of the soil air indicate that the amounts of CO_2 and O_2 do not vary much within the immediate surface. There may be some question as to the applicability of these data to all soil conditions, especially to impervious soils. Nevertheless, in light of the intensity of CO_2 production through plant growth and microbiological activities, the apparent lack of any large accumulation of CO_2 in the surface soil (often true of the subsoil) suggests a rather rapid exchange of gases with the atmosphere. The renewal of the soil air is brought about by diffusion and by mass flow resulting from the meteorological factors: soil temperature changes, barometer variations, action of the wind, and changes in the amount of pore space occupied by air as a result of the entrance of rain or irrigation water.

Mass Flow

Soil-Temperature Effects

Temperature may influence the renewal of soil air in two ways. First, there may be temperature differences within the soil between the different layers. It is possible that the contraction and expansion of the air within the pore spaces as well as the tendency for warm air to move upward may cause some exchange between the various horizons and perhaps with atmosphere. Second, the soil and the atmosphere usually have different temperatures. This temperature differential should permit an exchange between the atmosphere and the soil air in the immediate surface.

It is difficult to estimate the significance of temperature effects on gas exchange in soils. Romell (1922) suggested that daily variations of temperature within the soil are responsible for less than $\frac{1}{800}$ of the normal aeration. Temperature differences between soil and atmosphere were considered to be responsible for not more than $\frac{1}{240}$ to $\frac{1}{480}$ of the normal aeration. Thus it appears that temperature is a minor factor in soil aeration.

Barometric-Pressure Effects

Theoretically, according to Boyle's law, any increase in the barometric pressure of the atmosphere should cause a decrease in the volume of the soil air. This diminution in volume should permit an equivalent amount of atmospheric air to penetrate the soil pores. On the other hand, any decrease in barometric pressure should produce an expansion of the soil air and cause part of it to enter the atmosphere above the soil. Thus a rinsing of the soil with air should occur from time to time. These statements are predicated upon the fact that any change in barometric pressure of the atmosphere is readily reflected within the soil pores.

Buckingham (1904) calculated the possible rinsing action due to barometric changes and showed that penetration of atmospheric air within a permeable soil column 10 ft deep would amount only to about 0.12 to 0.22 in., dependent upon the magnitude of the barometric change. Thus it is seen that fluctuations in atmospheric pressure have little influence upon soil aeration, even where there is ready access between the soil air and atmosphere. Romell estimated that not more than about $\frac{1}{100}$ of the normal aeration of soils can be attributed to variations in barometric pressure.

Wind Action

One might expect that the pressure and suction effects of high winds would exert some influence upon the renewal of the soil air. Although Romell gave some attention to this phase of soil aeration and concluded, from calculations based upon wind velocities, that wind action could not be responsible for more than about $\frac{1}{1000}$ of the normal aeration on vegetated soils; it is possible that this value is higher for bare, unprotected soils that are extremely porous. More recent experiments on the effect of air turbulence on the transfer of vapor in the soil suggest that mass air flow may be much greater than normally assumed (Farrell, Greacen, and Gurr, 1966). Air with a wind speed of 15 mph can penetrate coarse sand and mulches to a depth of several centimeters. Even though there is no net mass flow, the fluctuations in air pressure at the soil surface result in a mixing of air within the surface which enhances the transport beyond that due to diffusion (Scotter, Thurtell, and Raats, 1967).

Rainfall Effects

The infiltration of rainfall into the soil may cause a renewal of the soil air in two ways: the displacement of air in the pores by the water, which is subsequently displaced again with air; and the carrying in of dissolved oxygen in the water. One can visualize a rather complete renewal of the soil air following a rain, especially if the water is able to displace the major portion of the air within the pores. In many instances, however, a considerable amount of entrapped air remains, not forced out of the soil by infiltered water. Renewal of the soil air through rainfall effects is periodical, depending on the distribution of the rains. Romell estimated that rainfall accounts for only about $\frac{1}{12}$ to $\frac{1}{16}$ of the normal aeration.

Diffusion

Diffusion is the molecular transfer of gases through porous media. According to the kinetic theory of gases, the molecules of gases are in a state of movement in all directions. Two gases will readily mix as the molecules of each gas move into the space occupied by the other. Inasmuch as the soil air tends to contain more CO_2 and less O_2 than the atmosphere, the diffusion process in soils consists primarily in the movement of CO_2 out of the soil into the atmosphere and O_2 from the atmosphere into the soil. If this action is complete when equilibrium is attained, there will be a tendency for the soil air to approach the composition of the atmosphere.

Fick's Law

According to Fick's law, diffusion is a function of the concentration gradient, the diffusion coefficient of the medium, and the cross-sectional area participating in the diffusion:

$$dQ = DA \left(\frac{dc}{dx}\right) dt \qquad (6\text{-}1)$$

where dQ is the mass flow (moles) diffusing during time dt across area A (cm^2), dc/dx the concentration gradient [mols/(cm^3)(cm)], and D the proportionality constant or diffusion coefficient (cm^2/sec). The concentration at a distance, x, is dependent upon the concentration in the air atmosphere: $c_x = c_0 + ax$, where a is a constant. D depends upon the property of the medium as well as the gas. It varies directly with the square of the temperature (absolute) and inversely with the total pressure.

The diffusion coefficient of O_2 is about 1.25 times that of CO_2. The rate of diffusion in air of both gases is nearly 10,000 times greater than in water (Grable, 1966). The greater solubility of CO_2 in water increases its concentration gradient and will be transferred in water at a higher rate than O_2. When dispersion is significant, equation 6-1 is employed but the dispersion coefficient is used instead of the diffusion coefficient (Scotter et al., 1967).

Diffusion Through Porous Media

Hannen (1892) was the first to show that the sum of the cross-sectional areas of the effective pore volume was the most important factor affecting the diffusion of CO_2 in soils. In other words, A in equation 6-1 requires an adjustment for the soil pore space. Buckingham (1904) observed that the diffusion rate increased with the square of the free pore space; that is, the rate of diffusion is reduced one-fourth as the free pore space is reduced one-half.

Penman (1940a, 1940b) studied the diffusion of gases through porous solids and concluded that the reduced rate of diffusion of a vapor through a porous body as compared through free air is due in part to the reduced cross-sectional area available for the movement of gaseous molecules and in part to the increased length of path the molecules must follow, since the effective channels are tortuous in nature. Theoretically, the rate of diffusion in a steady state through a solid is expressed by the equation

$$\frac{dq}{dt} = \frac{D}{\beta} A \frac{\rho_1 - \rho_2}{l} \qquad (6\text{-}2)$$

where A = the cross-sectional area of the solid,
l = length of the solid
D = the diffusion constant,
β = a proportionality constant, and
ρ_1 and ρ_2 = partial pressures of the vapor on each side of length l of the solid.

However, when one takes into consideration the available cross-sectional area for diffusion and the effective path length the molecules must travel, this equation becomes

$$\frac{dq}{dt} = \frac{D_0}{\beta} AS \frac{\rho_1 - \rho_2}{l_e} \qquad (6\text{-}3)$$

where D_0 = coefficient of diffusion in air,
S = pore space or available cross-sectional area, and
l_e = effective path length through the solid.

From equations 6-2 and 6-3, it is seen that

$$D = D_0 S \frac{l}{l_e} \qquad (6\text{-}4)$$

A study of diffusion rates through different media showed that within limits a curve can be drawn denoting the relationship between D/D_0 and S. Such a curve indicated that $D/D_0 = 0.66S$.

Penman called attention to the fact that the true steady state would rarely be attained in soils, owing to centers of microbiological activity releasing gases as well as to adsorption of gases in other areas.

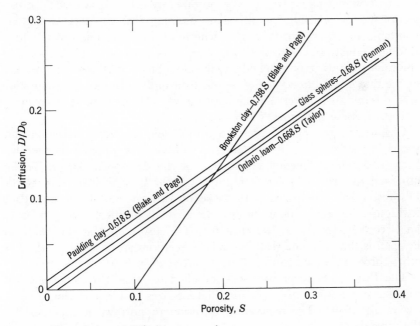

FIG. 6-4. The relation of diffusion to porosity.

Blake and Page (1948) made direct measurements of gaseous diffusion in soils, using carbon bisulfide. For 22 samples of Paulding clay, they found that D/D_0 was equal to $0.618S$; for 15 samples of Brookston clay, this value was $0.798S$. Their data are shown in Figure 6-4. It is significant to note that the curve for the Brookston clay does not approach the origin but extrapolates to zero at about 10 percent porosity. This behavior is attributed to the nature of the granules, which contain many blocked pores that are not available for diffusion but contribute to the total air porosity.

Taylor (1949) modified Penman's equations slightly and arrived at the expression

$$D = \frac{1}{\lambda^2} D_0 \qquad (6\text{-}5)$$

where λ is a parameter called the *equivalent diffusion distance*. It has the dimensions of length and characterizes the soil or other materials which restrict regular gaseous diffusion. It represents the length of an equivalent tube which has a unit cross section and just the right length so that free diffusion through the tube is exactly equivalent to the diffusion that would take place in any porous media having a specific cross-sectional area available for diffusion, assuming that all other factors are constant. In other words, it is a length of tube of unit cross section through which free diffusion will supply oxygen to a given point at the same rate as it is supplied by the soil. Since it is a direct measure of the rate at which oxygen is supplied through diffusion, the suggestion was offered that it serve as a measure of soil aeration. The quantity $1/\lambda^2$ is the same as Penman's D/D_0.

Changes in the partial pressure of oxygen produces a straight line when $\log \rho_0/(\rho_0 - \rho)$ is plotted as a function of time. This fact has been used as experimental verification of the developed theory. Taylor observed that the value of D/D_0, or $1/\lambda^2$, for Ontario loam was $0.668S$. His results are compared with those of Penman and Blake and Page in Figure 6-4. It is important to note that all of these values refer to an effective porosity of less than 0.4. It is also seen that the curve of Penman is the only one that follows the relationship of $D/D_0 = aS$. The other three curves are expressed by $D/D_0 = aS + b$.

The Penman concept has been modified to stress the importance of the size distribution of pores of different sizes in the diffusion process (Marshall, 1959). The relation of diffusion to porosity is expressed by the formula

$$\frac{D}{D_0} = S^{3/2} \qquad (6\text{-}6)$$

Differences with the Penman equation are not great when S is greater than 0.7. Freedom from obstruction is considered to be a function of porosity rather than tortuosity. The higher the porosity, the greater are the chances of pore continuity.

Millington (1959) considered both mass flow and diffusive flow to be functions of the cross-sectional area available for flow and the path-length increase or tortuosity (characteristics of the pore shape). The

area available for flow would be dependent upon the pore-volume distribution within the soil. The following modification of the Penman equation
was suggested:

$$\frac{D}{D_0} = S^{4/3} \qquad (6\text{-}7)$$

The equations of Marshall and Millington produce a curvilinear relationship between D/D_0 and the gas-filled porosity, due to the nature of
the changes relating to the size distribution and the continuity of the
pores. These changes occur as the total porosity varies due to the dry
solids and (to be discussed later) as the gas-filled porosity is affected

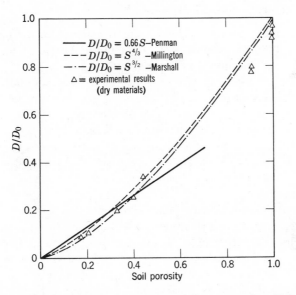

FIG. 6-5. The relation of diffusion in porous media to porosity. (After Papendick
and Runkles, 1965.)

by the moisture content. The results in Figure 6-5 show the close similarity between the curves of Marshall and Millington. They also point out
the deviations from the Penman formula at the higher and lower
porosities.

van Bavel (1951, 1952) developed a theory of soil aeration in which
he took into account not only the partial pressures of the vapors and
the rate of diffusion but also the rate at which gases were liberated
or consumed in the dynamic processes of respiration occurring in soils.
This rate of consumption or liberation was termed "activity" (after

Romell). For diffusion in the steady state the following equation was proposed:

$$\frac{\partial^2 p}{\partial x^2} = -\frac{\beta}{D}\alpha \qquad (6\text{-}8)$$

where x = the dimension in the direction of diffusion,

D = the diffusivity of the gas into air in the porous medium, equivalent to Penman's D/S,

β = a constant dependent upon temperature and inversely proportional to porosity,

α = the activity, expressed in mass units per unit volume of soil per unit time, and

β/D = specific diffusion impedance.

This equation indicates that the rate of change of partial pressures in the soil is directly proportional to gaseous activity and inversely proportional to "specific diffusion impedance."

In defining all units on the basis of the whole medium rather than solely the gaseous phase, the diffusion of a gas through the soil is considered dependent upon the nature of the gas, temperature, and pressure only and not upon soil characteristics. Diffusion must depend primarily upon the ratio between mass changes and partial pressure changes. In this statement the role of effective porosity is expressed in the mass changes that take place as a result of variations in moisture content and compaction.

One of the most significant points developed from this theoretical approach is the relative unimportance of the condition of the surface layer in influencing the magnitude of the diffusion process. The significant factor is the total depth of the "active" soil and the properties of the deepest part of this layer. Only when the depth of the surface layer, either compacted or loose, is large in comparison with the total depth of the active soil does the specific diffusion impedance of this layer play an important part in diffusion. In other words, one should not expect much hindrance to gaseous exchange through the surface crusting of a deep, porous soil. Neither should one anticipate much physical improvement from the surface tillage of just a few inches of a deep, densely compacted soil. The surface compaction of a shallow layer of soil which restricts the depth of roots to several inches will cause poor aeration conditions. Diffusion studies of alcohol through quartz sand and a soil-sand mixture showed that D/D_0 was more nearly equal to $0.6S$ (van Bavel, 1952).

Currie (1962), with the help of the van Bavel equation, calculated the O_2 concentraton gradients in the soil profile as affected by varying the diffusion resistance and the soil respiratory activity. His results are shown in Figure 6-6. The largest decrease in O_2 with depth occurs in the soil profile that has the higher resistance and the greater respiratory activity. If one takes the 50-cm depth for comparison of oxygen values, the data point out that doubling the activity under conditions of the higher resistance decreases the O_2 content by 10 percent; at the lower diffusion resistance, this decrease is only 2 percent. On the other hand, reducing the diffusion rate by ¼ at the higher activity decreases the O_2 content by 14.4 percent; this decrease is seven percent at the lower

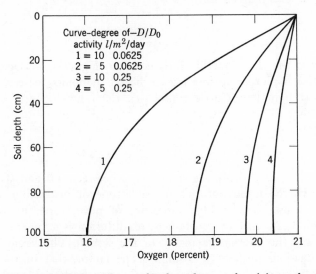

Fig. 6-6. Oxygen profiles in soils as related to degree of activity and resistance to diffusion. (After Currie, 1962.)

activity level. The curves portray quite clearly that the respiration rate has only a small effect on the O_2 concentrations at the lower depths when there are low resistances to diffusion. However, both must be evaluated for obtaining a clear picture of the aeration status of the soil with depth.

Impact of Particle Nature

PARTICLE SHAPE. Particle shape affects diffusion according to the formula (DeVries, 1950)

$$\frac{D}{D_0} = \frac{S}{1 + (k - 1)(1 - S)} \qquad (6\text{-}9)$$

where k is the shape factor. If the value for k for spherical particles was considered to be $\frac{1}{5}$, the data of Penman agreed fairly well with the theoretical curve. The relationship between diffusion and porosity becomes more curvilinear as the particles deviate more from a spherical geometry.

Currie (1960) stressed the significance of particle shape on diffusion by elaborating on the work of Bruggeman and Burger, who stated that diffusion depended not only upon porosity but also on particle shape. The Burger concept has been given in equation 6-9. The Bruggeman equation, in its simplified form, is

$$\frac{D}{D_0} = S^m \qquad (6\text{-}10)$$

where m is the shape factor. It is equal to $\log\,(D/D_0)/\log S$. When $\log\,(D/D_0)$ is plotted against $\log S$, the set of points for each material fall on a straight line. Currie found that the following relationship fits all materials:

$$\frac{D}{D_0} = \gamma S^\mu \qquad (6\text{-}11)$$

where $\gamma = 1$ and $\mu = m$; γ generally lies between 1.0 and 0.8 and increases with the mean porosity of each type of material; μ is usually less than m and is probably a measure of pore shape. Equations 6-9 and 6-10 indicate that the coefficient of diffusion is a function of both porosity and the internal geometry of the system. The Bruggeman and Burger equations both have a single-shape factor, that of the particle. However, a single-shape factor cannot define both the particle shape and the spatial distribution of particles. Therefore empirical equation 6-11 is suggested to serve for all dry granular materials.

TYPE OF PARTICLE. The curves in Figure 6-1 clearly showed that the type of particle determines the nature of the porosity in soils. The porosity of sands is entirely interparticle. That of crumbs or soil aggregates consists of both intercrumb and crumb pores. Great significance is attached to the effect of the soil microstructure on aeration (Currie, 1961a, 1961b, 1962, 1965). The heterogeneous pore-space distribution in a soil with a well-developed structure is visualized as having distinct zones of crumb pores that are separated by a more continuous system of intercrumb pores. When the intercrumb pores are drained, gaseous diffusion takes place between the crumb surfaces and the soil surface. The soil crumbs of the cultivated layer, or the peds of the natural soil structure, are the sites of microorganisms and most of the active root system (root hairs). Respiration occurs here and available water is held in these pores. Diffusion must take place between the crumb surface and the active root surface. The steady-state equation for a crumb in equilibrium with its surrounding is

$$P_r = P_a + \frac{\alpha_c(a^2 - r^2)}{6D_c} \tag{6-12}$$

where P_r is the concentration of gas at a distance r from the center of the crumb, P_a the concentration of gas at the crumb surface, a the crumb radius, α_c the respiratory activity within the crumb, and D_c the effective diffusion coefficient within the crumb. The O_2 concentration at any point within the crumb depends upon the crumb size and the total O_2 deficit between the surface and the center; it is proportional to the square of the radius. Gaseous exchange, therefore, in an aggregated soil involves two factors, D_v/D_o in intercrumb pores and the diffusion between the crumb surface and the respiring surface, D_c/D_o.

Impact of Soil Moisture

Buckingham showed that diffusion decreased with increasing moisture content; this was the result of changes in the water-free pore space. Taylor (1949) observed that very little diffusion took place at tensions lower than 20 cm of water. There was very little effect of moisture at tensions higher than 30 cm. The curves broke very abruptly between 20 and 30 cm of water tension. This abrupt break should be expected in light of the fact that there is little change in tension as the free pores fill with water, although there is a large decrease in the area of pores available for diffusion. When a liquid is included in a porous solid, the effective area for diffusive flow of a sparingly soluble gas is determined by the number of pores drained, that is, the gas-filled pore space (Millington, 1959). The diffusion equation, $D/D_0 = S^{4/3}$, then becomes

$$\frac{D}{D_0} = n^2 \left(\frac{S_1^{4/3}}{m^2}\right) \tag{6-13}$$

where S_1 is the air-filled porosity and m is equal-volume pore-size groups that make up the porosity when n of them are drained. The air-filled porosity $S_1 = n(S/m)$.

Adding water decreases the coefficient of diffusion because of a diminution of the air-filled porosity. Since the pore space is modified by the water, particle shape has much less effect upon diffusion in wet materials. Equation 6-11 is not applicable in wetted materials because of the alteration of the gas-filled pores by the liquid phase (Currie, 1960). The diffusion of a gas in the liquid phase takes place according to the relationship shown in equation 6-12, except that the concentrations refer to the amount of gas in solution and D_c becomes D, the coefficient of diffusion of the gas in the liquid. The latter will be about $\frac{1}{10,000}$ the value of the coefficient of diffusion in air. For example, Letey, Stolzy, and Kemper (1967) have quoted D_a and D_w for air and water to be 1.89×10^{-1} and 2.56×10^{-5}, respectively.

Root-Soil Relationships

One must know the O_2 relations exterior to the root surface to understand soil aeration from the point of view of the plant. This involves the O_2 demand characteristics of the plant root as well as the O_2-supplying power of the soil. In other words, root respiration is important in affecting the effective diffusion coefficient since it determines the amount of O_2 that has to diffuse to the root surface (Lemon, 1962; Currie, 1962). The supply of O_2 that reaches the root surface is controlled by the rate of gaseous exchange between the soil air and the atmosphere and the transfer of O_2 from the soil pores to the root surface. This occurs through the moisture films that exist around the plant roots and the soil particles; this means that O_2 diffusion must take place through the liquid phase. Wiegand and Lemon (1958) proposed the following equation to explain the O_2 supply to plant roots:

$$C_R = C_p + \frac{qR^2}{2D_l} \ln \frac{R}{r_e} \qquad (6\text{-}14)$$

where C_R is the concentration of the oxygen at the root surface (g/cm^3) C_p the O_2 concentration at the liquid-gas interface in equilibrium with the partial pressure of O_2 in the gaseous phase (g/cm^3), q the O_2 consumption of the root $[(g/(cm^3)(sec)]$, R the root radius (cm), r_e the radius of the root plus the moisture film (cm), and D_l the diffusion coefficient in the liquid-solid matrix around the root (cm^2/sec). Kristensen and Lemon (1964) elaborated further on this concept and suggested that the phase geometry around a root consists of solid particles at random surrounded by water films. Diffusion takes place in pathways of different lengths and tortuosities through these films. The parameters D_l, r_e, and R in equation 6-14 are "apparent" or "effective" values and are related to the moisture content and the porosity. The value of D_l is determined by the volume of the porosity occupied by water, and the shape, tortuosity, and cross-sectional area of the pathways. Even though these water films represent only a small fraction of the total length of the diffusion path between the soil surface and the root surface, the "effective" length is increased nearly 10,000 times because of the smaller coefficient of diffusion in solution (Currie, 1962). This is illustrated by the fact that the O_2-concentration difference across a water film 1 mm thick may be 10 times that across the same distance of air-filled pore space.

Summary of Basic Concepts

The diffusion of gases through soils, as expressed by the ratio D/D_0, is dependent upon five independent variables: (1) the intercrumb pore space, (2) the shape factor of the crumbs acting as individual units, (3) the porosity of the crumb, (4) the shape of the particles making up the crumbs, and (5) the fraction of the crumb pore space occupied by water. The water films around the soil particles in the crumbs and the root hairs govern the true aeration status of the respiring surface. This status is dependent upon the soil parameters: (1) the apparent liquid-path length between the liquid-gas interface at the pore walls and the root surfaces, (2) the effective diffusion coefficient for the liquid-solid matrix surrounding the root, and (3) the O_2 concentration at the liquid-gas interface at the pore walls.

Measurements of O_2 Diffusion

One of the earliest methods used to determine the O_2 concentration in the soil pore space is illustrated by the technique of Raney (1949). It consisted of allowing the free diffusion of O_2 into a diffusion chamber that was inserted into the soil. Analyses of the gas were made at 10-min intervals. The calculated partial pressure of O_2 in the chamber was used to evaluate the diffusion rate.

Willey and Tanner (1963) determined the O_2 concentration in the soil air with a membrane-covered polarographic electrode. The probe is enclosed in a plastic membrane that is permeable to O_2. When a small voltage is applied between the polarographic cathode and the nonpolarizable anode, almost all the O_2 which diffuses to the cathode is reduced and the concentration of O_2 at the electrode surface is very close to zero. The resulting current is proportional to the rate of reduction of O_2, which in turn is limited by the rate of diffusion of O_2 to the electrode. The measurement is applicable to measurements in situ.

The platinum electrode was introduced by Lemon and Erickson (1952) to measure the oxygen-diffusion rate (ODR) in soils. Its use is predicated upon the concept that the limiting factor in the O_2 supply to plant roots is the rate of diffusion of O_2 through the moisture film around the root. The ODR therefore characterizes the soil O_2 conditions. The method is based upon the principle that the electric current resulting from the reduction of O_2 at the platinum surface is governed (within limits) solely by the rate at which O_2 diffuses to the electrode surface. A voltage of 0.8 V between the platinum electrode and a saturated calomel cell was used. Calculations were made by knowing the electrode area and the amperes of current flowing between the two electrodes.

According to Fick's Law, the ODR for a linear system should reflect the O_2 concentration gradient at the electrode surface and the O_2 diffusion coefficient. The measured ODR is an integrated effect of (1) the O_2 concentration in solution, (2) the O_2 diffusion coefficient, (3) the thickness of the moisture film around the electrode, (4) the length of the diffusion path to the electrode, and (5) the electrode radius (Stolzy and Letey, 1964; Letey and Stolzy, 1964). The method fails in relatively dry soils since the electrode must be wetted so that the whole electrode surface is covered by a moisture film. The rupturing of moisture films at higher moisture tensions makes the technique more applicable to fine-textured soils over a wider range of soil-moisture tensions than to coarse-textured soils. The ODR decreases as the moisture content increases because of a diminution in the volume of air-filled pores and a lengthening of the diffusion path to the electrode.

Temperature affects the ODR by increasing the rate of respiration and the diffusion coefficient of O_2 in water; the solubility of O_2 in the liquid phase is decreased. These combined effects result in about a 1.8 percent increase in the ODR per degree Centigrade (Stolzy and Letey, 1964).

There is a great deal of empiricism in the ODR measurements with the platinum electrode. This requires standardization of techniques so that results from different investigators will be comparable. Measurements must be made during the steady state, which takes 3 or 4 min to achieve. The applied potential should be between 0.55 and 0.75 V. Since the measured current varies with the radius of the electrode, a 25-gage wire is recommended as the electrode size.

Bertrand and Kohnke (1957) obtained a good correlation between the ODR values measured at 8-in. depths with the platinum electrode and D/D_0 determined at the same depth with an oxygen analyzer. However, the correlation was not as good at the 16-in. depth. Wiersum (1960) found that visual observations of the depth of root penetration correlated very well with measurements by the platinum electrode. Since the boundary conditions at the platinum electrode are quite similar to those around the root in the soil, and since individual readings can be made quickly using simple equipment, this technique offers possibilities for wider usage for evaluating the aeration status of soils. One must exercise caution in its use and make sure that the conditions previously referred to are taken into consideration. Measuring O_2 changes in a diffusion well with the use of an oxygen analyzer can provide reliable data under a wide range of conditions.

O_2-Diffusion Rates

BIOLOGICAL IMPLICATIONS. The statement was made in Chapter 5 that aeration was a prime factor in root development. Aeration was expressed

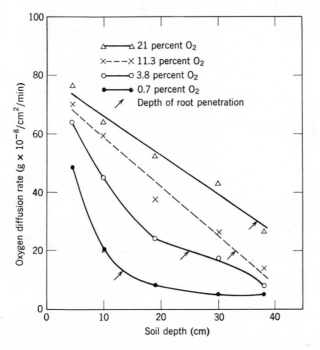

FIG. 6-7. Oxygen diffusion rate as a function of depth for various oxygen concentrations at the soil surface. (After Stolzy and Letey, 1964.)

in terms of the air-filled porosity or the ODR. Although there is a relationship between the ODR and aeration porosity, the concentration of O_2 at the plant root surface is the important factor in root development. The ODR therefore becomes a significant parameter in root-soil relationships. The results in Figure 6-7 depict the ODR values with depth as the O_2 concentration at the surface of the soil varies (Stolzy and Letey, 1964). It is obvious that the ODR decreases with depth. The extent of this decrease depends upon the O_2 concentration at the soil surface. The ODR at a single depth does not represent the aeration status for the whole soil volume. Root initiation is reduced or stopped at ODR values ranging from 18 to 23×10^{-8} g/(cm^2)(min). The value of 20×10^{-8} g/(cm^2)(min) is recommended as an approximate figure for a wide variety of plants. Barley is one crop where the critical value is 15×10^{-8} g/(cm^2)(min). Roots do respond to ODR values between 20 and 30×10^{-8} g/(cm^2)(min).

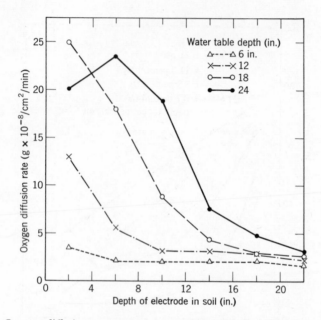

Fig. 6-8. Oxygen diffusion rates at a given soil depth as a function of depth of water table. (After Williamson and van Schilfgaarde, 1965.)

The ODR for optimum top growth appears to be greater than 40×10^{-8} g/(cm^2)(min). The reader is referred to the excellent review of Stolzy and Letey (1964) for detailed evaluations of the impact of the ODR on plant responses.

The importance of drainage on the ODR is illustrated in Figure 6-8. These data are self-explanatory and emphasize the necessity of lowering the water table to provide the proper aeration status for plant roots. The data indicate that aeration is below optimum 8 to 10 in. above the water table, irrespective of its depth.

References

Bertrand, A. R., and H. Kohnke (1957). Subsoil conditions and their effects on oxygen supply and the growth of corn roots. *Soil Sci. Soc. Am. Proc.*, 21:135–139.

Blake, George R., and J. B. Page (1948). Direct measurement of gaseous diffusion in soils. *Soil Sci. Soc. Am. Proc.*, 13:37–42.

Boynton, D., and W. Reuther (1938). A way of sampling soil gases in dense subsoil and some of its advantages and limitations. *Soil Sci. Soc. Am. Proc.*, 3:37–42.

Brown, N. J., E. R. Fountaine, and M. R. Holden (1965). The oxygen requirement of crop roots and soils under near field conditions. *J. Agr. Sci.*, 64:195–203.

Buckingham, E. (1904). Contributions to our knowledge of the aeration of soils. *U.S. Bur. Soils Bull. 25*.

Currie, J. A. (1960). Gaseous diffusion in porous media. Part II. Dry granular materials. *Br. J. Appl. Physics*, 11:318–324.

Currie, J. A. (1961a). Gaseous diffusion in the aeration of aggregated soils. *Soil Sci.*, 92:40–45.

Currie, J. A. (1961b). Gaseous diffusion in porous media. Part 3. Wet granular materials. *Br. J. Appl. Phys.*, 12:275–281.

Currie, J. A. (1962). The importance of aeration in providing the right conditions for plant growth. *J. Sci. Food Agr.*, 13:380–385.

Currie, J. A. (1965). Diffusion within the soil microstructure—A structural parameter for soils. *J. Soil Sci.*, 16:279–289.

DeVries, D. A. (1950). Some remarks on gaseous diffusion in soils. *Trans. 4th Int. Cong. Soil Sci.*, 4:41–43.

Epstein, Eliot, and Helmut Kohnke (1957). Soil aeration as affected by organic matter application. *Soil Sci. Soc. Am. Proc.*, 21:585–588.

Farrell, D. A., E. L. Greacen, and C. G. Gurr (1966). Vapor transfer in soil due to air turbulence. *Soil Sci.*, 102:305–313.

Grable, Albert R. (1966). Soil aeration and plant growth. *Advances in Agronomy*, 18:57–106.

Hannen, F. (1892). Untersuchungen über den Einfluss der physikalischen Beschaffenheit des Bodens auf die Diffusion wer Kohlensäure. *Forsch. Gebiete Agr.-Phys.*, 15:6–25.

Kristensen, K. J., and E. R. Lemon (1964). Soil aeration and plant-root relations. III. Physical aspects of oxygen diffusion in the liquid phase of the soil. *Agron. J.*, 56:295–301.

Lemon, E. R., and A. E. Erickson (1952). The measurement of oxygen diffusion in the soil with a platinum electrode. *Soil Sci. Soc. Am. Proc.*, 16:160–163.

Lemon, E. R. (1962). Soil aeration and plant root relations. I. Theory. *Agron. J.*, 54:167–170.

Letey, J., and L. H. Stolzy (1964). Measurment of oxygen diffusion rates with the platinum electrode. I. Theory and equipment. *Hilgardia*, 35:545–554.

Letey, John, Jr., L. H. Stolzy, and W. D. Kemper (1967). Soil aeration in irrigation of agricultural lands. *Drainage of Agricultural Lands, Agronomy Monograph*, No. 11, Academic Press pp. 941–949.

Marshall, T. J. (1959). The diffusion of gas through porous media. *J. Soil Sci.*, 10:79–82.

Millington, R. J. (1959). Gas diffusion in porous media. *Science*, 130:100–102.

Nelson, W. L., and L. D. Baver (1940). Movement of water through soils in relation to the nature of the pores. *Soil Sci. Soc. Am. Proc.*, 5:69–76.

Papendick, R. I., and J. R. Runkles (1965). Transient-state oxygen diffusion in soil. I. The case when rate of oxygen consumption is constant. *Soil Sci.*, 100:251–261.

Papendick, R. I., and J. R. Runkles (1966). Transient-state oxygen diffusion in soil. II. A case when the rate of oxygen consumption varies with time. *Soil Sci.*, 102:223–230.

Penman, H. L. (1940a). Gas and vapor movements in the soil. I. The diffusion of vapors through porous solids. *J. Agr. Sci.*, 30:437–462.

Penman, H. L. (1940b). Gas and vapor movements in the soil. II. The diffusion of carbon dioxide through porous solids. *J. Agr. Sci.*, 30:570–581.

Raney, W. A. (1949). Field measurement of oxygen diffusion through soil. *Soil Sci. Soc. Am. Proc.*, 14:61–65.

Romell, L. G. (1922). Luftväxlingen i marken som ekologisk faktor. *Medd. Statens Skogsfarsöks-anstalt*, 19, No. 2.

Russell, E. J., and A. Appleyard (1915). The atmosphere of the soil, its composition and causes of variation. *J. Agr. Sci.*, 7:1–48.

Scotter, D. R., G. W. Thurtell, and P. A. C. Raats (1967). Dispersion resulting from sinusoidal gas flow in porous materials. *Soil Sci.*, 104:306–308.

Stolzy, L. H., and J. Letey (1964). Characterizing soil oxygen conditions with a platinum electrode. *Advances in Agronomy*, 16:249–279.

Tamm, E., and G. Krzysch (1964). Changes in the profile of a loamy sand due to long-term differential cultivation and manuring of the soil. *Z. Acker-u. Pflanzenbau*, 121:1–28. Also (1965) *Soils and Fert.*, 28:107(693).

Taylor, Sterling A. (1949). Oxygen diffusion in porous media as a measure of soil aeration. *Soil Sci. Soc. Am. Proc.*, 14:55–61.

van Bavel, C. H. M. (1951). A soil aeration theory based on diffusion. *Soil Sci.*, 72:33–46.

van Bavel, C. H. M. (1952). Gaseous diffusion in porous media. *Soil Sci.*, 73:91–104.

Wiegand, C. L., and E. R. Lemon (1958). A field study of some plant-soil relations in aeration. *Soil Sci. Soc. Am. Proc.*, 22:216–221.

Wiersum, L. K. (1960). Some experiences in soil aeration measurements and relationships to the depth of rooting. *Neth. J. Agr. Sci.*, 8:245–252.

Willey, C. R., and C. B. Tanner (1963). Membrane-covered electrode for measurement of oxygen concentration in soil. *Soil Sci. Soc. Am. Proc.*, 27:511–515.

Williamson, R. E., and Jan van Schilfgaarde (1965). Studies of crop response to drainage: II. Lysimeters. *Trans. Am. Soc. Agr. Engr.*, 8:98–100, 102.

CHAPTER 7

The Thermal Regime of Soils

Soil temperature is one of the more important factors that control microbiological activity and the processes involved in the production of plants. It is a well-established fact that the rate of organic-matter decomposition and the mineralization of organic forms of nitrogen increase with temperature. Consequently, the amount of organic matter that remains in soils is greater under lower than under higher temperatures. Other important microbiological processes also vary in their intensity with temperature; there seem to be certain optimum soil temperature ranges for these processes.

Soil temperature affects plant growth first during the germination of seeds. Seeds of different plants vary in their ability to germinate at low temperatures. Germination is a slow process in a cold soil. Practical experience has shown that germination becomes more rapid as the temperature of the soil rises, up to a certain optimum temperature. The quicker the germination of the seeds, the earlier will be the crop. Thus cold or warm soils in the spring have different agricultural significance.

The growth of the plant is also influenced by soil temperatures. The functional activity of plant roots, such as the absorption of water and nutrients, can be affected at both low and high soil temperatures. The rate of growth and subsequent crop yields are related to the temperature status. For example, corn yields in Iowa were found to increase almost linearly with soil temperature at the 4-in. depth between the range of 60° and 81.3° F. (Allmaras et al., 1964). The yields decreased as the soil temperature rose above 81.3° F.

The thermal regime of the soil usually includes heat flux into the soil, the thermal characteristics of the soil, and the heat exchange between soil and air. It is generally expressed in terms of soil temperature.

SOURCE AND AMOUNT OF HEAT

Radiation

Radiant energy from the sun is the power source that determines the thermal regime of the soil and the growth of plants. Agriculture is the exploitation of solar energy in the presence of an adequate water supply and sufficient plant nutrients to maintain plant growth (Monteith, 1958). Only about $\frac{1}{200,000}$ of the total energy radiated by the sun is received by the earth. The solar constant, which is the rate at which radiation is received at the top of the earth's atmosphere, has been estimated to be about 2.0 langleys per minute (1 langley = 1 g cal/cm^2). Over 99 percent of this solar energy is contained in the wavelengths between 0.3 and 4 μ (Chang, 1968) and is known as "short-wave radiation."

Partitioning of Solar Radiation

Much of the short-wave radiation is dissipated and distributed by components of the earth's atmosphere as it passes downward toward the earth. A portion is reflected by clouds into outer space; some is absorbed by the molecules of carbon dioxide, oxygen, ozone, and water vapor present. Part is scattered and diffused by the molecules and small particles in the air. A portion of this diffused radiation is returned to outer space. The remainder is transmitted to the earth's surface as "sky radiation." That fraction of the solar radiation that is not affected by the earth's atmosphere reaches the surface as a direct beam.

Geiger (1965) estimated the distribution of incoming radiation during a summer day to be as follows: (1) global radiation, which is the sum of the direct beam penetration to the earth's surface (19 percent) and the diffuse sky radiation (26 percent); (2) scattered radiation returned to outer space (11 percent); (3) reflected by clouds (28 percent); and (4) absorbed by the atmosphere (16 percent). This distribution is illustrated in the upper portion of Figure 7-1. It is seen that the short-wave radiation reaching the ground, or global radiation, represents only 45 percent of the incoming energy.

The global radiation is further partitioned into reflected radiation and absorbed energy, which is utilized in heating the soil and the air above the soil (sensible heat), as latent heat of evaporation, and as effective long-wave radiation. This distribution over a mature cane field in Hawaii is depicted in the lower half of Figure 7-1.

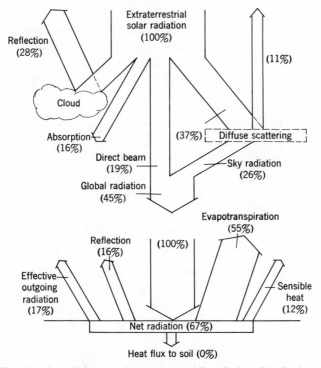

FIG. 7-1. Daytime heat balances. (Extracterrestrial radiation distribution after Geiger, 1965; global radiation after Chang, 1961.)

Physical Factors Affecting Radiation Distribution

LATITUDE. The angle at which the sun's rays meet the earth greatly influences the amount of radiation received per unit area. Radiation reaching the earth at an angle is scattered over a wider area than the same radiation striking the earth's surface perpendicularly. Consequently, in the former case the amount of heat received per unit area is decreased in proportion to the increase in area covered. The amount of radiation reaching the earth per unit area is proportional to the cosine of the angle made between the perpendicular to the surface and the direction of the received heat rays. Therefore the radiation received per unit area decreases with an increase in this angle.

The radiation distribution in the Northern Hemisphere in relation to latitude is illustrated in Figure 7-2 (Houghton, 1954). Both the incoming global and outgoing long-wave daily radiation are highest at the lower latitudes. The global radiation decreases rapidly at latitudes above 30°

and reaches a constant amount above 60°. The albedo is lowest (<10 percent) in tropical regions, increases slightly in the middle latitudes, and rises sharply in the polar areas. The relatively constant daily global radiation and the tremendous increase in the percentage reflection at the higher latitudes are the result of the lower solar elevation. It is impor-

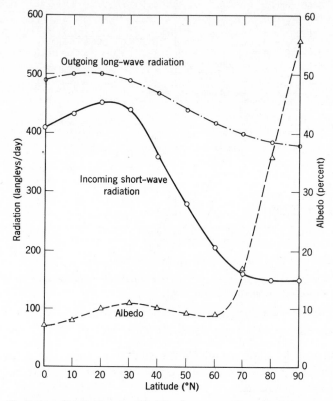

Fig. 7-2. Radiation distribution in the Northern hemisphere in relation to latitude. (After Houghton, 1954.)

tant to remember that not only is the daily global radiation high and the albedo low in the tropical latitudes but these values exist 12 months of the year.

The outgoing long-wave radiation is highest in the tropics because of the greater temperatures. It decreases gradually with increasing latitudes, as expected.

EXPOSURE. The effects of latitude may be simulated on a small scale within certain latitudes by changes in the direction of exposure and the degree of slope of the land. For example, the angle at which the rays of the sun strike a steep south slope is entirely different from that on a steep north slope; the southern slope receives more solar radiation per unit area. These differences in exposures have great ecological and agricultural significance, inasmuch as the temperature of the soil is always higher on southern exposures than on northern. Wollny (1878) pioneered studying the effect of exposure on the warmth of soils. He found that the southern exposures were always several degrees warmer than the north slopes. Because this increased warmth was due entirely to the direct action of the sun's rays, greater temperature variations from night to day were observed on the south exposure. He also found that the temperature differences between exposures were greater the steeper the slope. Direction of exposure, however, was of greater significance than the degree of slope.

Exposure is of little importance in the tropics because of the high elevation of the sun. It is of significance in the middle latitudes where the elevation is lower. The greater the percentage of diffuse sky radiation in the global radiation, the smaller is the difference in the incoming solar energy per unit area for slopes of different exposures. This is because the amount of sky radiation diffusing to north and south slopes is the same. The high percentage of sky radiation in polar regions makes exposure less significant in high than in middle latitudes.

The impact of the degree of slope on the solar climate is well illustrated by the work of Alter (1912). He stated that a 1° slope in Idaho facing north had the same solar climate as a level field at a latitude 1° farther north. A 5° slope in Idaho with a southern exposure had the same solar climate as a level field in southern Utah, 350 miles to the south; this was equivalent to a 5° change in latitude.

DISTRIBUTION OF LAND AND WATER. In general, island climates are more uniform than continental climates. The presence of large bodies of water tends to stabilize the temperature because of the high specific heat of water, which is responsible for the absorption of large amounts of heat. In addition the atmosphere surrounding these bodies of water is highly saturated with water vapor, which reduces the amount of radiant energy reaching the earth. The effect of the absence of these two factors is well illustrated in the temperature changes within a continent. For example, the central plains region and corn belt of the United States are characterized by hot summers and cold winters. Even the nights are warm as a result of the radiation from the earth. It should be emphasized, however, that the beneficial effects of water on the equalization of temperature are associated only with large masses of water. Little can be achieved in this direction through artificial ponds and lakes. This fact is completely overlooked in many of the statements associated with water conservation projects.

The influence of water currents on air and soil temperatures is well known. The Gulf current along the coast of Great Britain and the Japanese current along the shores of northwestern United States are examples of warm currents that materially affect the climate of the land they touch.

VEGETATION. Vegetation plays a significant role on soil temperature because of the insulating properties of plant cover. Bare soil is unprotected from the direct rays of the sun and becomes very warm during the hottest part of the day. When cold seasons arrive, such an unprotected soil rapidly loses its heat to the atmosphere. On the other hand, a good vegetative cover intercepts a considerable portion of the sun's radiant energy, which prevents the soil beneath from becoming as warm as bare soil during the summer. In winter the vegetation acts as an insulating blanket that reduces the rate of heat loss from the soil. A vegetative cover reduces the daily variations in soil temperature as compared with bare soil. This effect is related primarily to the extent of the protective canopy rather than to other special characteristics of the crop itself. Frost penetration is more rapid and the depth of freezing is greater with bare soils than under a vegetative cover.

The major impacts of vegetation are associated with (1) the albedo effect, (2) decreasing the depth of penetration of global radiation through the canopy (Chang et al., 1965), (3) increasing the latent heat in evapotranspiration, and (4) decreasing the rate of heat loss from the soil through its insulating influence.

Mulches can affect the thermal regime of the soil in several ways. Light-colored plastic mulches will increase the albedo of the surface. Transparent plastic mulches will transmit solar energy to the soil and produce a green-house effect. Mulches with low thermal conductivities will decrease both the positive and negative heat flux in the soil. The soil will be cooler during the day and warmer during the night. These mulches protect the soil from excessive cooling during the winter. They also will prevent the soil from warming up early in the spring.

Net Radiation and Heat Balance

Net radiation represents the difference between the total downward and upward radiation flux (Chang, 1968). It is a measure of the energy available at the soil surface. Geiger (1965) expressed net radiation by the equation

$$S = I + H + G - \sigma T^4 - R \tag{7-1}$$

where S is the net radiation, $I + H$ the global radiation, G the counter-radiation, σT^4 the radiation emitted by the soil surface, and R the albedo. $I + H$ and R are zero at night and net radiation becomes negative. This is shown in Figure 7-3b.

Table 7-1 summarizes the available data on the fraction of global radiation that is retained as net radiation (Chang, 1968). Since the global

TABLE 7-1
The Ratio of Net Radiation to Global Radiation during 24-Hour Periods
(after Chang, 1968)

Location	Type of cover	Global radiation (langleys/day)	Net radiation / global radiation
Copenhagen, Denmark 55° 44′	Grass	463 July)	0.51
Rothamsted, England 51° 48′N	Grass	550	0.41
	Tall crops	550	0.46
Davis, California 38° 30′N	Grass	750 (June)	0.56
		175 (December)	0.33
Aspendale, Australia 38° 2′S	Grass	181 (July)	0.18
		689 January)	0.69
Hawaii 21° 18′N	Sugar cane	725 (Summer)	0.69
		400 (Winter)	0.65
	Pineapple	710 (May–June)	0.66
		500 (December)	0.53

radiation is much less during the winter months in the middle and higher latitudes, there is a greater difference in this ratio in these latitudes than in subtropical regions. Net radiation becomes negative during the colder months. This is illustrated in Figure 7-4. There is a linear relationship between net and global radiation in the tropics and during the summer months in middle and high latitudes.

The earth's surface as a boundary surface cannot absorb heat (Geiger, 1965). Consequently, according to the law of conservation of energy, the heat balance at this surface is governed by the fundamental equation

$$S + B + L + V = 0 \tag{7-2}$$

where S is the net radiation, B the heat flux upward through the ground $(+)$ or downward from the surface $(-)$, L the heat flux in the air either to or from the ground as sensible heat, and V the evaporation effect [units are $cal(cm^2)(min)$]. The heat balance during the day and night is illustrated in Figures 7-1 and 7-4, respectively.

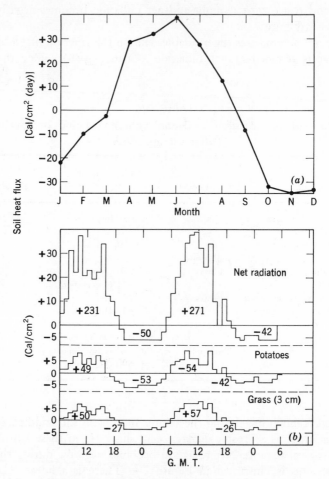

FIG. 7-3. (a) yearly and (b) daily heat flux cycles in soils. (a) After Carson and Moses, 1963; (b) after Monteith, 1958.)

The radiation balance throughout the year at Hamburg, Germany, is illustrated in Figure 7-5. The long-wave balance is depicted in part *a*. It is seen that the maximum radiation loss per day of about 14 percent of the terrestrial radiation occurred around August 1. Minimum loss (6 percent) took place about January 1. The short-wave balance is portrayed in Figure 7-5b. Both the global radiation $(I + H)$ and the albedo (R) were at a maximum about June 1; approximately 20 percent of the global radiation was reflected. The albedo became practically zero by the end of December. The net radiation balance is illustrated in Figure 7-5c. This represents the difference between the short-wave radiation balance 2 and the long-wave radiation loss 1. This balance was positive between the

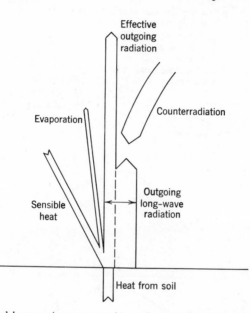

FIG. 7-4. Nocturnal heat exchange at earth's surface. (Adapted from Rose, 1966.)

middle of February to the end of October; it was negative during the winter months.

Parameters of Net Radiation

ALBEDO. The percentage reflection, or albedo, is dependent upon the nature of the surface, the angle of the sun, and the solar elevation. The higher the solar elevation, the lower is the albedo of the surface and vice versa. This is illustrated in Figure 7-2 where the reflection of incoming radiation at the equator is only about 7 percent; that near the North Pole is 56 percent (Houghton, 1954). The albedo percentages of various surfaces are given in Table 7-2. Water surfaces have a low albedo; forests have lower albedos than grasses or cultivated crops. Mature crops in the middle and high latitudes that provide complete ground cover have maximum albedo values of about 25 percent (Chang, 1968). Such crops in the tropics, however, have much lower albedos; this is undoubtedly due to the greater solar elevations. Chang's data in Figure 7-1 show that the reflecton from a mature sugar cane field is about 16 percent. This

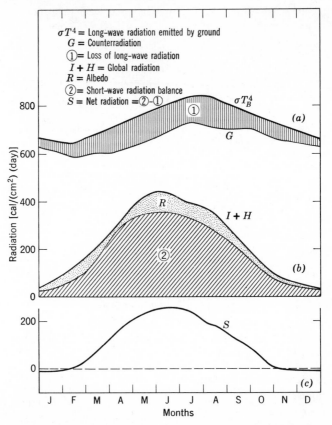

Fig. 7-5. Radiation balances throughout the year at Hamburg, Germany. (After Fleischer according to Geiger, 1965.)

value for a newly planted field of cane is only about 7 percent. The albedo is higher in the early morning and late afternoon because of the lower angle of the sun (Chang, 1961).

The type of soil and its moisture content determine the albedo of the bare ground surface. This is primarily related to color (Table 7-2). For example, dry, light-colored desert soils (serozems) have twice the albedo of dry, dark-colored chernozems. Adding water to light soils darkens the surface and decreases the reflection about 50 percent. Application of white $MgCO_3$ to the surface of a bare soil doubles the albedo from 195 to 390 g cal(cm^2)(day) (Stanhill, 1965). There is a 10° C decrease in soil temperature.

TABLE 7-2
The Relation of the Nature of the Surface on
the Reflection of Short-Wave Radiation

Nature of surface	Albedo (percent)	Reference
Fresh snow cover	75–95	Geiger, 1965
Light sand dunes	30–60	Geiger, 1965
Meadows and fields	12–30	Geiger, 1965
Forests	5–20	Geiger, 1965
Water surfaces	3–10	Geiger, 1965
Soils		
serozem, dry	25–30	Chudnovskii, 1966
wet	10–12	Chudnovskii, 1966
clay, dry	23	Chudnovskii, 1966
wet	16	Chudnovskii, 1966
chernozem, dry	14	Chudnovskii, 1966
wet	8	Chudnovskii, 1966
Crops		
Corn (New York)	23.5	Chang, 1968
Sugar Cane (Hawaii)	5–18	Chang, 1968
Pineapple (Hawaii)	5–8	Chang, 1968
Potatoes (U.S.S.R.)	15–25	Chang, 1968

EVAPOTRANSPIRATION. The major portion of the global radiation in humid climates is used in the process of evapotranspiration. This process consumes 580 cal/g of water that is changed from the liquid to the gaseous phase. This heat energy is lost to the soil and returned to the atmosphere. Of course, any condensation as dew will release this energy back to the soil. Chang's data in Figure 7-1 indicate that 55 percent of the global radiation is used in evapotranspiration over a mature sugar cane field. This represents 82 percent of the net radiation (incoming global radiation minus the sum of the reflected short-wave and the net outgoing long-wave radiation). This is about an average value for the tropics and for the summer season in middle latitudes. The daily and monthly variations in evapotranspiration correlates with fluctuations in the amount of incoming solar radiation (Baver, 1964; Chang et al., 1965). The total

solar energy available at the surface for evaporation has been used successfully to calculate the potential evaporation (Penman, 1948). Potential
evapotranspiration is defined as the amount of water that will be lost
by evaporation and transpiration from a surface that is completely
covered with vegetation if there is sufficient water in the soil at all times
for the use of the vegetation. Penman developed an expression for the
total amount of energy available for evaporation and heating of the air
and then divided the energy so obtained between evaporation and heating. His final equation was

$$E_0 = \frac{H + 0.27E_a}{\Delta + 0.27} \text{ mm/day} \tag{7-3}$$

where E_a is the evaporation when the saturation vapor pressure is that
at air temperature, H is the net radiant energy available at the surface,
and Δ is the slope of the evaporation-temperature curve when T_a is
at air temperature. Mean air temperature, mean air vapor pressure, mean
wind velocity, and mean duration of sunshine are required to calculate
the evaporation.

E_a is calculated from the equation

$$E_a = 0.35(1 + 9.8 \times 10^{-3}v)(e_a - e_d) \tag{7-4}$$

One can also calculate E_0 from the equation

$$E_0 = \frac{H}{1 + 0.27(T_s - T_a)/(e_s - e_d)} \tag{7-5}$$

where e_s is the mean saturation vapor pressure at the water surface,
determined by the mean surface temperature T_s; e_d is the mean vapor
pressure in the air at the dewpoint temperature; v is the wind velocity
in miles per day measured at 2 meters above the ground level. Penman
found that evaporation from a free water surface approximated that from
the soil or from vegetation.

HEAT FLUXES TO SOIL AND AIR. The incoming global radiation that is
not reflected, used in evapotranspiration, or back-radiated is responsible
for the flux of heat to the soil or to the air as sensible heat. The radiation
distribution over a mature sugar cane crop as shown in Figure 7-1 indicated no heat flux to the soil. This was because of the low degree of
penetration of radiation through the cane canopy. Chang et al. (1965)
showed that the subcanopy radiation in sugar cane decreased from about
91 to 8 percent from March 16 to July 27. The leaf area increased
21 times during this same period. Similar effects were observed with
corn in Missouri where the heat flux into moist soil decreased from 15

percent in the early stages of corn growth to 4 percent when the canopy was fully developed (Decker, 1959). More than 50 percent of the available energy at the surface of a bare Plainfield sand in Wisconsin goes into the soil heat flux (Fuchs and Tanner, 1968). About 11 percent of the net radiation in the summer months in Denmark is used to heat the soil (Aysling, 1960).

Monteith (1958) studied the diurnal variations of soil heat flux under grass, potatoes, and wheat. Part of his results for the 24-hr period beginning at 0600 GMT on June 12, 1956, are shown in Figure 7-3b. It is seen that approximately 20 percent of the net incoming radiation was stored in the soil between 6 AM and 6 PM under English conditions. The net radiation day was longer than the heat flux day. For example, net radiation on June 13 was positive between about 4:30 AM and 6:45 PM; this was 1½ hr before heat flux into the soil began and ¾ hr after the heat flux from the soil to the air started. All three crops stored approximately the same amount of heat during the day. The heat loss at night from potatoes and wheat and the outgoing net radiation were about the same; the nocturnal heat loss under grass was only slightly more than one-half of the outgoing radiation. Carson and Moses (1963) reported data on the annual cycle of heat flux in a yellow clay glacial till under a pasture cover for the 3-yr period 1953–1955 (Figure 7-3a). The heat flux was positive from March to September, with a maximum in June. Minimum negative values were registered during November.

About 12 percent of the global radiation over a cane field is expended in heating the air (Figure 7-1) (Chang, 1961); this is equal to 18 percent of the net radiation.

A greater percentage of the incoming global radiation in arid climates is used to heat the soil and the air since there is little opportunity to consume energy as latent heat of evaporation.

LONG-WAVE RADIATION. Long-wave (4 to 100 μ) radiation is emitted by the earth's surface to the sky. It is referred to as "terrestrial" or "outgoing" radiation. Terrestrial radiation is proportional to σT^4, where σ is the Stefan-Boltzmann constant and T is the absolute temperature of the surface. It takes place both day and night. Approximately 90 percent of the terrestrial radiation is absorbed by the earth's atmosphere, primarily by the water vapor present. There is considerable reradiation of this absorbed radiation back to the surface. This is known as "counterradiation." This counterradiation plays the major role in preventing excessive cooling of the earth's surface at night. The difference between the outgoing terrestrial radiation and the atmospheric counterradiation is referred to as "effective outgoing radiation."

The effective outgoing radiation during the day over a canefield is shown by the data of Chang in Figure 7-1. Although the effective outgoing radiation is about the same during the day as by night, it dominates the heat exchange picture at night because of a lack of short-wave radiation and the smaller turbulence. Rose (1966) portrayed the nighttime heat exchange at the earth's surface according to the illustration in Figure 7-4. It is observed that the effective outgoing radiation is less than the long-wave radiation emitted by the surface because of the counterradiation effect. Some heat is convected downward from the warmer air layers by the wind. If there is no wind, the heat flux upward from the soil is responsible for the effective outgoing radiation and any latent heat of evaporation. Chang (1961) observed that about 17 percent of the incoming global radiation was distributed as effective outgoing radiation.

THERMAL PROPERTIES OF SOILS

Heat Capacity

The specific heat of any substance is defined as the calories of heat required to raise one gram one degree on the centigrade scale. The heat capacity of a given material is equal to its specific heat times its mass. The specific heat of water is 1.0. All other constituents of soils have much lower specific heats.

The contributions of Lang (1878) and Ulrich (1894) on the specific heat of various soil constituents undoubtedly have been the most outstanding of all investigations. Some of their data along with more recent values are given in Table 7-3. It is seen that quartz has the lowest specific heat of the major soil constituents and humus the highest, excepting water. The aluminosilicate kaolin has a slightly higher specific heat than quartz. Since the major constituents in most soils are quartz, aluminosilicates, water, and humus, it is evident that humus and water will affect the heat capacity considerably.

One can calculate the specific heat of a soil, c_s, from the summation of the specific heat times the mass of the individual constituents:

$$c_s = c_1 M_1 + c_2 M_2 + c_3 M_3 \cdots c_n M_n \quad [\text{cal}/(\text{gm})(°\text{C})] \quad (7\text{-}6)$$

The heat capacity of a soil constituent is equal to its specific heat times its density. The heat capacity of the soil per unit volume can be calculated by the equation (de Vries, 1963)

$$C_s = x_s C_s + x_w C_w + x_a C_a \quad [\text{cal}/(\text{cm}^3)(°\text{C})] \quad (7\text{-}7)$$

where C_s is the heat capacity of the soil, x_s, x_w, and x_a the volume fraction of the solid material, water and air, respectively, and C_s, C_w, C_a their respective heat capacities. Since the solid material consists of mineral and organic matter whose heat capacities per unit volume are approximately 0.46 and 0.60, respectively, and since the air component

TABLE 7-3

The Specific Heat of Various Soil Constituents

Material	Specific heat [cal/(g)(°C)]			
	Lang (1878)	Ulrich (1894)	Kersten (1949)	Bowers and Hanks (1962)
Coarse quartz sand	0.198	0.191	0.190	0.19
Fine quartz sand	0.194	0.192	0.197	—
Quartz powder	0.209	0.189	—	—
Kaolin	0.233	0.224	—	—
Feldspar	—	0.194–0.205	0.190	0.210–0.220
Mica	—	0.206–0.208	—	—
Apatite	—	0.183	—	0.22
Dolomite	—	0.222	—	0.23
Al_2O_3	0.217	—	—	—
Fe_2O_3	0.163	0.165	—	—
Humus	0.477	0.443	—	—
Calcareous sandy soil	0.249	—	—	—
Humus calcareous sandy soil	0.257	—	—	—
Garden soil	0.267	—	—	—
Clay	—	—	—	0.27
Silty clay	—	—	—	0.26
Silt loam	—	—	0.164–0.194	—

in equation 7-7 is too small to be significant, this equation can be simplified as follows:

$$C_s = 0.46x_m + 0.60x_0 + x_w \qquad (7\text{-}8)$$

where x_m and x_o are the volume fractions of mineral and organic matter, respectively.

Thermal Conductivity and Diffusivity

The Theory of Heat Flow

In discussing the flow of heat through a given substance, let us follow the general ideas used by Patten (1909). Consider a metal bar, 100 cm long, placed in contact with a heat source so that under conditions of steady flow of heat the one end is at a temperature of 100° C and the other at 0° C (see curve 1, Figure 7-6). In order to calculate the amount of heat that flows through any given part of this bar, visualize a small rectangular section within the bar at a distance of x cm from the heated end. Let the cross-sectional area of this section be A and the thickness of an infinitesimal distance Δx.

If Θ is the temperature of the face of this section at x, the flow of heat per unit time through the surface A will be equal to $-KA(d\Theta/dx)$, where $d\Theta/dx$ is the temperature gradient, or the change in Θ with distance from the heat source, and K is the heat conductivity of the material. The negative sign indicates that the temperature decreases as the distance x increases. This expression simply states that the flow of heat from the hot to cold regions of a bar increases directly with the conductivity of the material, the cross-sectional area through which heat flows, and the greater the difference in temperature between the hot and cold ends. Heat conductivity is generally defined as the quantity of heat that flows through a unit area of unit thickness in unit time under a unit temperature gradient.

The temperature of the surface of this rectangular section at a distance $x + \Delta x$ from the source of heat will be lower than at x and will be equal to $\Theta - (d\Theta/dx)\,\Delta x$. Since $d\Theta/dx$ is the rate at which the temperature decreases as the distance x from the source of heat increases, then this rate of decrease times the distance Δx over which this fall in temperature occurs will give the total drop in temperature in passing from x to $x + \Delta x$. In this case Δx is so infinitesimally small that the rate of temperature fall may be considered constant over this distance. Therefore the flow of heat through the surface of the section at $x + \Delta x$ will be equal to $-KA[d/dx(\Theta - (d\Theta/dx)\,\Delta x)]$. The expression within the outer brackets is simply the rate of change of the temperature at $x + \Delta x$.

Since the temperature at $x + \Delta x$ is lower than at x, there will be less heat leaving the rectangular section than entering. This difference is given by the expression $-KA(d\Theta/dx) - \{-KA[d/dx(\Theta - (d\Theta/dx)\,\Delta x)]\}$, which reduces to

$$-KA\,\frac{d^2\Theta}{dx^2}\,\Delta x \qquad (7\text{-}9)$$

The term $d^2\Theta/dx^2$ is the rate of change of the temperature gradient, or the acceleration of the temperature change with distance. Under a constant gradient in a steady state this acceleration is zero, and there is no difference between the amounts of heat entering and leaving the section.

Consider now that Figure 7-6 represents a box of soil. Before reaching the steady state, the temperature at each point in the soil is changing according to curve 2. Under this condition the heat that flows into one side of the rectangular section is not the same as that which leaves it on the other side. Let the mean temperature of the section rise by a small amount $d\Theta$ in a small unit of time dt. The quantity of heat that is necessary per unit time to raise the temperature of this section will be equal to $AC(d\Theta/dt)\,\Delta x$, where C is the heat capacity of the soil

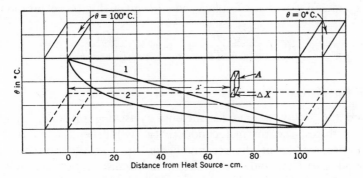

FIG. 7-6. Sketch illustrating the flow of heat through soils.

(C = effective specific heat of soil $\times$ apparent density). This expression states that the amount of heat required is equal to the volume of the section $A\,\Delta x$ times its heat capacity C times the change in temperature $d\Theta/dt$.

If the soil is thoroughly insulated so that there is no heat loss during the experiment, then

$$AK\left(\frac{d^2\Theta}{dx^2}\right)\Delta x = AC\left(\frac{d\Theta}{dt}\right)\Delta x \qquad (7\text{-}10)$$

or

$$\frac{K}{C}\left(\frac{d^2\Theta}{dx^2}\right) = \frac{d\Theta}{dt} \qquad (7\text{-}11)$$

This is the fundamental equation for calculating the heat conductivity K, from data on the effective heat capacity C, the rate of change of the temperature gradient $d^2\Theta/dx^2$, and the change in temperature at a given point with time $d\Theta/dt$.

The expression K/C, termed diffusivity, denotes the temperature change that takes place in any portion of the soil as the heat flows into it from an adjacent layer. It is the change in temperature (in °C) in 1 sec when the temperature gradient changes 1° C/cm³.

The theory of heat conductivity may be visualized somewhat more simply by considering the rectangular section in Figure 7-6 from a slightly different point of view. Let the temperature on each side of this section be equal to T_1 and T_2, the thickness of the section equal to d, and the amount of heat that flows across in a given time equal to Q. The rate of heat flow per unit area will be Q/At. The temperature gradient will be $(T_1 - T_2)d$. Therefore, according to definition, the heat conductivity of a substance will be given by the expression

$$K = \frac{Q/At}{(T_1 - T_2)/d} \quad \text{or} \quad \frac{Qd}{At(T_1 - T_2)} \tag{7-12}$$

Soil Parameters

Since the soil is a granular medium consisting of solid, liquid and gaseous phases, the thermal conductivity will depend upon the volumetric proportions of these components, the size and arrangement of the solid particles, and the interface relationships between the solid and liquid phases. The thermal conductivity of quartz is 26.3×10^{-3} cal/(cm)(sec)(°C) when measured parallel to the axis of the crystal and 16.0×10^{-3} when determined perpendicular to this axis. These values for water and air are 1.4 and 0.06×10^{-3}, respectively. This makes the ratio of thermal conductivity $333:23:1$ for quartz, water, and air.

It is obvious that the thermal conductivity of a granular soil will depend upon the intimacy of the contact of the solid particles and the extent to which air is displaced by water in the pore spaces between the particles.

SOIL TYPE AND POROSITY. The thermal conductivity of different soils follows the order of sand > loam > clay > peat. The results of several investigators are given in Table 7-4. Since the conductivity of the mineral components of the solid phase is of the same order of magnitude (Smith and Byers, 1938), these conductivity differences are related to the degree of packing and porosity of the system. Thermal conductivity diminishes with decreasing particle size due to reduced surface contact between the particles through which heat will readily flow. Patten (1909) observed that the conductivity of carborundum was lowered about 70 percent as the size of particles decreased from 450 to 6 mμ. He also found that the conductivity of quartz particles was only $\frac{1}{15}$ to $\frac{1}{20}$ that of a solid quartz block.

TABLE 7-4
Thermal Properties of Different Soils

Type of soil	Thermal conductivity [cal/(cm)(sec)(°C) × 10⁻³]				Thermal diffusivity (cm²/sec × 10⁻³)		
	Relative (moist sand = 100) von Schwarz (1879)	Geiger (1965)	Nakshabandi and Kohnke (1965)	van Duin (1963)	Geiger (1965)	Nakshabandi and Kohnke (1965)	van Duin (1963)
Sand, wet	100	4.00	4.35*	3.70§	7.0	12.6*	4.4§
Sand, dry	85.5	0.55	0.35	0.37	3.5	1.5	1.7
Clay, wet	90.3	3.50	1.40†	3.18§	11.0	3.2†	3.7§
Clay, dry	74.3	0.17	0.25	0.37	1.2	1.5	1.8
Loam, moist	99.3	—	4.25‡	—	—	6.0‡	—
Loam, dry	83.3	—	0.45	—	—	1.8	—
Peat, wet	58.8	0.85	—	0.82¶	1.2	—	1.2¶
Peat, dry	51.5	0.20	—	0.11	2.0	—	1.3

* 23 percent saturation with water.

† 38 percent saturation with water.

‡ 75 percent saturation with water.

§ 100 percent saturation with water.

¶ 66⅔ percent saturation with water.

Increasing the bulk density of soils lowers the porosity and improves the thermal contacts between the solid particles. The amount of low-conducting air is reduced and the overall thermal conductivity is increased. The impact of soil porosity on thermal conductivity is shown in Figures 7-7 and 7-8a. Van Rooyen and Winterkorn (1959) evaluated the conductivity data from Russian investigations on a chernozem soil (Figure 7-7).

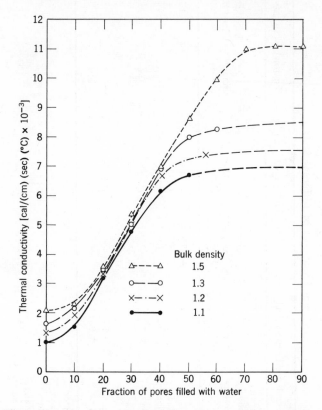

FIG. 7-7. Thermal conductivity as affected by density-moisture relations. (Adapted from data of van Rooyen and Winterkorn, 1957.)

As the bulk density of the dry soil increased from 1.1 to 1.5 (porosity decrease from about 59 to 43 percent), the thermal conductivity increased from 1.0 to 2.1 $\times$ 10^{-3} cal/(cm)(sec)(°C). In other words, a 27 percent decrease in porosity resulted in a doubling of the thermal conductivity. The data of van Duin (1963) in Figure 7-8a indicate similar relationships. A 50 percent decrease in the porosity of sand (as well as clay) caused a doubling of the thermal conductivity; van Duin's data in Figure 7-8b show that a 50 percent decrease in porosity caused a $33\frac{1}{3}$ percent increase in the thermal diffusivity of dry sand.

SOIL MOISTURE CONTENT. The increase in thermal conductivity as a result of raising the density is small compared with the impact of adding water to the soil. The presence of water films at the points of contacts between particles not only improves the thermal contact between particles

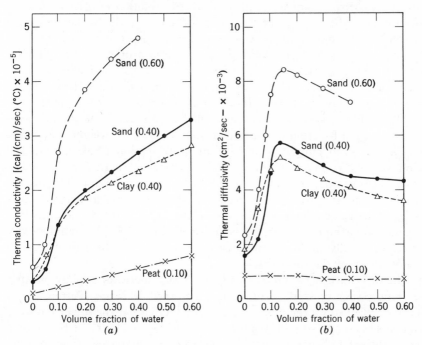

FIG. 7-8. The effect of water on thermal conductivity (*a*) and diffusivity (*b*). (After van Duin, 1963.) Note: (0.10), (0.40), and (0.60) = volume fraction of solid phase.

but also replaces air in the soil pore space with water, which has about 20 times the thermal conductivity of air. The curves in Figures 7-7 and 7-8 point out quite clearly the rapid increases in thermal conductivity and diffusivity as the percentage of water in the soil pore space rises. The greatest rate of increase in conductivity occurs at the lower moisture contents. If the thermal conductivity of the dry soil (Figure 7-7) with a bulk density of 1.1 is considered equal to 100, the conductivity at densities of 1.2, 1.3, and 1.5 is 130, 165, and 210, respectively. When the pore spaces are 25 percent filled with water, the relative thermal conductivities (1.1 clay = 100) are 400, 420, 420, and 440, respectively.

When they are 50 percent filled, these values are 670, 725, 800, and 860, respectively. In other words, the impact of density on thermal conductivity is greater at the higher moisture contents. These same relationships hold in the curves of van Duin in Figure 7-8a.

In order to visualize the impact of water on thermal conductivity, consider two dry quartz spheres that are in contact with each other. The major conduction takes place through a relatively small cross-sectional area at the point of contact. With the addition of a small amount of water in the wedges at the point of contact, the surface through which heat is conducted increases greatly; the distance of flow through the water wedge is small. With further additions of water, the distance of flow through water (volume of water) increases faster than the augmentation of surface. Therefore the rate of heat conduction should increase more slowly. This effect is clearly shown in the figures.

Although fine sand, silt loam, and clay exhibit different thermal conductivities as a function of the percentage of water by weight, the curves are similar when plotted against soil-moisture tension (Nakshabandi and Kohnke, 1965). There is only a slight increase in thermal conductivity from the oven-dry condition to pF 5 because the moisture films are not sufficient to provide the necessary thermal contacts between particles. Thermal conductivity begins to rise at a pF of 4.5 and reaches a value about equal to that of water at a pF of 3.8. The conductivity increases rapidly at lower moisture tensions due to the larger thermal conductivity of the soil particles.

Thermal diffusivity (K/C) (Figure 7-8b) increases rapidly at first with increasing moisture to a maximum and then decreases (Patten, 1909). This is explained by the greater rise in conductivity at the lower moisture contents as compared with the increase in the heat capacity of the system. The value of C becomes larger as the moisture increases, K approaches that of water, and diffusivity decreases. Jackson and Kirkham (1958) have suggested that the diffusivity curves, such as those shown in Figure 7-8b, represent apparent diffusivity and do not take into account the thermal transfer of water. Their technique of measurement showed no decrease in diffusivity after the maximum was reached.

Measurement of Thermal Conductivity

The steady-state method for determining the thermal conductivity of moist soils has two major weaknesses. It is subject to the redistribution of water under the influence of a temperature gradient and it cannot be used *in situ*. The transient-state cylindrical probe (Jackson and Taylor 1965; de Vries and Peck, 1968) overcomes these difficulties, although there is some temperature-induced moisture flow. The method basically consists of a thin metal wire that is heated electrically to serve as the heat source and a thermocouple to measure the temperature rise. There is radial flow of heat from the wire into the soil. One can calculate the conductivity from the equation

$$T - T_0 = \frac{q}{(4\pi K)d} + \ln (t + t_c) \qquad (t < t_1) \qquad (7\text{-}13)$$

where T_0 is the temperature at t_0, $T - T_0$ the temperature rise, q the heat developed per unit time and unit length of the source, d a constant, t_1 the time at the end of the heating period, and t_c a correction constant that depends upon the dimensions of the probe as well as the thermal properties of both the probe and the soil. If $T - T_0$ is plotted against ln t, a straight line is obtained for the higher t values. When this straight line is extrapolated to smaller values of t and the appropriate t_c chosen as the correction factor, the adjusted values fall on the extrapolated line.

The thermal conductivity is then calculated by the revised equation

$$K = \frac{2.303q}{4\pi S} \qquad (7\text{-}14)$$

where S is the measured slope and $q/4\pi K$ is obtained from equation 7-13. The value q is obtained from the current (amperes) applied to the wire and the measured resistance (ohms per centimeter of probe).

VARIATIONS IN SOIL TEMPERATURE

Global Patterns

The data in Figure 7-2 showed that global radiation was highest in low latitudes; reflection of solar energy was lowest in these latitudes. These differences in solar radiation are responsible for different soil temperatures at the various latitudes. Soil temperatures in both summer and winter in the Western hemisphere are depicted in Figure 7-9 in relation to latitude and soil depth. These data are approximate values estimated from the temperature isotherms of Chang (1957). There are several significant facts in these values:

1. Soil temperatures at 10-, 30-, and 120-cm depths are approximately the same at $0°$ latitude throughout the year.

2. Soil temperatures at $20°$ N and S during the summer are about the same at all depths.

3. Soil temperatures at 10 and 30 cm are approximately 5° C higher in the summer at 40° N than at 40° S.

4. Winter soil temperatures at all depths increase from high latitudes to a maximum at the equator.

5. Winter soil temperatures at 40° N are about 5° C colder at all depths than at 40° S.

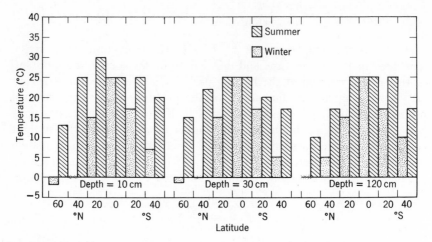

FIG. 7-9. Variations in soil temperature with latitude in the Western hemisphere. (After Chang, 1957.)

Chang also observed that the soil in July at 60° N was warmer than the air at a depth of 10 cm; this depth varied between 20 and 40 cm in the middle latitudes; it was 80 cm in the arid tropics. The 0° isotherm at 10 cm followed the air temperature except in western Europe and eastern North America where the soil temperature was higher because of snowfall.

Diurnal Variations

The early investigations of Wollny (1883) and Bouyoucos (1913) pointed out the diurnal variations in soils as affected by the nature of the soil, type of surface cover, and incoming radiation. The data of Yakuwa (1945) in Figures 7-10 and 7-11 are used to illustrate these variations. In the morning before sunrise, the minimum temperature of the soil was lowest at the surface and increased with depth. For example, at about 4:30 AM on August 15, 1929, when the surface temperature was approximately 13° C, the temperature at 20 cm was 24° C. This minimum temperature travels as a wave and lowers the temperature at different depths as it reaches them. This wave continues downward after sunrise even as the surface soil begins to warm up. Soil temperatures reached a minimum at 5, 10, 20, and 30 cm (Figures 7-10 and 7-11) approximately at 5.5, 7.0, 9.5, and 13 hours sidereal time, respectively, for loam and clay soils. Owing to its greater thermal conductivity, sand exhibited these minimums at about 5.75, 6.5, 8.75, and 11, respectively. Peat, which has a lower conductivity, required a longer time for the minimum temperature wave to reach the lower depths. The wave reached these depths at 6.0, 7.5, 15, and 22.5 hr, respectively.

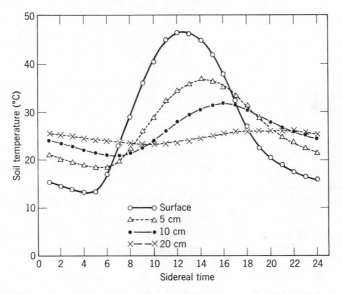

Fig. 7-10. Diurnal variations in temperature in a loam soil. (After Yakuwa, 1945.)

Soon after sunrise, the temperature of the surface rose and at about 7 AM was higher than that at the 20-cm depth, where it was still decreasing. Maximum surface temperature occurred at 12.5 (12:30 PM). This increase in surface temperature caused a maximum temperature wave to move downward. It reached 5, 10, and 20 cm at 14.5, 16.25, and 20.5 hr, respectively, in the loam soil. It took until 1 AM the next day to reach 30 cm. The wave in the clay soil was the same as that of the loam. The maximum temperature wave in the sand reached these same depths at 14.0, 16.0, 19.0, and 22.0 hr, respectively. The peat soil reached its maximum surface temperature at 12 M because of its higher absorptivity. Maximum values at 5, 10, 20, and 30 cm were obtained at 15.0, 17.0, 23.5, and 10.0 (next day), respectively. There was a rever-

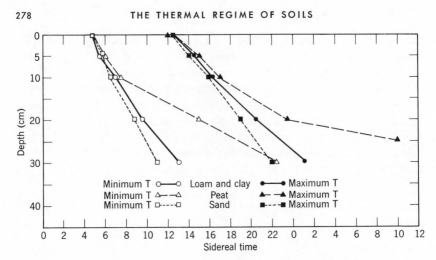

FIG. 7-11. The relation of type of soil to the rate of penetration of maximum and minimum temperature waves. (From data of Yakuwa, 1945; 0 = midnight; 12 = 12:00 M; 22 = 10:00 PM.)

sion of soil temperature at the surface after sundown. It became cooler than that at 5 cm at about 16.5; 2 hr later, it was cooler than the soil at 20 cm.

Yakuwa (1945) (Table 7-5) observed that the amplitude of soil temperature in different soils in the surface 5 cm followed the series sand > loam > peat > clay. The order of peat and clay was reversed below 20 cm. The relative depths where the amplitude was nearly 0.1° C for

TABLE 7-5

The Effect of Type of Soil on the Rate and Depth of Soil Temperature Changes (from data of Yakuwa, 1945)

Amplitude of soil	Sand (°C)	Loam (°C)	Clay (°C)	Peat (°C)
Temperature at depth (cm):				
0	40.0	33.6	21.5	23.2
5	19.4	18.5	13.7	13.9
10	12.3	10.7	7.7	5.4
20	4.8	3.0	2.2	0.7
30	1.6	0.7	0.6	0.3
Depth (in cm) where amplitude equals 0.1° C	56.9	46.6	46.8	39.5
Rate of movement (in hr/dm) of maximum temperature wave	3.30	3.95	3.94	5.31

sand, loam, clay, and peat were 100, 82, 82, and 70, respectively. This reflects the relative thermal conductivities of the four soil types. Yakuwa found that the rate of movement of the maximum temperature wave varied from 3.30 to 5.31 hr/decimeter for sand and peat, respectively.

Seasonal Variations

The seasonal variations in soil temperature with depth are similar in character to the diurnal changes. The summer months (June and July

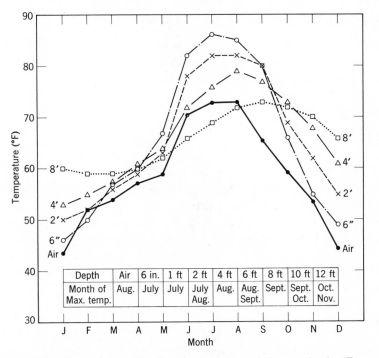

Depth	Air	6 in.	1 ft	2 ft	4 ft	6 ft	8 ft	10 ft	12 ft
Month of Max. temp.	Aug.	July	July	July Aug.	Aug.	Aug. Sept.	Sept.	Sept. Oct.	Oct. Nov.

FIG. 7-12. Monthly variation of soil temperature in relation to depth. (From data of Smith, 1932.)

in the Northern hemisphere), like midday, represent the peak of the global radiation and the maximum temperatures. The winter months have an effect similar to nocturnal daily temperatures. The California data of Smith (1932) in Figure 7-12 are typical of the variations observed. The soil from 6 in. to 8 ft was warmer than the air, except for the 8-ft depth, from the middle of May until the first of August. The deeper layers were warmer during the winter months than at the 6-in. depth;

there was a temperature gradient upward. The reversal of the temperature gradient took place from about May 1 to the middle of September when there was a downward gradient. There was a seasonal wave of soil temperature. Maximum temperatures at 6 in. and 2, 4, 6, 8, 10, and 12 ft occurred approximately on July 1, July 15, August 1, August 15, September 1, September 15, and October 15, respectively. These seasonal differences were associated with the incoming global radiation and the thermal properties of the soil profile, as related to changes in moisture content and temperature gradients.

MODIFYING THE THERMAL REGIMES OF SOIL

The thermal regime of the soil can be modified either by regulating the incoming and outgoing radiation or by changing the thermal properties of the soil.

Vegetation Effects

The basic impacts of a vegetative canopy on the thermal regime was discussed earlier. It suffices at this time to call attention to the effect of shelter belts on soil temperature changes. Chudnovskii (1966b) evaluated the influence of shelter belts on the microclimate of the soil environment between trees. He found that the wind velocity between the trees was reduced to 20 to 40 percent of that in the open. The relative humidity was 2 to 4 percent larger. Soil temperature at 50 cm was higher. The energy balance between the trees depended upon the cropping system used. The effective radiation and the thermal flux into the soil was higher under fallow than when wheat was grown on the land.

Mulches

Mulching effects on the thermal regime have been discussed previously. Mulches reduce the diurnal and seasonal fluctuations of soil temperature (Kohnke and Werkhoven, 1963). There is little difference between the mulched and bare plots in the middle of the summer. The temperature of the mulched soil at the 1-in. depth was about the same as that of the bare soil at 4 in. The mulched soil was cooler in the spring, summer, and fall; it warmed up more slowly from January to March. The bare soil reached the minimum temperature for good corn germination (61° C) 2 weeks sooner than the mulched. Chudnovskii recommended white paper mulches in hot, arid regions to increase the reflection of incoming radiation and to combat excessive heating and evaporation during the day.

Black paper should be used in the colder regions to absorb radiant energy during the day and to reduce heat loss at night.

Irrigation and Drainage

Irrigation increases the heat capacity of the soil, raises the humidity of the air, lowers the air temperature over the soil, and increases thermal conductivity. This reduces the daily soil temperature variations. Chudnovskii observed that the turbulent heat flux to the soil air over a nonirrigated field was -590 cal/(cm^2)(24 hr); that under sprinkler irrigation was reduced to -119 cal/(cm^2)(24 hr).

Drainage decreases the heat capacity of wet soils, which raises the soil temperature. This plays an important role in warming up the soil in the spring.

Changing the Physical Characteristics of the Soil Surface

Compaction of the soil surface increases the density and the thermal conductivity. Tillage, on the other hand, creates a surface mulch which reduces the heat flux from the surface to the subsurface layers. The amplitude of the diurnal wave in a cultivated soil is much larger than that in an untilled soil. A loosened soil is colder at night than a compacted one; this makes the loose one more susceptible to frost.

The ridging of fields causes increased evaporation. The albedo is decreased, which means that the effective incoming radiation is greater. Consequently, the temperature of the ridged fields is higher than those that are level.

References

Allmaras, R. R., W. C. Burrows, and W. E. Larson (1964). Early growth of corn as affected by soil temperature. *Soil Sci. Soc. Am. Proc.*, **28**:271–275.

Alter, J. Cecil (1912). Crop safety on mountain slopes. *U.S. D. A. Yearbook*, 1912, pp. 309–318.

Ayslung, H. C. (1960). Radiation energy balance recorded at soil surface. *Trans. 7th Int. Cong. Soil Sci.*, **1**:179–187.

Baver, L. D. (1954). The meteorological approach to irrigation control. *Hawaiian Planters' Rec.*, **54**:291–298.

Bouyoucos, G. J. (1913). An investigation of soil temperature and some of the most important factors influencing it. *Mich. Agr. Exp. Sta. Tech. Bull.* 17. (Also *Tech. Bull. 26*, 1916.)

Bowers, S. A., and R. J. Hanks (1962). Specific heat capacity of soils and minerals as determined with a radiation calorimeter. *Soil Sci.*, **94**:392–396.

Carson, J. E., and H. Moses (1963). The annual and diurnal heat-exchange cycles in the upper layers of soil. *J. Appl. Met.*, 2:397.

Chang, Jen-Hu (1957). World patterns of monthly soil temperature distribution. *Ann. Assoc. Am. Geographers*, 47:241–249.

Chang, Jen-hu (1961). Microclimate of sugar cane. *Hawaiian Planters' Rec.*, 56:195–225.

Chang, Jen-hu (1968). *Climate and Agriculture.* Aldine Publishing Co., Chicago.

Chang, Jen-Hu, R. B. Campbell, H. W. Brodie, and L. D. Baver (1965). Evapotranspiration research at the HSPA Experiment Station, *Proc. 12th Cong. Int. Soc. Sugarcane Technologists*, pp. 10–24.

Chudnovskii, A. F. (1966a). Plants and light. I. Radiant energy. *Fundamentals of Agrophysics, Israel Program for Scientific Translations, Jerusalem*, pp. 1–51.

Chudnovskii, A. F. (1966b). Transformation of radiant energy on an active surface and its thermal balance. *Fundamentals of Agrophysics for Scientific Translations, Jerusalem*, pp. 413–504.

Decker, W. L. (1959). Variations in the net exchange of radiation from vegetation of different heights. *J. Geog. Res.*, 64:1617–1619.

DeVries, D. A., and A. J. Peck (1958). On the cylindrical probe method of measuring thermal conductivity with special reference to soils. *Aust. J. Phys.*, 11:255–271.

Fuchs, S. M., and C. B. Tanner (1968). Calibration and field test of soil heat flux plates. *Soil Sci. Soc. Am. Proc.*, 32:326–328.

Geiger, Rudolf (1965). *The Climate Near the Ground*, Harvard University Press, Cambridge, Mass.

Houghton, Henry G. (1954). On the annual heat balance of the Northern Hemisphere. *J. Met.*, 11:1–9.

Jackson, Ray D., and Don Kirkham (1958). Method of measurement of the real thermal diffusivity of moist soil. *Soil Sci. Soc. Am. Proc.*, 22:479–482.

Jackson, Ray D., and Sterling A. Taylor (1965). Heat transfer. *Methods of Soil Analysis, Agronomy Monograph*, No. 9, Part 1, Academic Press pp. 349–360.

Kersten, Miles S. (1949). Thermal properties of soils. *Univ. Minn. Inst. Technology, Bull.* 28.

Kohnke, Helmut, and C. H. Werkhoven (1963). Soil temperature and soil freezing as affected by an organic mulch. *Soil Sci. Soc. Am. Proc.*, 27:13–17.

Lang, C. (1878). Über Wärmecapacität der Bodenconstituenten. *Forsch. Gebiete Agr.-Phys.*, 1:109–147.

Monteith, J. L. (1958). The heat balance of soil beneath crops. *Climatology and Microclimatology, Proc. Canberra Symposium, UNESCO*, pp. 123–128.

Nakshabandi, G. A., and H. Kohnke (1965). Thermal conductivity and diffusivity of soils as related to moisture tension and other physical properties. *Agr. Met.*, 2:271–279.

Patten, H. E. (1909). Heat transference in soils. *U.S. D. A. Bur. Soils Bull.* 59.

Penman, H. L. (1948). Natural evaporation from open water, bare soil and grass. *Proc. Roy. Soc. A*, **190**:120–145.

Rose, C. W. (1966). *Agricultural Physics*, Pergamon Press, London.

Smith, Alfred (1932). Seasonal subsoil temperature variations. *J. Agr. Research*, **44**:421–428.

Smith, W. O., and H. G. Byers (1938). The thermal conductivity of dry soil of certain of the great soil groups. *Soil Sci. Soc. Am. Proc.*, **3**:13–19.

Stanhill, G. (1965). Observations on the reduction of soil temperature. *Agr. Met.*, **2**:197–203.

Ulrich, R. (1894). Untersuchungen über die Wärmecapacität der Bodenconstituenten. *Forsch. Gebiete Agr.-Phys.*, **17**:1–31.

Van Duin, R. H. A. (1963). The influence of soil management on the temperature wave near the surface. *Inst. for Land and Water Mgt. Res., Wageningen, Tech. Bull.* 29.

Van Rooyen, Martinus, and Hans F. Winterkorn (1959). Structural and textural influences on thermal conductivity of soils. *Highway Res. Bd. Proc.*, **38**:576–621.

Von Schwarz, A. R. (1879). Vergleichende Versuche über die physikalischen Eigenschaften verschiedener Bodenarten. *Forsch. Gebiete Agr.-Phys.*, **2**:164–169.

Wollny, E. (1878). Untersuchungen über den Einfluss der Exposition auf der Erwärmung des Bodens. *Forsch. Gebiete Agr.-Phys.*, **1**:263–294.

Wollny, E. (1883). Untersuchungen über den Einfluss der Pflanzendecke und der Beschattung auf die physikalischen Eigenschaften des Bodens. *Forsch. Gebiete Agr.-Phys.*, **6**:197–256.

Yakuwa, R. (1945). Über die Bodentemperaturen in dem verschiedenen Bodenarten in Hokkaido. *Geophys. Mag. (Tokyo)*, **14**:1–12.

CHAPTER 8

Soil Water Retention

The amount and energy status of the water within the soil influence the soil physical properties more than any other factor. It is proper that soil water relations occupy a major portion of any discussion of soil physics.

WATER CONTENT

Water and air occupy the nonsolid porosity of the soil. All of the pores are filled with liquid water at saturation. However, the term water in this context includes dissolved substances such as soluble salts and gases which are slightly soluble in water. Unless directly involved, substances dissolved in water are not mentioned explicitly where the term water is used. As water content decreases, the large pores empty first; water is held tightest in the small pores, the interstices between particles, and on particle surfaces. Water is present also in the crystal structure but is so tightly bound that elevated temperatures are necessary to remove it. Most of the crystalline water may be considered a component of the crystal and is not thought of as soil water. However, it is not always possible to distinguish precisely between such water and tightly bound soil water.

For most practical purposes soil water is taken to be that which can be removed by drying to constant weight in an oven at 110° C. However, there exists no unique temperature at which different soil minerals can be heated to drive off all adsorbed water and leave behind only the water of crystallization. Such drying temperatures range from as little as 100 to over 400° C, depending upon the minerals present. Furthermore, organic materials oxidize at temperatures as low as 50° C. Hence a state of absolute dryness for soil that consists of one or more different minerals and some organic matter cannot be defined. However, in most cases, the relative water content between two different energy states is more important than absolute values. Thus the problem of water content measurement becomes more one of reproducibility than one involving absolute accuracy. With reasonable care in controlling drying-oven conditions, and losses due to evaporation during handling and weighing, water content values reproducible to within $\pm 0.5\%$ can be achieved. For greater precision, as in determining the formula weights of particular minerals, considerable care must be taken in defining the dry state and in making measurements (Gardner, 1965; Kittrick, 1970).

Water content is based upon oven-dry weight and is usually expressed in units of mass per unit mass or volume per unit volume. The relationship between mass and volume basis figures is easily derived as

$$\Theta_V = \frac{\Theta_m \rho_b}{\rho_w} \tag{8-1}$$

where Θ with an appropriate subscript represents water content either as a ratio or a percentage, and ρ_b and ρ_w are the densities of the bulk soil and water. Where water content on a volume per unit volume basis is available, it is easy to determine the depth at which water would stand if accumulated from a specified profile depth interval, or the volume of water per unit area per indicated depth interval. This is comparable to indicating the amount of rainfall in length units (e.g., inches or centimeters) or the root-zone water content in inches or centimeters of water. The unit of measurement is reported as a length, but the proper interpretation must always be a volume per area per length, or the depth of water per indicated depth interval, for example, in the root-zone depth or in the topmost 100 cm or 8 in. of soil. To get such water depth figures, one must multiply the volume of water per unit volume of soil by the depth interval d, considered

$$\Theta_{\text{depth}} = \Theta_V d \tag{8-2}$$

Water content of porous materials may be described in several other ways such as on a wet rather than a dry weight basis as is sometimes used in mining engineering. Also, some water content determination equipment is calibrated on a wet-weight basis. The relationship between wet and dry mass water contents is

$$\Theta_{dm} = \frac{\Theta_{wm}}{1 - \Theta_{wm}} \tag{8-3}$$

Water contents for some purposes are based upon the proportion of the total porosity which is water filled and referred to as relative or percentage saturation. If all pores were water-filled, a soil would have a relative saturation of unity or would be 100 percent saturated. To determine such values, it is necessary to evaluate the total porosity or the ratio of pore volume to bulk volume of a soil. This can be done in terms of the bulk and particle densities of soil, ρ_b and ρ_s:

$$\text{pore space ratio} = \text{PSR} = \frac{\text{vol. of pores}}{\text{bulk volume}} = 1 - \frac{\rho_b}{\rho_s} \qquad (8\text{-}4)$$

Bulk densities range from values of the order of 1 g/cm³ to 1.6 or 1.7, whereas particle density of soils for many purposes may be taken as 2.65 g/cm³ without appreciable error. Relative saturation may be obtained by dividing the volumetric water content by the pore space ratio:

$$\text{relative saturation} = \Theta_{\text{RS}} = \Theta_V/\text{PSR} \qquad (8\text{-}5)$$

This becomes the saturation percentage when multiplied by 100.

WATER CLOSE TO PARTICLE SURFACES

Before dealing with the energy state of soil water as it relates to water retention and water flow, it is appropriate to consider how water behaves in thin films close to particle surfaces. Ordinary water is the most common and the most important liquid found in nature. However, it has extraordinary properties which make its behavior unlike that of most other liquids. These properties largely arise from complicated intermolecular forces which are described superficially as hydrogen bonding and short-range London and van der Waals forces (Adamson, 1967; Adam, 1952).

The hydrogen atom, consisting of a single proton with a single electron, is well suited to interact electrically with other charged particles. The electrons of two hydrogen atoms, when chemically or covalently shared with an oxygen atom in two vacancies of oxygen's outer ring, form a single water molecule. The individual molecule has a polar tetrahedral structure with two hydrogen protons that protrude at an angle of 105° from the oxygen nucleus. These protons form positive poles in the tetrahedron, whereas the two electrons in internal orbits of the oxygen atom form negative poles at right angles to the plane of the hydrogen protons (Bernal and Fowler, 1933; Buswell and Rodebush, 1956; Robinson and Stokes, 1959; and Pauling, 1960). These latter electrons are referred to as "lone-pair" electrons. The center of electronegative charge for the molecule is near the oxygen nucleus, but offset toward the hydrogen protons. Oxygen and hydrogen protons in the water molecule are 0.97

Å apart, whereas the hydrogen protons are 1.54 Å apart. Several water molecules are held together in turn by weaker forces (hydrogen bonds) that exist between a positive proton at a corner of one tetrahedron and a lone-pair electron of an adjacent tetrahedron corner. The nuclei of oxygen in two adjoining hydrogen-bonded molecules of solid water (ice) are 2.72 Å apart. Each molecule thus may be associated with four addi-

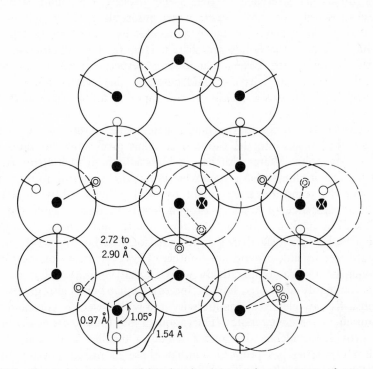

2.72 to 2.90 Å

0.97 Å   1.05°

1.54 Å

Fig. 8-1. Suggestive structure of ice showing hexagonal arrangement of water molecules and their orientation. The straight lines indicate hydrogen bonds, open circles hydrogen protons, and the filled circles oxygen nuclei. When ice melts the bonds are stretched so that the distance between oxygen nuclei increases. (Bernal and Fowler, 1933.)

tional molecules. A hexagonal array of water molecules suggesting the nature of ice structure is shown in Figure 8-1. The straight lines represent hydrogen bonds, the small open circles represent hydrogen protons, and the filled circles represent oxygen nuclei. Hydrogen proton pairs oriented perpendicular to the plane of the paper are shown as concentric circles. Three adjoining molecules associated with one hexagonal ring beneath the plane of the paper are shown by means of dashed lines. These lines show the position of the fourth molecule associated with each molecule in the structure.

When ice crystals melt, the hydrogen bond length increases to about 2.90 Å. Water density actually increases with increasing temperature up to a temperature of nearly 4° C, where water has its maximum density. The reason for this apparent anomaly is that some hydrogen bonds break as the bonds are stretched so that a less ordered crystalline structure of higher density develops. Each water molecule may be surrounded at any one time by five or more molecules in a fluctuating structure in which some of the bonds are shared. Despite these fluctuations, the structure is sufficiently strong that in some respects aggregates of water molecules themselves behave as molecules, so that water exhibits some polymerlike characteristics. Only in the vapor state does intermolecular structure disappear.

Intermolecular forces in the liquid state make it difficult for water molecules to escape into the vapor state. This gives water its unusually high boiling point. Water also has an unusually high melting or fusion point because it already has a somewhat crystalline structure in the liquid state and the transition from liquid to solid is achieved with relative ease. The existence of weak hydrogen bonds in liquid water requires that bonds be broken in order for a molecule to move about. This gives water a higher viscosity than that for unassociated liquids, where molecules are free to slide around one another. However, many of the shared hydrogen bonds are comparatively weak so that viscosity is not as great as it is for highly associated liquids. Intermolecular bonds also give water its unusually low electrical conductivity and high dielectric constant.

A number of recent reports (Deryaguin, 1966; Lippincott et al., 1969) have provided evidence that water which is distilled into small Pyrex or silica capillaries has properties unlike those of ordinary water, even after removal from the capillary. Such water, referred to as "anomalous water" by Deryaguin and "polywater" by Lippincott et al., is reported to be considerably more dense than ordinary water (with densities up to 1.4 g/cm^3) and to show appreciable differences in infrared and Raman spectra. The existence of such water would be of considerable interest in connection with studies of water associated with colloidal silicates such as clays. However, recent studies strongly suggest that the properties ascribed to anomalous water or polywater are similar to those that would be observed in water solutions containing impurities such as silica, silicate sols, or certain organic compounds (Bascom et al., 1970; Pethica et al., 1971; Davis et al., 1971). The infrared spectrum of such water also has been shown to be similar to that of sodium lactate, the primary constituent of human perspiration (Rousseau, 1971). Nonetheless, the fact that minute quantities of impurities in water that are removed from close proximity to a solid surface could so markedly influence the properties of this "water" provides some insight into problems associated with water properties and behavior in thin films.

In the same way that water molecules associate with other water molecules through hydrogen bonding, water molecules are able to associate with charged ions in the electric double layer and with charged solid surfaces. Positive ions may be surrounded by water molecules that are attracted through lone-pair electrons and through negative ions attracted to positive protons of the molecules. Water "hulls" which form about charged ions impart sluggishness to the movement of such ions, and the attractive force between water molecules and ions increases the energy required for removal of water. Constraint of water by charged ions in the electrical double layer is analogous to the effect of a semipermeable membrane, and gives rise to what are known as "osmotic" forces and to the "osmotic potential" discussed later in this chapter. Double layer effects have been considered in Chapter 2 since they are particularly important in disperse systems. Detailed discussions of the theory of this phenomenon are to be found in Bolt and Peech (1953) and Bolt (1955).

Colloidal clays are characterized by the presence of oxygen atoms and hydroxyl units on their surfaces, particularly at broken edges. Hydrogen bonds readily form in association with the lone-pair electrons of oxygen and protons of hydrogen thus exposed. Bonding to oxygen in the lattice approaches the strength of covalent bonding because of the fact that lone-pair electrons of the oxygen atom are somewhat repelled by excess electrons in the negatively charged crystal lattice; this makes possible a closer tie with a hydrogen proton of the water molecule.

Bonding of water molecules to surfaces leads to "structured" water close to such surfaces. Evidence exists that water close to particle surfaces is less dense than normal water because of the increased structural organization (Low, 1961). It has been demonstrated also that the viscosity of water in the first few molecular layers may be several times greater than that in water unaffected by such surfaces (Kemper, 1961a, 1961b; Low, 1959).

Evidence of strong attractive forces at particle surfaces is obtained through experimental measurements such as those performed by Goates and Bennett (1957), who measured the heat necessary to remove or absorb the first molecular layer of water from particle surfaces to vapor form in the surrounding atmosphere. They found the energy of adsorption of water on a kaolinite clay with about 50 percent of the surface covered to be about -14.5 kcal/mol and with the entire surface covered by a monolayer to be about -13 kcal/mol. (The negative sign indicates that work must be done to remove the adsorbed water.) Energy required for similar removal from a free water surface is of the order of -10.4 kcal/mol (heat of vaporization); therefore 3 or 4 kcal/mol additional energy was required to remove water from particle surfaces of the clay used. Removal of successive layers of water requires less and less energy, approaching that for vaporization of free water after the first few layers are removed. However, adsorption energy is influenced by the nature of the particle surface and by the kind of exchangeable cations present.

Montmorillonite with its high cation-exchange capacity per unit mass and its large specific surface has a much greater capacity to adsorb water than illite, whereas kaolinite has the lowest adsorption capacity of the three common soil minerals. At 50 percent relative humidity (at which the soil is very dry), water adsorption on a mass basis is of the order of 21, 4.5, and 0.5 percent for montmorillonite, illite, and kaolinite, respectively. These differences arise from the makeup of the crystal lattices and the sites of the negative charges. As shown by Thomas (1928) and Baver and Horner (1933), and by numerous investigators more recently, the nature of the exchangeable cation has a sizable impact upon water adsorption. Generally, clays saturated with divalent cations exhibit greater water adsorption than do those saturated with monovalent cations, with the exception of the $Li+$ ion, which behaves more like a divalent ion, Adsorption of water vapor decreases with increasing size of the ion within both the monovalent and divalent groups.

Inasmuch as the freezing temperature of water is affected by its structure, the presence of ions and the proximity of surfaces, both of which affect water structure, also will depress the freezing-point temperature. Water at considerable distance from soil particles freezes first as temperature is reduced and the highly structured water adjacent to particle surfaces freezes last. Freezing-point depression has been used as a measure of the energy state of soil water (Day, 1942; Babcock and Overstreet, 1957). However, water can be appreciably supercooled before freezing; the degree of supercooling depends heavily upon the presence or absence of freezing nuclei. The complicated nature of the mineral constituents of soil and the associated soil solution make the degree of supercooling difficult to predict, hence interpretation of measurements sometimes is difficult (Low, 1961).

As a consequence of water structuring near particle surfaces and near hydrated ions, problems of water or ion movement close to particle surfaces are considerably more complicated than flow farther from such surfaces. Water film thicknesses, ranging from a few molecular layers (e.g., about 8 Å in kaolinite) to many layers (e.g., up to 68 Å in montmorillonite), have been reported to be affected by clay surfaces. The extent of this effect depends also upon the nature of the exchangeable cations on the surfaces. As porous materials dry out, increasingly greater proportions of the water are associated in thin films with strong attractive forces at particle surfaces. This results in steep water potential-water content curves in relatively dry soil systems.

Problems of ion movement and exchange near colloid surfaces and problems of ion uptake by roots are intimately associated with water and its properties in thin films. Gross water flow and water uptake by plants over a large part of the water content range involve water in which the structure is not greatly affected by particle surfaces. However, as soil dries out the properties of water in thin films that are dominated by surface forces become increasingly important. In much of the following discussion the presence of these forces, though often of dominating and crucial importance, appears only implicitly as the underlying causative agent of such gross properties of water as viscosity or its potential.

ENERGY STATE OF WATER IN SOIL

Next to water content, the energy state of the water is probably the most important single physical soil characteristic. Understanding the relationship between soil water content, its energy state, and processes that involve energy gradients in soil-water systems is still far from perfect. Enough is known, however, to make soil water phenomena constitute the subject area in soil physics for which the largest body of mathematical theory is now available.

Both kinetic and potential energy concepts have importance in dealing with soil water. Kinetic energy is that energy which matter has by virtue of its motion; quantitatively it is $\frac{1}{2} mv^2$ (dimensions ML^2/T^2), where m is the mass of a body that has a velocity of v. On atomic and molecular scales, all matter above absolute zero of temperature is in motion and has kinetic energy. Bulk water in motion has kinetic energy by virtue of its motion. Additionally, changes of state and variations in temperature involve changes in kinetic energy. However, a great many processes that involve water in soil and plant systems may be dealt with through consideration only of potential energy and potential energy changes; kinetic energy enters into equations only implicitly. These systems will be dealt with first; those where kinetic energy changes play the dominant roles will be the subject of later discussions. Use of potential energy concepts is restricted to those situations where temperature is uniform throughout the system or where the temperature change may be regarded to have a negligible effect upon the process under consideration.

Potential energy is the energy which a body has by virtue of its position in a force field. Thus potential energy is measured by the force required to move a body directly against the force field and is the product of the force and the distance moved, Fd (dimensions ML^2/T^2, the same as for kinetic energy). Work must be done to move a body against the force field, or work is done by the body if it is allowed to move with the field. Of course, potential energy is converted to kinetic energy when a body moves so as to reduce its potential energy or when work is done to move the body which increases its potential energy. However, kinetic energy terms need not always appear in equations that describe such motion.

Numerous forces act upon water in a porous material like soil. The earth's gravitational field pulls vertically upon the water. Force fields that are caused by the attraction of solid surfaces for water pull water in various directions. The weight of water, and sometimes the additional weight of soil particles which are not constrained in the soil matrix, also act upon water below due to the pull of gravity. Ions dissolved in water have an attractive force for water and resist the pulling away of water. An especially important force existing at water surfaces is associated with the attraction of water molecules for each other and the unbalance of these forces that exist at an air-water interface.

The variety of forces and the directions in which they act make description of force networks in soil water difficult. However, it is possible to assess the potential energy associated with an increment of water as a consequence of the forces acting upon it. Potential energy differences from point to point in isothermal systems determine the direction of flow, the amount of work available for causing flow, or the amount of work that must be done from the outside to cause flow. This is the merit of describing soil water in terms of the potential energy associated with it.

Quantitative description of the potential energy of soil water requires its expression in terms of the mass or volume of water with which it is associated. Potential energy per unit quantity of water is defined as the *potential*. Potential may be defined on either a mass or volume basis, but, since water has a density close to unity in the cgs system of units, numerical values are almost the same for either system. Correction for nonunity water density would be needed only where extraordinary precision is required. However, the dimensions are different. The dimensions of potential on a mass basis are $FL/M = L^2/T^2$; on a volume basis they are the same as those for a pressure, or $FL/L^3 = F/L^2 = M/(LT^2)$.

POTENTIAL TERMINOLOGY AND POTENTIAL MEASUREMENTS

Various potentials or combinations of potentials are involved that depend upon the phenomenon under consideration. Theoretically, a potential is associated with each force acting on water. However, some of the separate potentials are combined into a single potential for practical convenience. Absolute potential energies or potentials would be difficult to define; hence each potential must be defined with respect to an arbitrary reference level. For water at rest in a system with respect to the earth's surface, the sum of all the potentials, or the total potential at a point, must be either constant or zero; this depends of course upon the reference taken. Total water potential may be defined as

$$\psi_T = \psi_M + \psi_g + \psi_p + \psi_\pi + \psi_\Omega \tag{8-6}$$

where ψ_M, ψ_g, ψ_p, ψ_π, and ψ_Ω are the matric, gravity, pressure, osmotic and overburden potentials, respectively.

Matric Potential, ψ_M

Matric potential is the amount of work that must be done per unit quantity of pure water in order to transport reversibly and isothermally an infinitesimal quantity of water from a pool containing a solution identical in composition to the soil water at the elevation and the external gas pressure of the point under consideration to the soil water (Commission I, ISSS, 1963). It may be seen that matric potentials are negative inasmuch as water from the reference pool would flow readily into dry soil with a release of energy in the form of heat. This means that work would be done in the wetting process rather than that work is required according to the foregoing definition. The matric potential of water above the water table always is negative and the matric potential becomes zero below a free water table.

Matric potential is associated with the attraction of solid surfaces for water as well as the attraction of water molecules for each other; it includes the unbalanced forces across air-water interfaces which give rise to the phenomenon of surface tension. This potential has been referred to historically as the capillary potential inasmuch as it is analogous over a significant part of its range to the situation that exists where water rises in small capillary tubes. However, phenomena that are described by the term matric potential extend beyond those associated with air-water interfaces in small pores. As water content decreases in a porous material, the water that is contained in pores with clear-cut negative air-water interfaces, as in capillary rise, becomes negligibly small when compared to the water held directly on particle surfaces. The term matric potential therefore covers phenomena beyond those for which a capillary analogy is appropriate.

Matric potential may be measured by balancing the weight of a hanging water column (or column of water plus mercury) against matric forces in the soil across a porous ceramic membrane (Figure 8-2). A negative pressure or positive tension is created in the water-filled pores of the ceramic by means of the hanging liquid column or manometer. A ceramic membrane with small pores, all of which are small enough to retain water at the negative pressures involved, must be used. Large,

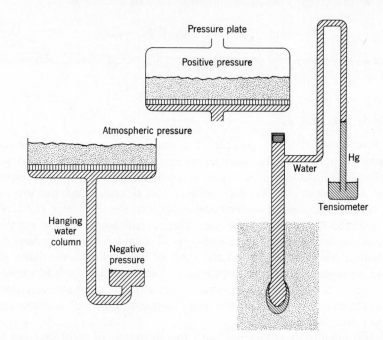

FIG. 8-2. Apparatuses for application of a pressure differential across a porous plate for measurement of matric potential, and a tensiometer which operates on the same principle and may be used in the field.

air-filled pores in the soil or in the ceramic membrane will not permit development of the negative pressures needed to balance those that exist in the liquid-filled pores of the soil. Water pressure in the manometer comes into equilibrium with the adjacent soil through flow across the ceramic membrane; the height of the liquid column at this time is an index of the matric potential. The force per unit area, or negative pressure of the water in the porous membrane, is the weight per unit cross section of the hanging column. This is the volume of the column divided by the area multiplied by the density of the liquid water and the acceleration of gravity, g:

$$\frac{mg}{A} = V \times \rho_w \times \frac{g}{A} = \frac{hA\rho_w g}{A} = h\rho_w g \qquad (8\text{-}7)$$

where ρ_w is density of water and the potential is in units of potential energy per unit volume. This expression becomes

$$h\rho_w g \times \frac{V}{\text{mass}} = \frac{h\rho_w g}{\rho_w} = hg \qquad (8\text{-}8)$$

after division by the density of water to get it into units of potential energy per unit mass. In cgs units, 1020 cm of water, or the equivalent of 1020 cm of water when water and mercury are used, would exert a negative pressure of $1020 \times 1 \times 980$ D/cm^2 = 10^6 D/cm^2; since 10^6 D/cm^2 = 1 bar in cgs pressure units, this would amount to 1 bar of pressure. Converting to units of potential energy per unit mass requires division by the density of water, or approximately by 1 g/cm^3. One bar or 1020 cm of water would be the equivalent of $1020 \times 1 \times 980$ D/cm^2, where the density of water is taken as unity; division by 1 g/cm^3 would yield (1020 cm) (1 g/cm^3) (980 cm/sec^2)/(1 g/cm^3) = 10^6 D-cm/g = 10^6 ergs/g = $10^6 \times 10^{-7}$ joule/10^{-3} kg = 100 joules/kg. This is the potential on a mass basis in commonly used units. The practical limit of the tensiometer shown in Figure 8-2 is about -0.85 bar potential (or 0.85 bar tension) because of the problem of cavitation when air comes out of solution or out of the walls of the tubing used. However, tensions or negative pressures considerably in excess of this are possible in fine pores in the soil or in some of the conducting tissues of plants.

Matric potentials of much larger magnitude (larger negative values) may be measured by using a positive gas pressure across a porous ceramic membrane rather than negative pressures developed by hanging liquid columns. Soil is placed on a porous ceramic plate arranged in a pressure chamber (Figure 8-2) so that gas pressure forces water through the plate to atmospheric pressure on the other side until equilibrium is achieved between the pressure and matric forces in the soil. The gas pressure in appropriate units is the equivalent of the matric potential. When such equipment is used, it is possible to construct curves that show the relationship of matric potential to water content. These curves are not unique; wetting and drying curves differ because of the effect of *hysteresis*, as is discussed later. Typical drying, or desorption curves for several soils are shown in Figure 8-3. Such curves are sometimes referred to as *moisture characteristic curves* for a soil.

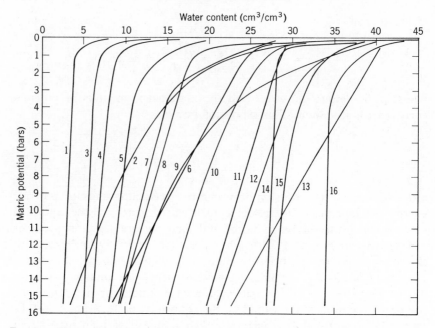

FIG. 8-3. Desorption curves for various soils sketched from data at 0.1, 0.3, 0.6, 3, and 15 bars tension given by Holtan et al., 1968. 1. Continental gravelly sandy loam, Arizona, 2. Sassafras sandy loam, Maryland. 3. Progresso fine sandy loam, New Mexico. 4. Vaucluse sandy loam, Georgia. 5. Albion loam, Oklahoma. 6. Abilene clay loam, Texas. 7. Hartsells loam, Ohio. 8. Palouse silt loam, Washington. 9. Fayette silt loam, Wisconsin. 10. Nellis gravelly loam, New York. 11. Larb-like silty clay loam, South Dakota. 12. Memphis silt loam, Mississippi. 13. Drummer silty clay loam, Illinois. 14. Auston silty clay, Texas. 15. Marshall silty clay loam, Iowa. 16. Bascom-like clay, South Dakota.

Gravity Potential, ψ_g

Gravitational potential is the amount of work that must be done per unit quantity of pure water in order to transport reversibly and isothermally an infinitesimal quantity of water from a pool containing a solution identical in composition to the soil water at a specified elevation of the point under consideration (Commission I, ISSS, 1963). The elevation that is chosen as a reference determines whether gravity potentials are negative or positive. If the reference pool is below the point in question, work must be done on the water and the gravity potential is positive; if the pool is above, work is done by the water and the gravity potential is negative. If the reference pool is below the point in question, such as the level of an underlying free-water table, and water is free to flow, the gravity potential and the matric potential would be equal and opposite

w/. H₂O pushing down

in sign at equilibrium so that their sum would be zero; this is a condition for equilibrium. Such a system, if tall enough to cover the range of interest, can be used to determine the water content-water potential relationship for both sorption and desorption conditions. This is due to the fact that the gravity potential, which can be measured with a ruler, is equal and opposite in sign to the matric potential; the column therefore can be sectioned and water content can be determined. The range can be extended considerably by means of a centrifuge where centripetal acceleration replaces gravitational acceleration in the equations used.

If the reference pool is chosen above the point in question and flow is permitted, the negative gravity potential is balanced at equilibrium by the weight of the column of water (a positive pressure potential due to the weight of liquid, as is described later) and the sum of the two is zero. The matric potential is zero in this case. Pressures due to the weight of solid particles, if free to move, also may be involved.

Equilibrium conditions may exist for total potentials other than zero, which would depend upon the choice of reference for gravity potential.

Pressure Potential, ψ_p

A potential that is due either to the weight of water at a point under consideration or to gas pressure which is different from that which exists at a reference position may be referred to as a "pressure potential." If the point is beneath the water table, the potential is equal and opposite to the gravity potential that is measured from the free water surface; it sometimes is referred to as the "submergence potential." It could be referred to as a "piezometric potential," which corresponds to a "piezometric head." It sometimes is referred to as a "pneumatic potential" in soil above the water table where the gas pressure is different from that at the reference point. The gas pressure involved in the measurement of the matric potential with the use of a porous plate apparatus as previously described would be a pneumatic or pressure potential. The exact definition of a pressure potential depends on whether water or gas pressures are involved; otherwise, it is completely analogous to the definition of a matric potential.

Pressure potentials due to gas pressure may be measured with an ordinary manometer. Pressure potentials due to water pressure may be measured manometrically or by using a piezometer, which is a tube placed in the soil with its end open; it may have openings in the wall at the point where the pressure is desired. The level of water in the tube, measured from some suitable reference, is the piezometer reading, which may be converted to potential units.

Osmotic Potential, ψ_π

Osmotic potential is the amount of work that must be done per unit quantity of pure water in order to transport reversibly and isothermally an infinitesimal quantity of water from a pool of pure water at a specified elevation at atmospheric pressure to a pool containing a solution identical in composition with the soil water (at the point under consideration) but in all other respects identical to the reference pool (Commission I, ISSS, 1963). Osmotic potentials result from the hydration of ions in the soil solution. The polar nature of water, with two sites that are electropositive and two that are electronegative, causes water molecules to be attracted to ions in the soil solution. A negative pole of the water molecule is attracted to positive ions and the hydrogen proton positive poles of water are attracted to negative ions. These attractive forces tend to orient water around ions and the osmotic potential refers to the work required to pull water away from these ions.

The most common example of the presence of osmotic forces is the rise of water in one arm of a U-tube that contains dissolved substances in solution and which is separated from pure water in the other arm by a semipermeable membrane at the bottom. The membrane must have openings that are large enough to permit passage of water molecules but too small for ions of the dissolved substance to pass. Because of the attraction of the ions for water, pure water passes through the membrane into the solution until the hydrostatic pressure difference between the two tubes balances the effect of the ion-water attraction forces. This pressure difference is referred to as the "osmotic pressure."

Osmotic forces that are important in dealing with soils and plants may be observed in another way. Most of the dissolved substances in the soil water and in plant tissues are nonvolatile at ordinary temperatures. Hence they do not evaporate and are left behind by evaporating water. More energy must be supplied to remove water from solutions that contain solutes than from pure water. The air-solution interface acts like a membrane and the vapor pressure in the surrounding air at equilibrium is reduced proportionately in a comparable manner to the pressure difference developed across a solution membrane. Therefore osmotic pressures are involved wherever a liquid and an air phase are free to interact, even in the absence of a solution membrane.

Clay particles suspended in a solution will cause the same phenomenon because of the attraction of clay for water. Whether or not this should be called an osmotic effect because the clay particle is behaving as an ion, or a matric effect because it concerns the attractive forces of a solid particle, might be argued; however, we consider this to be a matric effect.

Osmotic potentials behave similarly to the potentials already described wherever a membrane or membranelike situation exists and can be added to such potentials to form the total potential. However, in the absence of a membrane of some kind, soluble ions diffuse spontaneously into the soil solution by virtue of the kinetic energy they possess. Osmotic potentials under such conditions may not be balanced against any of the other potentials and cannot be treated as a potential in the same way. The net flow of water in response to a difference in matric potential from point to point may be in the opposite direction to flow of water brought about by an osmotic potential gradient. Osmotic effects under such conditions are sometimes better handled by using principles similar to those applied to water flow under a temperature gradient, as will be discussed later.

Overburden Potential, ψ_Ω

Overburden potentials exist wherever soil is free to move and its weight becomes involved as a force acting upon water at a point in question. It behaves similarly to a pressure potential that is caused by the weight of water above and sometimes is included as an implicit part of a pressure potential. However, when an overburden potential exists and is implicitly included as a part of a pressure potential, the gravity and pressure potentials are no longer equal and opposite; this fact suggests the desirability of recognizing the overburden potential explicitly as a separate potential.

Measurement of overburden pressures under saturated conditions requires the use of a piezometer which filters out soil particles at its entrance. Piezometer readings are compared with the level of the free-water surface within the soil system.

The weight of the overburden when particles are free to move sometimes becomes involved with water conditions above the water table. Its influence on the water potentials is difficult to separate and the overburden potential often is included as a part of the matric potential. Where an air phase exists in the soil, the effect of overburden pressure is to reduce the pore space which only indirectly affects the water potential. Whether or not the water potential is changed through the existence of an overburden pressure depends on water content in the compressed soil and whether the adjacent soil below or to the side has an appreciable buffering capacity to supply or absorb water as overburden pressure changes. In the latter case, the potential may quickly return to a previous equilibrium value as a consequence of water flow, although the water content in the soil would be changed. The loss of pore space that attends compression at high water contents may lead to increased water potentials. On the other hand, compression at lower water contents may reduce the size of some of the air-filled pores to the point where they fill with water from larger pores and the water potential is reduced. Overburden pressures and the overburden potential would be included implicitly in the matric potential from a practical point of view. However, if the matric potential is measured on a soil sample that is removed from a profile where overburden pressures exist, the release of overburden potential in the sample could lead to some error. No completely satisfactory method exists for measuring overburden potential directly.

Other potentials may be defined according to need. However, where a potential which is not zero is neglected, it must be assumed that it is implicity included in one of those which is explicit in the definition, e.g. where overburden potential is neglected, it becomes implicit in the pressure (or matric) potential. Some useful combinations of potentials relevant to certain practical situations are discussed in the following section.

COMBINATIONS OF POTENTIALS FOR PARTICULAR APPLICATIONS

Soil-Water-Plant Relations

The potential that has the greatest relevance in soil-water-plant relations is made up of a matric and an osmotic component:

$$\psi_s = \psi_M + \psi_\pi \tag{8-9}$$

where the subscript s refers to stress, which is the common term that is used to identify the force or potential associated with water availability to plants. This combination of potentials often is referred to as *soil-water potential*, *stress potential*, or without a negative sign and in appropriate units, as *suction* or *soil-water suction*. Other potentials influence water uptake by plants, but usually their effect is small compared to the effect of osmotic and matric forces or the errors associated with measurements. Gravity potential is one of these. Work always must be done to raise water from the root zone into the plant. However, the distance is of the order of 1 meter for the major rooting zone of many plants; this corresponds to about 0.1 bar potential. Gravity potential should be considered for precise work in wet soil or where the energy required to remove water from deep in the soil is significant.

Biological Materials

The potential of interest in biological materials usually is the sum of the osmotic potential, the matric potential, and a turgor potential:

$$\psi_b = \psi_\pi + \psi_M + \psi_t \qquad (8\text{-}10)$$

where the subscripts b and t refer to biological materials and to turgor. Turgor potential is a pressure potential that arises from the fact that physical pressures exist within the turgid cells. Where the biological potential is zero, the turgor potential is equal to the sum of the osmotic and matric potentials that exist external to the cell.

Unsaturated Flow

Matric and gravity potentials are of the greatest concern in the majority of work on unsaturated flow:

$$\psi_u = \psi_M + \psi_g \qquad (8\text{-}11)$$

where the subscript u refers to unsaturated conditions. There exists no commonly accepted term for the sum of these two potentials and most equations include both potentials explicitly in one form or another.

Saturated Flow

Pressure potential and gravity potential are of major interest in saturated flow:

$$\psi_h = \psi_p + \psi_g \qquad (8\text{-}12)$$

The sum of these two potentials often is referred to as the hydraulic potential and is designated by the symbol Φ. The pressure potential usually includes the overburden potential implicitly or else a separate term is included. The pressure here is due to the weight of water, but if the gas pressure differs appreciably from that which exists at the reference, a separate term might be required.

Flow in the Field

Head rather than potential units are often used in field work. Moreover, the hydraulic potential often is written so as to cover both conditions above and below the water table; thus in potential units

$$\psi_H = \psi_M + \psi_p + \psi_g \qquad (8\text{-}13)$$

and in head units

$$H = H_M + H_p + z$$
$$= \phi + z \qquad\qquad (8\text{-}14)$$

The sum $\psi_M + \psi_p$ sometimes is referred to as *pressure potential;* ψ_p is zero above the water table and ψ_M is zero below. The hydraulic head H is positive below the water table and negative above.

Other symbolism appears in soil and plant physics literature. It is important to avoid confusion to study carefully the definitions of the symbols used and the context in which various terms appear.

WATER IN THE VAPOR PHASE

The amount of water in soil as vapor in the gaseous part of the system is relatively small; it is of the order of 0.012 mg/cm^3 in a wet soil. Therefore the amount of water stored in the gaseous phase is negligible in quantity compared to that which the soil holds in the liquid state. However, the significance of the vapor phase water lies not in the amount present at a given time but in the processes which involve passage of water through the vapor phase and the energy that is required to vaporize water or that is given up when water vapor condenses. Energy involved in vaporization or condensation is 585 cal/g (2.45×10^6 joules/kg) at 20° C.

Evaporation of water from the soil and transpiration from plant leaves are important cooling mechanisms; the condensation of water at night often is an effective warming mechanism. Evaporative loss of water from surface soil and indirectly from the soil through transpiration by plants are major water loss processes. Water also may be transferred downward in the soil in the vapor phase when the surface soil is warmer than the subsoil; movement would be upward when the surface soil is cooler. Such water transfer in the vapor phase can have practical significance under some conditions, particularly in connection with supplying small amounts of water for germination of seeds which otherwise might not be available. Circulation of water downward in the vapor phase during warm days and upward in the liquid phase at night can often result in concentration of soluble materials in the soil surface.

The kinetic energy of the gaseous-phase water at equilibrium and at the same temperature is greater than that of the liquid-phase water by an amount equal to the heat of vaporization or condensation of water. However, the average energy difference with which they can interact with each other without change of state is zero. Nonetheless, because of an unequal distribution of the kinetic energy of the separate molecules in each state, some molecules of water have, or acquire through collision or absorption of energy from the surroundings, sufficient energy to escape the liquid state. Some gaseous molecules lose energy and return to the liquid state. Where temperature is uniform throughout the system, the number of molecules that leave the system at equilibrium is exactly balanced by the number that enter.

If water in the liquid state is constrained in some way, such as through binding by solid particles or soluble materials in solution, the escaping tendency for water molecules is reduced and the system becomes drier. When temperature is uniform in such a system, the quantity of water vapor in the atmosphere, compared to what would be present if the water were completely free of such forces, is a measure of the difference in energy in the two systems. Since these are equilibrium systems, the energy difference is appropriately described as a potential (energy) difference. It is a matric effect if the difference is due to the solid particle-water interaction and an osmotic effect if soluble materials in the water are the causative factors. Similarly, if the system is under pressure (negative or positive) with respect to the reference system, a potential due to that pressure would be evident. A slurry containing only solid particles and no dissolved substances would have a nonzero water potential. The situation is similar to that which exists with soluble ions and could be described as an osmotic potential. However, it is more aptly described as a matric potential since it involves attractive forces between water and a solid surface.

One additional potential is in evidence in the relationship between water and associated water vapor. Vapor is more concentrated near a free water surface than at greater elevations because of the pull of the earth's gravitational field. The amount of water vapor present in still air in a closed container at uniform temperature decreases with elevation above the surface. Therefore, apart from any other forces, the vapor content of the gas at various elevations in such a closed system depends upon the gravity potential. Of course, conditions are modified by the presence of soluble materials in the water. As noted earlier, the gravity potential is measured from an arbitrary reference which, on a mass basis, is zg, where z is the elevation and g is the acceleration of gravity. If the reference is taken at the free water surface, then the water potential at the surface is equal to the gravity potential by definition and both are zero. The water potential at equilibrium in the region above the free water surface is equal and opposite the gravity potential. Subsequent sections will show that these concepts have importance as they relate to water conditions near water films in soil and in connection with an important technique for measuring water potentials.

Temperature has been specified as uniform thus far in most of the discussion. Obviously, the energy content of substances change with temperature and temperature must be specified under some conditions. However, it often does not have to appear explicitly in equations when it is uniform, or when only small differences exist in a system. Temperature largely is an important factor in soil water problems where significant differences in temperature exist from point to point. This is discussed in later sections and chapters.

Vapor Pressure and Relative Humidity

The presence of water vapor in an atmosphere ordinarily is sensed by its pressure. The number of water vapor molecules present and their average kinetic energy determine how hard they push on the walls of a container. Thus the partial pressure of water vapor in a fixed volume, or *vapor pressure*, depends upon the temperature and upon the number of molecules present. This is expressed by the well-known combined gas law or the Boyle-Charles law,

$$pV = nRT \tag{8-15}$$

where p is the pressure, V the volume, n the number of moles of gas present, R the universal gas constant, and T the absolute temperature. The energy associated with this vapor may be obtained by considering the work, dw, involved in expanding or compressing isothermally a small volume increment, dV, of water vapor

$$dw = -p \, dV \tag{8-16}$$

The volume occupied by water vapor at constant temperature T is a function of the vapor pressure and is sufficiently approximated by the combined gas law (equation 8-15) even though water vapor may not behave quite as a perfect gas. Differentiating equation 8-15 with respect to V and p, substituting into equation 8-16, and integrating over suitable limits gives

$$\frac{w - w_0}{n} = \frac{\Delta w}{n} = RT \ln \frac{p}{p_0} \tag{8-17}$$

The ratio p/p_0 may be recognized as the relative humidity where p_0 is the pressure at saturation for the existing temperature and $\Delta w/n$ is the loss in energy per mole as the vapor pressure p decreases. It should be noted that the ratio of the two pressures is the same as the ratio of the absolute humidities or the ratio of water content per unit volume in a region compared to that which would be present at saturation.

If a system containing liquid water is free to interact with the adjacent atmosphere, the free energy content per mole at equilibrium (the adjective "free" implies that the energy content above or below that of the reference is available to do work or that work has been done to reach that state) will be equal to the free energy content per mole in the adjacent water. Thus a measure of free energy in the vapor phase at equilibrium also measures the free energy in the liquid state. Since potential usually is defined in soil-water relations as the potential energy per unit mass or unit volume of water, the relationship between water potential and p/p_0, or relative humidity, becomes

$$\psi_{\text{mass}} = \frac{\Delta w}{M} = \frac{RT}{M_w} \ln \frac{p}{p_0} \tag{8-18}$$

or

$$\psi_{\text{vol}} = \frac{\Delta w}{V} = \frac{RT}{V_m} \ln \frac{p}{p_0} \tag{8-19}$$

where the water potential is given both on a unit mass and unit volume basis and M_w and V_m are the molecular weight and molecular volume of liquid water. The potential indicated by equations 8-18 and 8-19 ordinarily would be the sum of the matric and osmotic potentials, but it depends upon the reference where p_0 is taken.

Relative Humidity as a Measure of Matric and Osmotic Potential

An important measuring technique makes use of the relationship given in equations 8-18 and 8-19. The water potential of a porous material containing water may be computed by using these equations if the relative humiditiy, p/p_0, can be measured. A small (0.025 mm diameter) chromel-constantan thermocouple that is contained in a thin-walled ceramic cup is buried in the soil so that the atmosphere which surrounds the thermocouple is at equilibrium with the soil solution. By using a small, direct current, the temperature of the thermocouple is reduced to the dew point by Peltier cooling so that water is condensed on the junction. Cooling is then stopped and the temperature of the junction is measured by employing a microvoltmeter as the junction is cooled by evaporation. The temperature of the junction depends upon the evaporation rate, which in turn depends upon the relative humidity of the atmosphere. Such units, when calibrated over osmotic solutions of known humidities, permit measurements of water potentials in a range from about $-\frac{1}{2}$ bar down to the order of -70 bars; even lower values are possible with special techniques. The technique also may be used in the laboratory on plant tissues and on soil and plant samples.

Vapor Pressure at Air-Water Interfaces

The curvature of an air-water interface at equilibrium is related to the pressure difference across the interface. If the water is pure and the interface flat, the pressure is exactly the same at an infinitesimal distance above and below the interface; the pressure difference is zero. However, if the interface is curved, the pressure is greater on the concave side of the interface by an amount which depends upon the degree of curvature and the surface tension of the liquid.

Surface tension is the force required per unit length to pull the surface of a liquid apart, or it may be conceived as the work required per unit area to expand a surface; these are equivalent expressions. It is brought about by a change in orientation of the molecular bonds which hold liquid molecules together as the surface, or interface with another substance, is approached. Such a surface has some "membranelike" properties. The influence of surface tension on curvature and pressure is best illustrated in a raindrop. A small raindrop is essentially spherical because of surface tension forces which act to make the surface a minimum. Deviations from the spherical shape are due only to resistance to the air, if the droplet is in motion, and to the pull of gravity. If the raindrop is treated as two hemispheres, the force acting inside that tends to push the hemispheres apart is $\pi r^2 p$, where p is the pressure within. They are held together by a force that is due to surface forces acting along the perimeter, or $2\pi r\sigma$, where σ is the surface tension. These forces are equal at equilibrium so that

$$\Delta p = \frac{2\sigma}{r} \qquad (8\text{-}20)$$

An equation of this form is applicable to all liquid-vapor interfaces, although it becomes a little more complicated where associated solid surfaces cause the liquid surface to be nonspherical. The curvature may be negative for water in a capillary tube and the pressure therefore will be negative with respect to the pressure of the atmosphere. Curvature, which is the inverse of radius of curvature, of liquid-vapor interfaces generally, may be shown to be

$$C = \frac{1}{R} = \frac{1}{r_1} + \frac{1}{r_2} \qquad (8\text{-}21)$$

so that

$$\Delta p = \frac{1}{r_1} + \frac{1}{r_2} \qquad (8\text{-}22)$$

where r_1 and r_2 are radii of the curved surface at a point measured in planes at right angles to each other. Equation 8-22 reduces to equation 8-20 where r_1 and r_2 are the same, as for a spherical surface.

Water in porous materials is held both on solid particle surfaces and in the interstices between particles. The geometrical distribution of the water depends upon a balance between forces that pull the water toward the solid surface and surface tension forces in the liquid-vapor interface

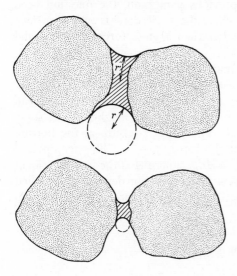

Fig. 8-4. Water films between soil particles. The negative curvature of the film at the top is the same as the positive curvature of the meniscus water at right angles to the page, thus the matric potential is zero. A negative matric potential exists in the water film at the bottom.

which act to keep the surface area of liquid water at a minimum. This may be illustrated by considering the intersticial space between two particles as shown in Figure 8-4. If considerable water is present, the air-water interface will be positive everywhere and, as with the raindrop, the pressure in the water will be positive. However, as water is withdrawn, by evaporation or through contact with a drier material, the curvature of the water surface between particles becomes zero (refer to equation 8-21) then negative in the region between the particles; the minimum surface areas are maintained consistent with the amount of water present and the pull of the solid surface for water. The pressure difference across the film, or the negative pressure in the vapor adjacent to the film, as measured with respect to the pressure which would exist over a flat surface of the liquid, is given by equation 8-22. Ultimately, a point is reached where surface tension no longer is great enough to maintain water between the particles in a configuration for which the curvature of the air-water interface is meaningfully associated with the pressure in the vapor phase. Water films usually are of molecular dimensions at this point and the vapor pressure depends upon the tightness with which water is held against the surface rather than upon the curvature of the air-water interface, which is definitely positive over particles with positive curvature.

It is worth noting that description of water retention in a porous system is considerably facilitated by specifying the energy required to remove water from the system rather than by specifying the mechanism involved. In the instance described in the previous paragraph, the question need not be raised as to the exact point where the work done is against surface binding forces near the particle rather than against forces in the liquid water.

Vapor pressure of a still atmosphere at equilibrium over a flat pure water surface (e.g., a closed, insulated air column above a free water surface) varies with elevation. This may be inferred by isolating a small increment of vapor of unit cross section and unit height, dz, at some elevation z above the water surface. The vapor pressure at the bottom of the vapor increment is p and at the top is $p - dp$. The net upward force then will be $p - (p - dp)A = A dp$, where A is the cross-sectional area. The mass of vapor in the increment is the vapor density ρ multiplied by the volume, or $\rho A dz$. One may equate these forces at equilibrium and rearrange them to give

$$\frac{dp}{dz} = \rho g = \frac{g n M_w}{V} \tag{8-23}$$

where density ρ has been replaced by the mass of vapor $n M_w$, n is the number of moles, and M_w is the molecular weight of the gas divided by its volume V. The volume of the gas V can be replaced by its equivalent from the combined gas law, assuming, as for equation 8-17, that water vapor behaves sufficiently like a perfect gas, $V = nRT/p$, to yield

$$\frac{dp}{p} = \frac{n M_w g}{nRT} dz = \frac{M_w g}{RT} dz \tag{8-24}$$

which yields upon integration between the limits p and p_0 and z and $z = 0$ for the pressure distribution above a free water surface where z is zero and where the vapor pressure is p_0

$$\ln \frac{p}{p_0} = \frac{M_w g}{RT} z \tag{8-25}$$

or

$$p = p_0 \exp\left(\frac{M_w g z}{RT}\right) \qquad (8\text{-}26)$$

Elevation z above the free water surface is taken as negative. Where potential on a mass basis is gz, it may be noted that this equation reduces to equation 8-18.

Equations 8-25 or 8-26 also give the distribution of water vapor pressure in the soil air space for a soil column with its base in a free water table; osmotic components are neglected. Both a saturated and a dry soil column held at constant temperature would come to the same equilibrium water vapor distribution although the time required for equilibrium and the water content at equilibrium would be different. The columns need not connect with the water table since water will move in the vapor phase if given sufficient time and condense in or evaporate from the column until equilibrium is reached. The unequal water contents of the two columns results from hysteresis, which is discussed in a later section.

Radii of Curvature of Air-Water Interfaces

The thermodynamic expression to show the effect of pressure change on free energy is

$$\Delta w = \int V \, dp \qquad (8\text{-}27)$$

This becomes $\Delta w = V \Delta p$ where volume is held constant. From equation 8-20 $\Delta p = 2\sigma/r$ for spherical curvature so that the energy per unit volume associated with pressure change for a pure water system is

$$\psi_V = \frac{\Delta w}{V} = \frac{2\sigma}{r} \qquad (8\text{-}28)$$

The free energy per unit volume in terms of the vapor pressure or relative humidity is given in equation 8-19. The relationship between vapor pressure, or relative humidity, and the radius of curvature of a spherical air-water interface is obtained by using

$$\frac{2\sigma}{r} = \frac{RT}{V_m} \ln \frac{p}{p_0} \qquad (8\text{-}29)$$

The radius r would represent the effective radius of the largest water-filled or smallest air-filled pore in a system that has a matric potential on the volume basis of ψ_V (equation 8-28) or that has a vapor pressure p or relative humidity p/p_0 as given by equation 8-29.

Capillary Rise

The rise of a liquid in a capillary tube may be obtained by computing the pressure exerted by the hanging water column. Water "hangs" around the perimeter of the tube by virtue of adsorption forces between the tube surface and the liquid and the cohesive forces in the liquid surface, or surface tension. The pressure is the mass of the liquid per unit cross section, $r^2 h \rho / r^2$, multiplied by the acceleration of gravity, g, or $h \rho g$. Putting this into equation 8-20 gives

$$h = \frac{2\sigma}{r\rho g} \tag{8-30}$$

or

$$h = \frac{2\sigma \cos \Theta}{r\rho g} \tag{8-31}$$

where σ is the surface tension, r is the radius of the capillary tube, ρ is the density of the liquid, and $\cos \Theta$ appears in the equation if the liquid does not wet the surface perfectly and the surface film at the point of contact hangs at an angle of Θ from the vertical wall, or if the wetting angle Θ is not zero.

Effect of Temperature on Vapor-Phase Water

Temperature changes will occur in the soil atmosphere if evaporation or condensation takes place or if heat is conducted or carried in from the outside. Soil temperature still may remain reasonably constant if heat flow is rapid into or out of the system. Temperature change is small enough to be neglected under real conditions in some systems where the heat capacity is sufficiently large, conduction sufficiently rapid, or evaporation and condensation tend to compensate each other. Temperature change may be large enough under other conditions that it cannot be neglected. Heat will flow either by conduction or by transport in flowing matter if a temperature difference exists in materials that are free to interact. Such flow is spontaneous and cannot be balanced. By virtue of kinetic energy imbalances, water vapor diffuses spontaneously as well. Wherever systems could be described on the basis of potential energy content, or by potential, it was possible to balance one potential against another to establish equilibrium, for example, matric potential against a pressure potential or against a gravity potential. Where the kinetic energy difference is the cause of flow, as is the case where a temperature difference exists, balance becomes impossible and potential concepts are inappropriate. Temperature may be held uniform at a point through which heat is flowing by adding and subtracting energy at the same rate. Similarly, water content can be held constant at a point under appropriate conditions while flow takes place; this assumes that appropriate energy is applied and removed from the system. These are steady-state systems and may be referred to as steady state equilibria.

The incompatability—at least at an elementary level—of potential considerations in problems where temperature differences are important may be illustrated by a simple flow problem. Horizontal water flow under uniform temperature conditions is from regions of high to regions of low matric potential. Water vapor may be made to flow in the opposite directon if a temperature difference of appropriate size and direction is superimposed on such a flow system. Circulation results when water vaporizes at the hot end of a column and flows toward the cold end where it condenses and flows back in the liquid state. Liquid flow occurs because of unbalanced matric forces that are associated with the adhesion of water to solid surfaces and cohesion in the liquid water; these can be appropriately described by the use of potential concepts. On the other hand, vapor flow occurs because of the difference in kinetic energy of the water molecules from point to point; this is a diffusional process. No simple method exists for treating the two flow problems together with the use of potential concepts. Flow is considered in greater detail in the following chapter.

HYSTERESIS IN WATER CONTENT—ENERGY RELATIONSHIPS

Water content and the energy status of soil water are not uniquely related because the energy state is determined by energy conditions at the air-water interfaces and the nature of surface films rather than by the quantity of water present in pores. Soil pores are highly variable in size and shape and interconnect with each other in a variety of ways. Inherent to such a porous material is the presence of pores larger than the average isolated from each other except for interconnecting smaller pores and passages. Water is held most tenaciously in small pores so that they generally fill first when water is admitted to a system. But they do not always empty again during drying in the same order as they filled.

The factors involved may be most clearly discussed by starting with a soil-water system in the absence of air so that problems of entrapped air are not involved. In such a system it may be seen that if the matric potential is increased by adding water, small pores fill first, followed by successively larger and larger pores until all pores are filled and the matric potential is zero. At potentials high enough so that vapor-water interfaces can exist between particles and in small pores, the curvature of such interfaces is given by equation 8–22, where the pressure difference is between that of the vapor and that of the free liquid water. Water content and water potential in such a system will follow the wetting curve in Figure 8-5. Some small pores could be isolated so that during the wetting process they might remain dry while larger pores are filled. However, this would not be the case at equilibrium in the absence of air inasmuch as vapor transfer would assure the wetting of all pores small enough to retain water at a particular matric potential.

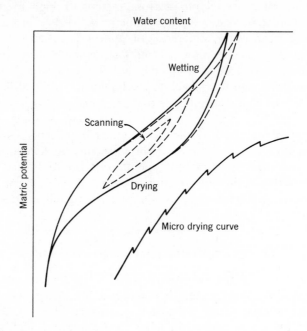

FIG. 8-5.   Wetting and drying curves and scanning curves.

If the system is dried either by evaporating water or by bringing the soil into contact with a dry, porous material which pulls water away from the system, pores will begin to empty, generally from large to small. However, water may now be trapped in large pores in such a way that they will not empty in the order that they filled. Water will be held in large pores until conditions are reached where at least one interconnecting smaller pore can empty; at this time the larger pore quickly empties. The sudden release of a relatively large amount of water from a large pore floods surrounding pores and increases the matric potential in them temporarily. If matric potential were followed in a small system having discrete differences in pore size, the matric potential-water content relationship for drying would be saw-toothed as indicated for a microsystem in Figure 8-5.

In real soil systems, because of the random nature of the pore-size distribution with many pores in all size ranges, the water content and potential distributions tend to average out so that a smooth curve is real-

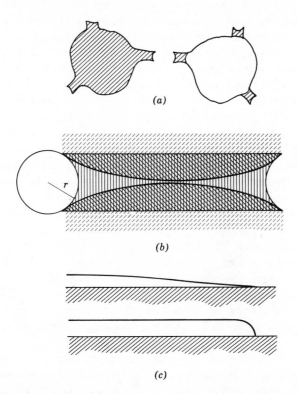

(a)

(b)

(c)

FIG. 8-6. Diagrammatic representation of three forms of water content-matric potential hysteresis.

ized; however, the water content for a given matric potential is higher than for the wetting system as is shown by the drying curve in Figure 8-5. This principle is illustrated by the pore-water system shown in Figure 8-6a. Here it may be observed that the curvature of the vapor-water interface in the small pores in two identical systems can be in equilibrium with each other even though their water contents are grossly different. The matric potential is determined by the curvature of the liquid interface (equation 8–22), which at equilibrium would be precisely the same in the small pores connecting with the large pore in each case. This idea commonly is referred to as the "ink bottle principle," which refers to the fact that an ink bottle has a small opening into a large cavity.

Large pores that are interconnected by smaller pores are not required for hysteresis. It is possible for a single pore to contain the same amount of water at two different water potentials, as is shown in Figure 8-6b. Water that condenses initially into such a capillary from a humid environment is shown by the diagonal cross-hatched area. However, as condensation proceeds, water at the center finally coalesces and a concave meniscus is formed as a consequence of surface tension forces (vertical cross-hatching). Whereas positive pressure existed in the system before coalescence, the system suddenly goes under negative pressure as a consequence of its new configuration. In the example in Figure 8-6b, water in a cylinderical pore about 1 cm in length and 0.1 cm in radius would have a slight positive potential of about 0.15 mbar immediately before coalescence and a potential of −1.5 mbar immediately afterward.

Dependence of water content-water potential relations upon wetting and drying further involves a surface wetting phenomena. Unless particle surfaces are meticulously clean, a wetting angle exists when the surface is wetted (Figure 8-6c). This results in thicker films than would be present in the drying phase where water films are drawn tightly over the surface by adsorptive forces.

Thus far the hysteresis phenomena have not involved the presence of air in the system. When air is involved, additional opportunities exist for hysteresis. As small pores and interstices between particles fill with water, air may become entrapped in large pores. Continued water entry into such pores must occur against a buildup of air pressure. Since air is slightly soluble in water, pressures in such pores may gradually be relieved, which will sometimes allow more complete pore filling. However, the order of filling and access to pores will be influenced by entrapped air so that water content still is permanently affected despite the fact that some air may go into solution and disappear.

In the absence of air, the water potential-water content relationship for complete wetting and complete drying will follow approximately the dashed line-solid line loop shown in Figure 8-5. However, such a loop is not exactly reproducible because of the inherent difficulty associated with repetition of the exact order of pore filling over each cycle. When air is present in the system, the curves are offset somewhat toward the dry side (solid curve, Figure 8-5). If a soil is completely wetted so that no air is present and then dried, it will follow the dashed-solid curve down; upon rewetting in the presence of air it will follow the solid wetting curve and will not return to the starting point because of the presence of entrapped air. If the process is reversed at any time during wetting or drying, curves like those in the interior (dotted curves) of the hysteretic envelope are produced. These interior curves have been called *scanning curves;* the curves that form the hysteretic envelope have been called *characteristic curves* or *soil moisture characteristic* (Childs, 1969). The wetting curve of the hysteretic envelope is commonly known as a *sorption* curve and the drying curve a *desorption* curve. Since air almost always is present when such curves are produced experimentally, the solid curves shown in Figure 8-5 are obtained. However, some ambiguity exists because the starting point for many measurements is a wet, and sometimes even puddled sample in which the degree of air removal is unknown.

Hysteretic phenomena also exist in soil materials as a consequence of shrinking and swelling. Shrinking and swelling affect pore size on a micro basis as well as on the basis of overall bulk density. Both factors would lead to altered volumetric water content for a given energy state from that which would exist if the soil matrix remained fixed. Shrinking and swelling often take place slowly and usually irreversibly, particularly when organic matter is involved; this complicates consideration of their contribution to hysteresis. Experimental observations do not always reveal which measurements involve true hysteresis and which involve permanent or semipermanent changes in the porous system.

References

Adam, Neil Kensington (1941, reprinted 1952). *The Physics and Chemistry of Surfaces*, 3rd ed. Oxford University Press, London.

Adamson, A. W. (1967). *Physical Chemistry of Surfaces*. Interscience, New York.

Babcock, K. L., and Roy Overstreet (1957). The extra-thermodynamics of soil moisture. *Soil Sci.*, **83**:455–464.

Bascom, Willard D., Edward J. Brooks, Bradford N. Worthington, III (1970). Evidence that polywater is a colloidal silicate sol. *Nature*, **228**:1290–1293.

Baver, L. D., and G. M. Horner (1933). Water content of soil colloids as related to their chemical composition. *Soil Sci.*, **30**:329–353.

Bernal, J. D., and R. H. Fowler (1933). A theory of water and ionic solution, with particular reference to hydrogen and hydroxyl ions. *J. Chem. Phys.*, **1**:515–548.

Bolt, G. H. (1955). Analysis of the validity of the Gouy-Chapman theory of the electron double layer. *J. Colloidal Sci.*, **10**:206–218.

Bolt, G. H., and M. Peech (1953). The application of Gouy theory to soil water systems. *Soil Sci. Soc. Am. Proc.*, **17**:210–213.

Buswell, Arthur M., and Worth H. Rodebush (1956). Water. *Sci. Am.*, **194**:Apr. 77–89.

Childs, E. C. (1969). *The Physical Basis of Soil Water Phenomena*. Wiley-Interscience, New York.

Davis, R. E., D. L. Rousseau, and R. D. Board (1971). "Polywater": Evidence from electron spectroscopy for chemical analysis (ESCA) of a complex salt mixture. *Science*, **171**:167–170.

Day, P. R. (1942). The moisture potential of soils. *Soil Sci.*, **54**:391–400.

Derjaguin (Deryaguin), B. V. (1966). Effect of lyophile surfaces on the properties of boundary liquid films. *Disc. Faraday Soc.*, **42**:109–119.

Gardner, W. H. (1965). Water content. *Methods of Soil Analysis. Agronomy Monograph*, No. 9 Part I, Academic Press, pp. 82–129.

Goates, J. Rex, and S. John Bennett (1957). Thermodynamic properties of water adsorbed on soil minerals: kaolinite. *Soil Sci.*, **83**:325–330.

Holtan, H. N., C. B. England, G. P. Lawless, and G. A. Schumaker (1968). Moisture-Tension Data for Selected Soils on Experimental Watersheds. ARS 41-144, USDA.

International Society of Soil Science Commission I (1963). *Soil Physics Terminology*, Bulletin No. 23:7–10.

Kemper, W. D. (1961a). Movement of water as effected by free energy and pressure gradients: I. Application of classic equations for viscous and diffusive movements to the liquid phase in finely porous media. *Soil Sci. Soc. Am. Proc.*, **25**:255–260.

Kemper, W. D. (1961b). Movement of water as effected by free energy and pressure gradients: II. Experimental analyses of porous systems in which free energy and pressure gradients act in opposite directions. *Soil Sci. Soc. Am. Proc.*, **25**:260–265.

Kittrick, J. A., and E. W. Hope (1970). Preventing water resorption in weight loss determinations. *Soil Sci. Soc. Am. Proc.*, **34**:536–537.

Lippincott, Ellis R., Robert R. Stromberg, Warren H. Grant, and Gerald L. Cessac (1969). Polywater. *Science*, **164**:1482–1487.

Low, P. F. (1959). The viscosity of water in clay systems. *Proc. 8th Nat. Clay Conf.*, Pergamon Press, New York.

Low, Philip F. (1961). Physical chemistry of clay-water interaction. *Advances in Agronomy*, **13**:269–327, Academic Press, New York.

Pauling, Linus (1960). *The Nature of the Chemical Bond*. Cornell University Press, Ithaca, N.Y.

Pethica, B. A., W. K. Thompson, and W. T. Pike (1971). Anomalous water not polywater. *Nature Physical Sci.*, **229**:20–22.

Robinson, R. A. and R. H. Stokes (1959). *Electrolyte Solutions*. Butterworths, London.

Rousseau, D. L. (1971). "Polywater" and sweat: similarities between the infrared spectra. *Science*, **171**:170–172.

Thomas, M. D. (1928). Aqueous vapor pressure in soils: IV. Influence of replaceable bases. *Soil Sci.*, **25**:485–493.

CHAPTER 9

Soil Water Movement

GENERAL FLOW PRINCIPLES

Water is seldom at rest in soil and the direction and rate of its movement is of fundamental importance to many processes that take place in the biosphere. Flow in soil is a special case of a larger problem of fluid flow in porous media (Childs, 1969; Crank, 1956; Morel-Seytoux, 1969; Kirkham and Powers, 1972). Only the basic principles of water movement in soil are dealt with in this chapter.

Flow Equations

Saturated Flow

It is possible to treat water flow under ordinary conditions in soil as a steady-state phenomenon and ignore inertial forces or accelerations. The flux of water under these conditions is proportional to the gradient of water potential and the conductivity in a way similar to the flux of electricity being proportional to the electrical potential difference in a circuit and electrical conductivity (Ohm's law). However, the nature of the forces that give rise to the potentials varies in soil-water systems and the kinds of forces involved have an important bearing upon the way that flow takes place. Additionally, the properties of the fluid as well as the nature of the flow channel are involved in the conductivity factor in complicated ways. Each of the components involved in flow equations requires careful discussion. However, such discussion will be more meaningful if the general nature of flow is discussed first.

Application of a force to a solid piston in a horizontal cylinder causes the piston to move. If the force applied is constant, the piston will accelerate until the opposing frictional force between the piston and the cylinder wall, which is proportional to the velocity, is equal to the applied force.

Back pressure from air compressed by the piston, which might be important in some situations, is ignored here. At this point, the "flow" of
the piston in the cylinder will be at a constant velocity or at steady
state. At steady state the energy added by application of a force will
be dissipated in heat at the cylinder wall. The flux of the substance
of the piston on a volume basis will be the product of the velocity v
and the cross-sectional area of the piston, a volume per unit time, or

$$F = \pi a^2 v \qquad (9\text{-}1)$$

or the flux per unit area is

$$v = \frac{F}{\pi a^2} = \text{constant} \qquad (9\text{-}2)$$

If a fluid like water replaces the solid piston, the frictional force that
resists flow at the solid surface now becomes distributed outward into
the water, inasmuch as shear is possible between water molecules. A
thin water layer at the solid surface is held tightly by molecular forces
and its velocity is zero. A shear stress is developed between each concentric ring of water outward from the surface which is proportional to
the differential velocity between the two layers. As a consequence, the
velocity of flow, which is zero at the surface, increases outwardly from
the surface to a maximum at the center of the cylinder. By summing
the forces that resist flow between each successive infinitesimal concentric
ring and equating the resulting differential expression to a like expression
that sums the forces that cause flow (i.e., a pressure difference between
two points along the cylinder at unit distance, multiplied by the area
of each of the infinitesimal concentric rings), a differential equation is
obtained that is the velocity of flow as a function of the radius in the
cylinder. Integration of this equation over the limits $r = 0$ to $r = a$,
where a is the radius of the cylinder, gives the parabolic equation for flow
velocity in any concentric ring of radius $r + dr$:

$$v = \frac{p_1 - p_2}{4\eta s} (a^2 - r^2) \qquad (9\text{-}3)$$

where p_1 and p_2 are the fluid pressures at two points along the cylinder
separated by the distance s, and η is the coefficient of viscosity as defined
by the equation

$$f = \eta \frac{\partial v}{\partial r} \qquad (9\text{-}4)$$

which says that the shear stress is proportional to the gradient of the
velocity at right angles to the direction of flow. It should be noted that

the viscosity may be considerably greater within several molecular depths from the solid surface than it is in bulk water; this can complicate detailed flow analyses but is ignored for the present.

Contrasted with the velocity of flow of the solid piston, which is constant across the area of the cylinder, the velocity of flow of a fluid in a cylinder depends upon the square of the cylindrical radius a and the square of the radius r to the point in question within the cylinder as defined by the parabolic equation 9-3 and illustrated in Figure 9–1. The flux of the substance of the solid given by equation 9-2 is propor-

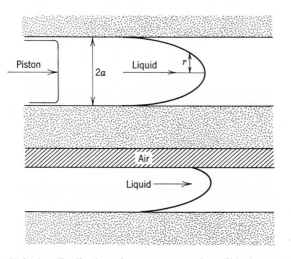

FIG. 9-1. Top: Velocity distributions for movement of a solid piston and movement of liquid water in a filled flow channel. Bottom: Velocity distribution for flow of liquid water in a channel not completely filled.

tional to the cross-sectional area of the piston, πa^2, whereas the flux of a fluid is proportional to the fourth power of the radius of the cylinder. This is seen if the velocity given by equation 9-3 is multiplied by the area of concentric cylinders, $2\pi r \, dr$, and summed by integration from $r = 0$ to $r = a$ to yield the Poiseuille equation:

$$F = \int_0^a 2\pi r v \, dr = \int_0^a \frac{(p_1 - p_2)}{(2\eta s)} \pi r \, dr = \frac{\pi a^4}{8\eta} \frac{p_1 - p_2}{s} \qquad (9\text{-}5)$$

The average flux per unit area is

$$v = \frac{F}{\pi a^2} = \frac{a^2}{8\eta} \frac{p_1 - p_2}{s} \qquad (9\text{-}6)$$

The preceding equation may be misleading, since the flux is not uniform over the cylindrical area of the tube, as might be implied. However, it does serve to point up the substantial difference between piston flow (equation 9-2) and flow where the velocity distribution is parabolic as given by equation 9-3. This has numerous implications to water flow and to transport of soluble materials in the soil solution. If the ratio of cross-sectional areas of two different capillary tubes is N, the ratio of fluxes at the same pressure gradient and for the same viscosity is N^2. Thus

$$N = \frac{A_1}{A_2} = \frac{a_1^2}{a_2^2}$$

$$\frac{F_1}{F_2} = \frac{a_1^4}{a_2^4} = \left(\frac{a_1^2}{a_2^2}\right)^2 = N^2 \tag{9-7}$$

By analogy, it is evident that the flux under saturated conditions for unit pressure gradient (or potential gradient) in a porous soil is not proportional to the cross-sectional area of the pores and that the flux per unit gradient in a tight clay with tiny pores and that in a loam soil with larger pores would be grossly different, even though the total porosity might be exactly the same.

A similar analysis for flow between parallel plates or in a slit spaced a distance d apart yields a flux of

$$F = \frac{wd^3}{12\eta}\frac{p_1 - p_2}{s} \tag{9-8}$$

where w is the width of the section through which flow takes place. The average flux per unit cross section, comparable to that for a cylindrical tube (equation 9-6), is

$$v = \frac{F}{wd} = \frac{d^2}{12\eta}\frac{p_1 - p_2}{s} \tag{9-9}$$

This again illustrates the heavy dependence of flux upon the depth of water on the surface of the confining solid.

The flow channels in a porous medium like soil will be considerably more complicated than cylinders and slits, so that rational flux equations would be more complicated. Also, the distance factor s, which up to the present has been the length of a cylinder or of a slit, now becomes the length of a flow channel which winds its way tortuously through the pores of the medium. The length of a path, and there are many different paths, cannot be measured directly and can only be inferred from other measurments, such as electrical conductivity, or by comparing empirical data with theoretical computations. Furthermore, not all of the pore space in a soil is involved in flow since many pores are oriented in such a way that little or no flow takes place through them.

Despite the complications associated with flow in a nonideal porous system like soil, the flux equations for cylinders and slits given above provide an excellent basis for understanding the nature of flow. An equation of the same form as equation 9-6 was used by Darcy (1856) to describe the flow of water in filter beds. The equation is written in many ways, but the form most nearly resembling equations 9-6 and 9-9 is most suitable for the present discussion:

$$v = \frac{Q}{tA} = k \frac{p_1 - p_2}{s} \qquad (9\text{-}10)$$

where v is the average velocity of flow over the gross cross section A of the porous body arranged so that flow is horizontal, Q is the quantity of flow obtained in time t, p_1 and p_2 are the pressures at the inlet and outlet faces of the porous body of length s. The parameter k is referred to as the *permeability* or *hydraulic conductivity*, depending, in part, upon the units used for the energy or force term and in part upon whether or not viscosity is implicit or explicit. This parameter, which usually is measured empirically, includes all of the factors pertaining to variation in velocity over the flow section, to tortuosity and size of the channels, and to properties of the flowing fluid, such as viscosity. With the objective of eliminating the properties of the fluid from the conductivity parameter, viscosity sometimes is included in the equation so that k'/η replaces k. However, since water interacts with the colloidal component in practically all soils to cause shrinking or swelling, the nature of the pores is altered, at least internally, and the overall porosity often is changed. Hence the permeability k usually is different with various liquids quite apart from viscosity considerations. The sensitivity of k' to structural changes has led to its use for evaluation of certain aspects of soil structure stability; the ratio of k' with air to k' with water is used as an index. From the point of view of water flow, viscosity generally is not included explicitly in the equation unless temperature is a major factor under consideration. However, recognition of the effect of temperature upon the viscosity of water sometimes is appropriate.

Unsaturated Flow

Under conditions where negative pressure or negative potentials exist in soil water, an air phase is present and the flow channel is grossly modified. The liquid boundary then consists of a solid surface in some configuration and an air-water interface. The forces that cause flow now are associated with the attraction of solid surfaces for water and the surface tension in the air-water interface. As the water content decreases, the path of water flow becomes more and more tortuous inasmuch as liquid flow occurs along surfaces and through small pores of such size that they retain water at the prevailing water potential.

As in capillary flow, described by the Poiseuille equation (9-5) given earlier, the velocity of flow in a system with air present is zero at the surface and increases outward from the surface. Viscosity plays precisely the same role in such flow, as described earlier. However, the velocity of flow at the air-water interface that is required for completing an analysis of the type given earlier for full capillaries is not easily defined. The velocity of flow at the surface in larger flow channels where it can be measured is known to be less than that at the point of maximum flow velocity which is about $\frac{2}{3}$ to $\frac{9}{10}$ of the distance from the solid surface to the air-water interface. Qualitatively, the velocity curve is believed to be something like that shown in Figure 9-1. Since part of the flow is in completely filled small channels and part in channels involving an air-water interface, the velocity distribution in water films may be expected to be somewhat parabolic.

With the introduction of an air phase in the porous system, the flow equation must be modified to take into account the fact that the cross section available for liquid flow no longer is equivalent to the total porosity. Furthermore, since large pores fill last on wetting and generally empty first on drying, the fraction of the total pore water which is close to a solid surface increases markedly with decreasing water content. This causes the conductivity to decrease markedly with a decrease in water content; the lowered conductivity is much more than would be predicted on the basis of a loss in water-filled cross section. Thus the conductivity factor in the flow equation must be modified in such a way as to take into account the reduced cross section and the added effect of friction close to particle surfaces. This could be done in a variety of ways, but the addition of a dimensionless factor λ^* with k in equation 9-10 makes it possible to preserve the generality of the flow equation. The factor λ varies from 0 to 1 and disappears when the soil is saturated.

The forces that cause flow are changed by the introduction of an air phase. Gravity continues to be involved, but the pressure in the water

---

* The symbol λ was used by Buckingham (1907) for the conductivity in an unsaturated flow equation. That use or its use here is not to be confused with the occasional use of λ to designate a characteristic dimension of soil particles or aggregates.

due to the weight of water now becomes negative with respect to the surrounding gas pressure. Attractive forces between solid surfaces and water and the surface tension at the air-water interface now become important. These are referred to as matric forces or as soil-moisture tension. To retain the generality of the flow equation, it is appropriate to deal with these forces or the tension as potentials; the gradient of the potential is a force per unit mass or per unit volume. On this basis, whatever potentials are involved may be added directly. Potentials may be expressed in head units as indicated by equation 8-14. If the potentials are given on a unit mass basis, division by the acceleration of gravity converts them to head units. If they are given on a unit volume basis, division by the acceleration of gravity and the density of water converts these potentials to head units. Gradients in head units then become dimensionless.

If equation 9-10 is modified by adding the dimensionless factor λ, to account for any reduction in the cross section for flow where an air phase is present, and by replacing the pressure drop per unit distance with a more general expression for the force term, that is, by the potential gradient in head units, grad H, it then becomes

$$v = -k\lambda \text{ grad } H \qquad (9\text{-}11)$$

The negative sign appears in order to make the flux positive, since flow is from regions of higher potential to regions of lower potential which gives rise to gradients with negative signs. As indicated in the discussion accompanying equation 8-14. $H = H_M + H_p + z = \phi + z$, where, if $z = 0$ at the water table, H is positive below and negative above the water table. It is composed of two parts: (1) the pressure potential ϕ, which above the water table is the matric potential H_M, and below the water table is a potential due to positive water pressure head H_p; and (2) the gravity potential H_z or merely z in head units, which accounts for potential changes associated with change in elevation. For present purposes overburden potential, where it exists, is implicit in either H_M or H_p and other potentials are neglected.

The symbol "grad" in equation 9-11 is a vector notation that denotes the gradient of H in all directions so that equation 9-11, with $\phi + z$ replacing H, may be written for three-dimensional flow as

$$v = -k\lambda \left[\frac{\partial \phi_x}{\partial x} + \frac{\partial \phi_y}{\partial y} + \frac{\partial}{\partial z}(\phi_z + z) \right]$$

$$= -k\lambda \left[\frac{\partial \phi_x}{\partial x} + \frac{\partial \phi_y}{\partial y} + \frac{\partial \phi_z}{\partial z} + 1 \right] \qquad (9\text{-}12a)$$

for vertical flow as

$$v = -k\lambda \left(\frac{\partial \phi_z}{\partial z} + 1 \right)$$ (9-12b)

and for one-dimensional horizontal flow along x as

$$v = -k\lambda \frac{\partial \phi_x}{\partial x}$$ (9-12c)

Equations 9-11 and 9-12a are written for the case where $k\lambda$ depends only upon the nature and water content of pores at the point where v is taken and is independent of direction of flow. Where $k\lambda$ varies with direction the flow equations become considerably more complicated. The signs in equations 9-12 are chosen so that v is positive when flow is in the positive x or positive z directions. Furthermore, z is taken as positive in the upward direction. Hence a negative v in equation 9-12b would denote downward flux. This is covered in more detail in the section on equilibrium conditions in soil later in this chapter.

DIFFUSION-TYPE EQUATIONS FOR UNSATURATED FLOW. Thus far equations of flow have given flux as a product of a force term and a conductivity. However, neither conductivity nor potentials associated with unsaturated flow are easily measured in all ranges of interest. It is possible to deal with flow so that flux is proportional to the water content gradient and a diffusivity term, as in problems involving diffusion. This permits the use of water content measurements and diffusivity, both of which sometimes are more easily measured than potential and conductivity. However, if the diffusion process is defined as the flow of matter as a consequence of kinetic energy differences of the flowing substance along a flow path, then liquid flow cannot be described as a diffusion process. Vapor flow is aptly described as a diffusion process, but liquid flow under unsaturated conditions takes place because of the attraction of solid surfaces for water, the pull of surface tension in air-water interfaces, and the transmission of these forces through water by virtue of molecular attraction between water molecules. Despite this, however, it is possible to write flow equations in diffusion form and to use diffusion mathematics in flow problems. In such cases, the gradient of the water concentration becomes an index of the moving force. This may be shown by introduction into the flux equations the term

$$d(\Theta) = k\lambda \frac{\partial \phi}{\partial \Theta}$$ (9-13)

and application of the chain rule of calculus to yield for three-dimensional flow

$$v = -k\lambda \left[\frac{\partial \phi_x}{\partial \Theta_x} \frac{\partial \Theta_x}{\partial x} + \frac{\partial \phi_y}{\partial y} + \frac{\partial \phi_z}{\partial \Theta_z} \frac{\partial \Theta_z}{\partial z} + 1 \right]$$

$$= -D(\Theta) \left[\frac{\partial \Theta_x}{\partial x} + \frac{\partial \Theta_y}{\partial y} + \frac{\partial \Theta_z}{\partial z} \right] - k\lambda \qquad (9\text{-}14\text{a})$$

for vertical flow,

$$v = -D(\Theta) \frac{d\Theta}{dz} - k\lambda \qquad (9\text{-}14\text{b})$$

and for horizontal one-dimensional flow in the x direction,

$$v = -D(\Theta) \frac{d\Theta}{dx} \qquad (9\text{-}14\text{c})$$

Multiplication of the unsaturated conductivity $k\lambda$ by the slope of the matric potential-water content curve $\partial \phi / \partial \Theta$, as shown by equation 9-13, produces a concentration-dependent diffusivity $D(\Theta)$ which is more tractable for many purposes than is the unsaturated conductivity $k\lambda$.

HYSTERESIS AND FLOW PROBLEMS. The matric potential has been referred to in these flow discussions without mention of the fact that its relationship to water content is not unique. Just as it is necessary to recognize in discussions of energy-water content relationships that water content at a given matric potential depends upon antecedent conditions, it is also essential to take into account in flow discussions the fact that the flow path may be different for the same potential and potential gradient if antecedent water conditions are different. In unsaturated flow, the relationships of $k\lambda$ or $D(\Theta)$ to water content may be quite different in wetting and drying systems or in systems of uncertain wetting history. Also, the amount of water present in *dead-end* pores which does not participate in flow varies appreciably with antecedent conditions. Thus hysteresis becomes an important factor in water flow as well as in water-energy relationships. In the practical application of flow theory it is necessary to obtain $k\lambda$ or $D(\Theta)$ relationships under conditions that are typical of the water flow under consideration.

Equation of Continuity and the Laplace Equation

The relationship between flux and the gradient of the hydraulic head, or the concentration gradient, is useful in itself. But it becomes a powerful mathematical tool for the description of flow in both saturated and unsaturated systems when combined with the equation of continuity. This equation will not be derived mathematically here; it says that the algebraic sum of the quantity of water that enters a specified volume of soil and that which leaves the same volume in a given unit of time is equal to the change (increase or decrease) of water in that volume. The equation is derived in books on physical mechanics or theoretical physics and is discussed by Childs (1969).

If Θ is the water content per unit volume of soil, the equation of continuity in three dimensions, using rectangular coordinates, may be stated mathematically as

$$\frac{\partial \Theta}{\partial t} = -\left[\frac{\partial v_x}{\partial x} + \frac{\partial v_y}{\partial y} + \frac{\partial v_z}{\partial z}\right] \tag{9-15a}$$

where v_x is the flux in the x direction, v_y and v_z the fluxes in the y and z directions. Comparable expressions in cylindrical coordinates and in polar coordinates, which fit some problems in the field better than rectangular coordinates, are

$$\frac{\partial \Theta}{\partial t} = -\frac{1}{r}\left[\frac{\partial}{\partial r}(r\Theta v_r) + \frac{\partial}{\partial \phi}(\Theta v_\phi) + \frac{\partial}{\partial z}(r\Theta v_z)\right] \tag{9-15b}$$

$$\frac{\partial \Theta}{\partial t} = -\frac{1}{r^2 \sin \phi}\left[\frac{\partial}{\partial r}(r^2 \sin \phi \Theta v_r) + \frac{\partial}{\partial \phi}(r \sin \phi \Theta v_\phi) + \frac{\partial}{\partial \alpha}(r\Theta v_\alpha)\right] \tag{9-15c}$$

where r, ϕ, and z are the cylindrical and r, ϕ, and α are spherical coordinates.

Differential equations, whose the solutions give the water content as a function of time and position in a wetting or drying system, are obtained by substitution of the appropriate flux from equations 9-12a, b, and c and 9-14a, b, and c into the equation of continuity 9-15a (or equivalent substitutions into equations 9-15b or c for other coordinate systems). Using flux in potential form as given by equations 9-12a, b, and c, the desired differential equations are, for three-dimensional flow,

$$\frac{\partial \Theta}{\partial t} = \frac{\partial}{\partial x}\left[k\lambda \frac{\partial \phi_x}{\partial x}\right] + \frac{\partial}{\partial y}\left[k\lambda \frac{\partial \phi_y}{\partial y}\right] + \frac{\partial}{\partial z}\left[k\lambda \frac{\partial \phi_z}{\partial z}\right] + \frac{\partial(k\lambda)}{\partial z} \tag{9-16a}$$

for vertical flow,

$$\frac{d\Theta}{dt} = \frac{d}{dz}\left[k\lambda \frac{d\phi}{dz}\right] + \frac{d}{dz}(k\lambda) \tag{9-16b}$$

and for horizontal one-dimensional flow in the x direction,

$$\frac{d\Theta}{dt} = \frac{d}{dx}\left[k\lambda \frac{d\phi}{dx}\right] \tag{9-16c}$$

Using flux as given by equations 9-14a, b, and c in diffusion form the equation is, for three-dimensional flow,

$$\frac{\partial \Theta}{\partial t} = \frac{\partial}{\partial x}\left[D(\Theta)\,\frac{\partial \Theta_x}{\partial x}\right] + \frac{d}{\partial y}\left[D(\Theta)\,\frac{\partial \Theta_y}{\partial y}\right] + \frac{\partial}{\partial z}\left[D(\Theta)\,\frac{\partial \Theta_z}{\partial z}\right] + \frac{\partial (k\lambda)}{\partial z}$$

(9-17a)

for vertical flow,

$$\frac{d\Theta}{dt} = \frac{d}{dz}\left[D(\Theta)\,\frac{d\Theta}{dz}\right] + \frac{d(k\lambda)}{dz}$$

(9-17b)

and for horizontal one-dimensional flow in the x direction,

$$\frac{d\Theta}{dt} = \frac{d}{dx}\left[D(\Theta)\,\frac{d\Theta}{dx}\right]$$

(9-17c)

It should be noted that for saturated flow water content Θ remains constant with time and therefore $\partial \Theta / \partial t = 0$. With flux written in terms of hydraulic head H, equation 9-11 becomes

$$v = -k\left[\frac{\partial H_x}{\partial x} + \frac{\partial H_y}{\partial y} + \frac{\partial H_z}{\partial z}\right]$$

(9-18)

where $\lambda = 1$ and therefore disappears and k is uniform in all directions. The more general case in which k is not uniform in all directions due to soil anisotropy is discussed in detail by Childs (1969). Putting equation 9-18 into the equation of continuity, equation 9-15a, yields

$$\frac{\partial \Theta}{\partial t} = k\left[\frac{\partial^2 H_x}{\partial x^2} + \frac{\partial^2 H_y}{\partial y^2} + \frac{\partial^2 H_z}{\partial z^2}\right]$$

(9-19)

Where $\partial \Theta / \partial t = 0$, this reduces to

$$\frac{\partial^2 H_x}{\partial x^2} + \frac{\partial^2 H_y}{\partial y^2} + \frac{\partial^2 H_z}{\partial z^2} = 0$$

(9-20)

which is known as the Laplace equation. It has numerous important uses in solving equations of saturated flow. There is an infinite number of mathematical expressions which relate the potential to the space coordinates x, y, and z which satisfy Laplace's equation. The major problem is then to find the one and only solution which satisfies the conditions specified at the boundaries of the system. Many solutions of this equation have been worked out and can be found in the literature.

APPLICATION OF THEORY TO FLOW PROBLEMS

Saturated Flow Problems

Application of the Laplace Equation

A simple example of the solution to a saturated flow problem is a cylindrical well that removes water from a layer of saturated gravel of thickness z that lies between impermeable clay layers. The solution to this problem is easiest if cylindrical coordinates are used. The last two terms of equation 9-15b become zero for this problem because the velocity does not vary with angular position nor with elevation, then

$$\frac{\partial \Theta}{\partial t} = -\frac{1}{r} \frac{\partial}{\partial r} (r \Theta v_r) = -\frac{\Theta}{r} \frac{\partial}{\partial r} (r v_r) \qquad (9\text{-}21)$$

Since water content Θ is constant with time, the left-hand term is zero and equation 9-21 becomes

$$\frac{\partial}{\partial r} (r v_r) = 0 \qquad (9\text{-}22)$$

which is the Laplace equation appropriate to this situation. By carrying out the indicated integration equation 9-22 becomes

$$r v_r + \text{constant} = 0 \qquad (9\text{-}23)$$

When $r = r_0$, $v_r = v_{r_0}$ so that the constant is $-r_0 v_{r_0}$, equation 9-23 becomes

$$r v_r - r_0 v_{r_0} = 0 \qquad (9\text{-}24)$$

which is the solution of the Laplace equation for the situation described.

At the surface of the permeable well casing r_0, the discharge of the well, Q, is the flow velocity v_{r_0} multiplied by the surface area of the casing, $2\pi r_0 z$, so that

$$v_{r_0} = -\frac{Q}{2\pi r_0 z} \qquad (9\text{-}25)$$

The negative sign indicates that the velocity is in the opposite direction to the increasing radius. Equation 9-24 then becomes

$$Q = -2\pi r z v_r \qquad (9\text{-}26)$$

The Darcy equation 9-11, with $\lambda = 1$ for saturated conditions, and written in cylindrical coordinates with variation in angle and elevation zero, is

$$v = -k \operatorname{grad} H = -k \frac{\partial H}{\partial r} \tag{9-27}$$

Substituting this equation into equation 9-26 gives

$$Q = 2\pi r z k \frac{\partial H}{\partial r} \tag{9-28}$$

Changing to total derivatives, rearranging and integrating over the limits r_0 to r and H_0 to H, gives

$$H - H_0 = \frac{Q}{2\pi z k} \ln \frac{r}{r_0} \tag{9-29}$$

Thus the hydraulic head is a logarithmic function of the distance from the well. Or if the head H at some distance r_1 from the well is H_1, then the discharge Q of the well is

$$Q = \frac{2\pi z k (H_1 - H_2)}{\ln r_1/r_0} \tag{9-30}$$

Hydraulic Conductivity Measurements

Laboratory measurements of hydraulic conductivity may be made on undisturbed cores that are taken from the field or on reconstituted samples of field soil. A fixed head of water is applied and the flux after saturation is measured. The hydraulic gradient is the difference in head at the inlet and outlet faces of the core divided by the length of core. Although the flux may be measured to any desired precision, the applicability of such measurements and the need for precision is severely limited by problems of assuring that field structure is maintained or attained in reconstitution. Shrinking and swelling during handling and microbial activity often lead to appreciable changes from field conditions. Furthermore, vertical and horizontal conductivity often differ. Since field flow rarely is restricted exactly to either the vertical or the horizontal, field hydraulic conductivity is not easy to obtain from laboratory measurements on reconstituted samples. Field methods often are much more useful.

If a well were pumped for a sufficient period so that the discharge was essentially constant, then equation 9-29 could be used to find an effective field value for the hydraulic conductivity. The hydraulic head H_0 is the depth of drawdown of the water at the casing wall, r_0, and H_1 is the hydraulic head at a distance r_1 from the well. With such data and the well discharge, k may be computed. However, simpler methods have more practical value. One of several simpler field methods involves the same general idea. It may be carried out through measurement of the piezometric head and the time required for the head to rise a short distance after the water level has been lowered by withdrawal from the piezometer. Kirkham (1945) has applied the Laplace equation to a small cylindrical cavity at the bottom of a piezometer tube (and for some variations in cavity geometry). Using the Darcy equation, he has derived an equation for hydraulic conductivity in terms of the time

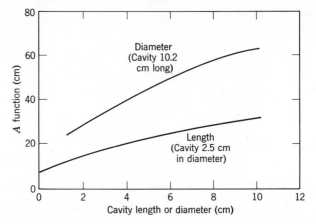

FIG. 9-2. A function for various cavity sizes for use in measurement of hydraulic conductivity by a piezometer method. (Abridged from Luthin and Kirkham, 1949.)

required for the hydraulic head in the piezometer tube to raise a certain distance after having been pumped to an appropriate starting level. Kirkham's equation, modified so as to use head terminology, is

$$k = \frac{\pi a^2 \ln h_1/h_2}{A(t_2 - t_1)} \qquad (9\text{-}31)$$

where a is the radius of the pipe, h_1 and h_2 are the distances below the water table at times t_1 and t_2, and A is a function of the geometry of the cavity that has length units. Luthin and Kirkham (1949) give some experimental curves for A obtained from an electrical analogue and from field measurements. Figure 9-2 shows two of these curves used for the estimation of A. The A values shown are reasonably constant for a cavity located approximately 15 to 45 cm below the water table and no closer than a few centimeters from an impermeable layer. Some useful relationships of flux, permeability, and driving force are shown in Table 9-1.

TABLE 9-1
Units and Dimensions of Terms in Saturated Flow Equation

Velocity or flux			Permeability factor				Driving force			
Symbol	Dimensions	Units	Symbol	Name of k	Dimensions of k	Units of k	Symbol	Name	Dimensions	Units
v, v_v	L/T	$cm^3/(sec)/(cm^2)$ $= cm/sec$	k	Permeability	T	sec	$\nabla\Phi$ grad Φ	Hydraulic potential gradient	L/T^2	D/g
			k/η	Intrinsic permeability	M/L	g/cm				
			k	Permeability	L^3T/M	$cm^3\ sec/g$	$\Delta p/\Delta L$	Pressure gradient	$M/(L^2T^2)$	$D/cm^2/cm$ $= D/cm^3$
			k/η	Intrinsic permeability	L^2	sec^2				
			k	Hydraulic conductivity	L/T	cm/sec	$h/L,$ $\Delta h/\Delta L$	Hydraulic head gradient	L/L	cm/cm $= unity$

Unsaturated Flow Problems

Simulation of Unsaturated Flow by Means of Capillary Tubes

As has been shown in previous sections in this chapter, the conductivity factor in unsaturated flow is highly dependent upon water content or upon the thickness of water films. This factor for most soils drops to a small fraction of its saturation value with the loss of only a small percentage of the water content. This conductivity property completely dominates important flow processes under some conditions. An understanding of the reason for such a great dependence upon water content or film thickness is essential to an understanding of unsaturated flow. Application of the Poiseuille equation (equation 9-5) to water flow in bundles of capillary tubes of different sizes simulates unsaturated flow and provides the needed perspective.

Bundles of capillary tubes with appropriate size distribution so that they drain in the same way as a soil are hypothesized. The water content-

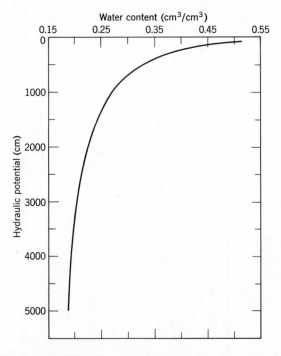

Fig. 9-3. Hydraulic potential-water content relationship for Salkum silty clay loam after which the size distribution of capillary tubes is modeled for use in producing the conductivity curve shown in Figure 9-5.

water potential curve for these tubes thus becomes the same as that shown in Figure 9-3 for Salkum silty clay loam for which experimental conductivity data are available (discussed in the next section). The flow system differs from that which would exist in soil in several ways: (1) each capillary is continuous and, although it may be considered to be meandering, it contains no place for temporarily nonparticipating or "dead" water; (2) capillaries are all of the same length; (3) flow boundaries are entirely solid-water interfaces with empty capillaries constituting the air space, whereas in soil the flow boundaries are a combination of solid-water and solid-air interfaces; (4) the size of the capillary entirely

Capillary tubes or soil

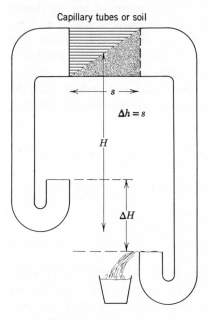

FIG. 9-4. Diagramatic picture of a system for measurement of water flux through capillary tubes or through soil at various hydraulic heads, H. The hydraulic gradient, $\triangle H/\triangle s$, can be adjusted to be unity.

governs the thickness of the films, whereas the film thickness on surfaces in soil is governed by the matric potential; and (5) flow is steady state, which is contrary to the usual situation in soil.

The flux through the system at the outset is considered to be the sum of the fluxes through all full capillaries; and the gradient is taken to be unity and is based upon the actual length of the capillary. The hydraulic conductivity per unit gradient is exactly equal to the flux so that a curve showing hydraulic conductivity as a function of water content may be produced. When this curve is compared with the curve for Salkum silty clay loam, the ratio of effective pore length to measured soil column length, or tortuosity, is taken into account.

An apparatus for measuring flux at different hydraulic heads, and thus at different average water contents, is shown in Figure 9-4. Such an apparatus will not work practically, except in the very wet range, because of the long hanging-water columns required; however, the principle is adequately illustrated.

The Poiseuille equation (9-5) used, written for head in centimeters rather than in pressure units, is

$$F = \frac{\pi a^4 \rho g}{8\eta} \frac{H_1 - H_2}{s_e} = \frac{\pi a^4 \rho g}{8\eta} \frac{\Delta H}{s_e} \qquad (9\text{-}32)$$

where pressure p has been replaced by $H\rho g$, ρ is the density of water, g is the acceleration of gravity, and s_e is the length of the capillaries that corresponds to the effective length of flow path in soil. The total flux at saturation is the sum of N_1, N_2 ... N_n capillary tubes that have radii of a_1, a_2, ... a_n. The flux at water contents less than saturation is obtained in successive increments by dropping successively smaller tube sizes, starting with the largest capillary tube and ending with the smallest that is of sufficient size to contribute significantly to the flux, thus:

$$F_{\text{total}} = F_{a_1, a_2, \ldots a_n} = \frac{\pi \rho g}{8\eta} \frac{\Delta H}{s_e} [N_1 a_1^4 + N_2 a_2^4 + \cdots N_n a_n^4]$$

$$F_{a_2, a_3, \ldots a_n} = \frac{\pi \rho g}{8\eta} \frac{\Delta H}{s_e} [N_2 a_2^4 + N_3 a_3^4 + \cdots N_n a_n^4]$$

$$\cdots \qquad \cdots$$

$$F_{a_n} = \frac{\pi \rho g}{8\eta} \frac{\Delta H}{s_e} N_n a_n^4 \qquad (9\text{-}33)$$

For these equations to simulate conditions in the Salkum silty clay loam, N must be the number of tubes of radius a in unit bulk cross section (pores plus solid). The sum of the pore cross section for all of the tubes in unit cross section would be the same as the volumetric saturation water content; this corresponds to the cross-sectional area of pores per unit bulk cross section. With N defined as pores of a size a per unit bulk cross section, the flux becomes the volume of flow per unit area per unit time or, dimensionally L/T; the latter corresponds to the common expression for water flux in soil. Sizes of capillary tubes that are emptied as the water content is decreased are determined from the Salkum sorption curve by taking the average hydraulic potential over each water content increment, $\Delta\Theta$, as water is removed and computing tube radius from the capillary-rise equation

$$a = \frac{2\sigma}{H\rho g} \cong \frac{0.147}{H} \text{ cm} \qquad (9\text{-}34)$$

where σ is the surface tension, ρ is the density of water, and g is the acceleration of gravity. This reduces to $a = 0.147/H$ cm at ordinary temperatures, where H is in centimeters of water. The number of pores, N, of each size then is obtained by dividing the water content loss over the range considered $\Delta\Theta$ (in cm^3/cm^3), as obtained from the sorption curve, by the volume of a unit length of capillary tube of radius a. Thus

$$N = \frac{\Delta\Theta \text{ cm}^3/\text{cm}^3}{\pi a^2 s_u \text{ cm}^3} = \frac{\Delta\Theta \text{ cm}^2/\text{cm}^2}{\pi a^2 \text{ cm}^2} = \frac{\Delta\Theta}{\pi a^2} \text{ cm}^{-2} \qquad (9\text{-}35)$$

where s_u is unit length. To provide flux in units of volume per unit area per unit time (or L/T), N is obtained on an area basis as shown in equation 9-35, or as the number of pores per unit of bulk area. Putting equation 9-35 into equation 9-33 yields

$$F_{\text{total}} = F_{a_1,a_2\ldots a_n} = \frac{\rho g}{8\eta} \frac{\Delta H}{s_e} [\Delta\Theta_1 a_1{}^2 + \Delta\Theta_2 a_2{}^2 + \cdots \Delta\Theta_n a_n{}^2]$$

$$F_{a_2,a_3\ldots a_n} = \frac{\rho g}{8\eta} \frac{\Delta H}{s_e} [\Delta\Theta_2 a_2{}^2 + \Delta\Theta_3 a_3{}^2 + \cdots \Delta\Theta_n a_n{}^2]$$

$$F_{a_n} = \frac{\rho g}{8\eta} \frac{\Delta H}{s_e} \Delta\Theta_n a_n{}^2 \qquad (9\text{-}36)$$

The factors outside of the brackets in equation 9-36 become 12,250 $cm^{-1}sec^{-1}$. Since $\Delta\Theta$ is dimensionless and a^2 is L^2, the flux has the appropriate dimensions of L/T. Where $\Delta H/s_e$ is unity, the fluxes are the hydraulic conductivities. Plotting these as a function of the average water content for each $\Delta\Theta_n$ gives the hydraulic conductivity curve for the capillary tube system, as shown in Figure 9-5, along with an experimental curve for Salkum silty clay loam.

The gradient for computations of fluxes and hence hydraulic conductivity has been based upon actual capillary length. Since the effective length of conducting paths in soil is considerably greater than the length of a soil core, or the length used in computing the gradient, direct comparison with the soil requires a correction so as to simulate core length rather than capillary length. Tortuosity usually is defined as the ratio of the effective or average length of flow paths to the length of the porous body:

$$\tau = \frac{\alpha}{a^\beta} \qquad (9\text{-}37)$$

where a is the radius of the largest water-filled pore and α and β are parameters. Fatt and Dykstra (1951), quoted by Morel-Seytoux (1969), indicate fairly good agreement with experimental data by using $\beta = 1$. For present purposes, the curve obtained by using equations 9-36 (Figure 9-5) is divided by $\tau = \alpha/a^\beta$ and the parameters α and β are determined so that a part of the curve is fit. The tortuosity factor obtained is

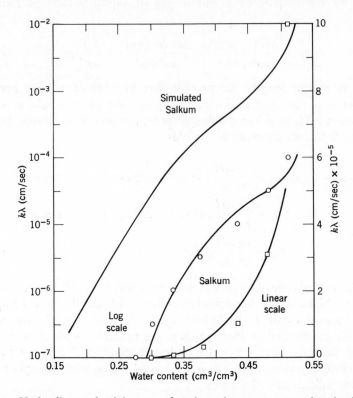

FIG. 9-5. Hydraulic conductivity as a function of water content for the Salkum silty clay loam (log and linear scale curves) and the capillary tube system simulating this soil (log curve only). The plotted points represent values for the simulated soil after division by the tortuosity or fitting factor as given by equation 9-37. Such curves may not be single-valued as noted later (Figure 9-9).

$\tau = 5.7/a^{0.33}$; the adjusted hydraulic conductivity values are the circled points that are plotted along the linear and logarithmic curves for Salkum silty clay loam (Figure 9-5). The tortuosity values that are obtained by using this equation are rather large. It must be recognized that other factors not taken into account in the analogy become involved in forcing the computed curve through a part of the Salkum curve. Thus the tortuosity equation becomes more of a fitting equation, which includes tortuosity, rather than a tortuosity equation alone.

If the tortuosity equation (9-37), with computed parameters, were included in the flow equation (9-31), it would become

$$F = \frac{\pi a^{5.3} \rho g}{.048 \eta} \frac{\Delta H}{s_e} \qquad (9\text{-}38)$$

This demonstrates that the flux in unsaturated flow depends upon the effective tube diameter a to a power higher than 4, as it is for capillary tubes in the Poiseuille equation, probably to a power of between 5 and 6.

Distribution and Flow in a Horizontal Soil Column

Water flux in soil under unsaturated conditions varies from point to point in the system and with time. Distribution of water in a soil system depends upon water content, its distribution antecedent to the addition of new water, and upon the flux into the system. Primary questions that relate to water flow usually are the infiltration rate at the surface, the depth of wetting, and the distribution of water in the soil column at various intervals of time. The latter includes the distribution after cessation of irrigation. Additional questions of interest involve flux of water within the soil at various times as caused by absorption of water by roots. The investigator can easily measure the infiltration rate under field conditions and with considerable effort can follow to a limited extent the spatial distribution of water in the profile as a function of time. He also can measure water potential in a limited number of locations. Analytical processes are required for the development of details of flow and water retention that are not easily measured. However, flow situations, particularly in nonuniform materials, often are so highly complicated that formal analyses become impractical and simplified approximations are used. The general approach in dealing with unsaturated water-flow problems is to consider particular situations in detail through use of experimental as well as analytical models. The findings then are extrapolated to other situations. Analyses of simple systems form a useful basis for understanding more complicated systems. An analysis of a simple system that involves flow into a dry horizontal column of Salkum silty clay loam is given here.

The distribution of water infiltrating into a dry soil column has been followed as a function of time by using gamma ray attenuation methods for sensing water content (Rawlins and Gardner, 1963; Stewart, 1962). The soil column and water-content distributions, after several time intervals, are shown in Figure 9-6. Using these and the sorption curve for Salkum silty clay loam shown in Figure 9-3, it is possible to obtain water potentials for the same time intervals. It is possible to obtain either the gradient of the hydraulic potential or the water content gradient from these data as shown in Figure 9-7, and required in the flow equations 9-11 or 9-12. Water fluxes could be computed for this

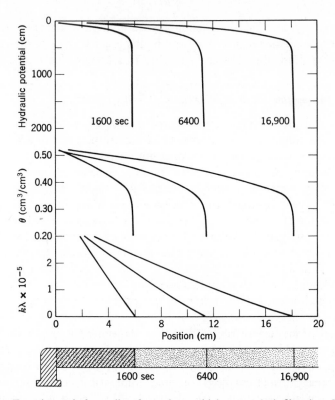

Fig. 9-6. Experimental dry soil column into which water is infiltrating. The associated curves show water content distribution, hydraulic potential, and hydraulic conductivity, $k\lambda$, at three different positions of the wetting front.

soil by using these gradients if hydraulic conductivity or diffusivity curves were available. Independent methods are available for measurement of such hydraulic conductivities or diffusivities (Gardner, 1956); however, for present purposes they are computed by step integration of the water-content distribution curves to obtain flux directly (Figure 9-8) and substitution in the equations for flow to give the following two functions:

$$k\lambda = -\frac{F(\Theta)}{dH/dx} \qquad (9\text{-}39)$$

$$D(\Theta) = -\frac{F(\Theta)}{d\Theta/dx} \qquad (9\text{-}40)$$

Curves for $k\lambda$ as a function of position in the soil column for Salkum silty clay loam are shown in Figure 9-6 and for $k\lambda$ and $D(\Theta)$ as a function

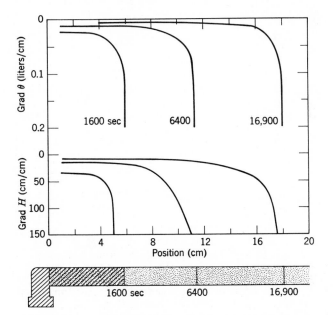

FIG. 9-7. Water content gradient and hydraulic potential gradient distributions for infiltrating water at three different positions of the wetting front as inferred from data given in Figure 9-6 for the same soil column.

of water content in Figure 9-9. Ideally, it is expected that these curves would be independent of time (or position of a wetting-front). However, multiple curves often are obtained in practice, which of course complicates the application of flow theory to practical flow problems. The reason for such multiple hydraulic conductivity and diffusivity curves is not precisely known. Nevertheless, a major assumption made in writing the flux equation is that flow is sufficiently slow that steady-state conditions dominate. Since flow of water into dry soil is not steady state, neglect of an inertial term may well be a part of the explanation.

The nature of water flux in a wetting soil column has been shown in the preceding paragraphs. It is possible to obtain infiltration rates and the advance of a wetting-front as a function of time (Figure 9-10) from the same data. Both the infiltration rate, or the flux at the water source, and the rate of advance of a wetting-front slow down markedly with

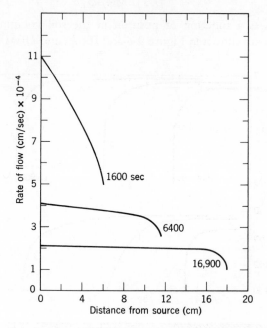

Fɪɢ. 9-8. Rate of water flow as a function of position in the soil column at the different times corresponding to the same positions of the wetting fronts as shown in Figures 9-6 and 9-7.

time. For time intervals of a few hours and for uniform materials, the equations are of the form

$$Q = At^B \tag{9-41}$$

$$S = Et^F \tag{9-42}$$

where B and F are close to but not always $\frac{1}{2}$. For flow longer than a few hours, particularly when flow is vertical so that gravitation is a factor, the infiltration equation has additional terms, as shown by Philip (1957). Philip, using a diffusion-type flux equation, has derived an equation for infiltration which is a time series, but where two terms usually are sufficient:

$$Q = St^{\frac{1}{2}} + At \tag{9-43}$$

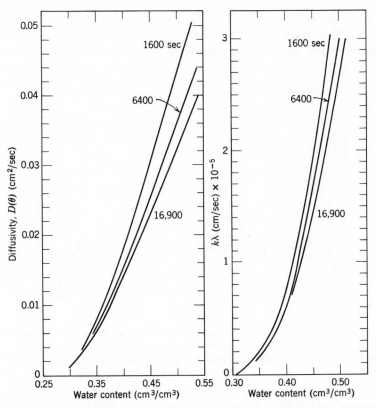

FIG. 9-9. Diffusivity, $D(\theta)$, and hydraulic conductivity, $k\lambda$, as a function of water content for the Salkum silty clay loam. The hydraulic conductivity curve for 1600 sec is the one shown in Figure 9-5.

He refers to the coefficient S as the sorptivity. Equations 9-41 and 9-42 may be written logarithmically with considerable advantage in working with data:

$$\log Q = \log A + B \log t \tag{9-44}$$

$$\log S = \log E + F \log t \tag{9-45}$$

Equation 9-41 produces a useful infiltration rate equation when differentiated against time, which also is convenient to use in logarithmic form:

$$\text{rate} = \frac{dQ}{dt} = ABt^{B-1} \tag{9-46}$$

$$\log \text{rate} = \log \frac{dQ}{dt} = \log (AB) + (B - 1) \log t \tag{9-47}$$

If hydraulic conductivity or diffusivity data are available, it is possible, with the aid of the equation of continuity, to write differential equations in water content, position, and time necessary for obtaining water content

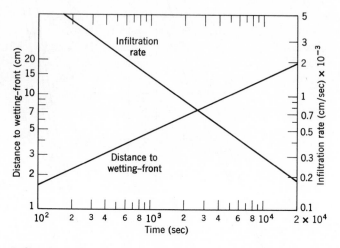

FIG. 9-10. Infiltration rate and distance from the water source to the wetting front plotted logarithmically as a function of time for infiltration into Salkum silty clay loam and for which data are given in the five previous figures.

distribution analytically. For a simple case of one-dimensional flow the equation of continuity (9-15a) becomes

$$\frac{\partial \Theta}{\partial t} = \frac{\partial v}{\partial z} \qquad (9\text{-}48)$$

Putting equation 9-12b for vertical flux into this equation yields the differential equation

$$\frac{\partial \Theta}{\partial t} = \frac{\partial}{\partial z}\left(k\lambda \frac{\partial \phi}{\partial z}\right) + \frac{\partial(k\lambda)}{\partial z}$$

$$= \frac{\partial(k\lambda)}{\partial z}\frac{\partial \phi}{\partial z} + \frac{\partial^2 \phi}{\partial z^2} + \frac{\partial(k\lambda)}{\partial z} \qquad (9\text{-}49)$$

Numerical methods usually are employed to solve this equation to give water content distribution as a function of position and time. If diffusion mathematics are used, the flux given by equation 9-14b is put into equation 9-48 to give

$$\frac{\partial \Theta}{\partial t} = \frac{\partial}{\partial z}\left[D(\Theta) \frac{\partial \Theta}{\partial z} \right] + \frac{\partial(k\lambda)}{\partial z} \qquad (9\text{-}50)$$

When flow is horizontal or when the gravitational gradient is negligible in comparison with the matric gradient the final term on the right-hand side of equation 9-50 can be omitted. For certain boundary conditions it then is possible to reduce the flow equation to an ordinary differential equation by use of a transformation due to Boltzmann and often used in heat flow theory. This transformation eliminates z and t through the introduction of a new variable,

$$\eta(\Theta) = \frac{z}{t^{1/2}} \qquad (9\text{-}51)$$

to give

$$-\frac{1}{2}\frac{d\Theta}{d\eta} = \frac{d}{d\eta}\left[D(\Theta) \frac{d\Theta}{d\eta} \right] \qquad (9\text{-}52)$$

where total derivatives replace the partials. Solutions to equation 9-52 for various boundary conditions are available in heat flow literature (Crank, 1956) after which z and t may be reintroduced. The Boltzmann transformation (equation 9-51) imposes the condition that the locus of points z at constant Θ in a flow system must be a linear function of $t^{1/2}$, which is not always found to be experimentally true. However, entirely reasonable profile water-content distributions for many situations are obtained through its use.

Flow in Stratified Soil Systems

Flow and distribution of water in stratified soil are difficult to analyze because of complicated boundary conditions. The effect of hydraulic conductivity upon flow is the greatest in such soils. Even though steep hydraulic potential gradients often exist, flow can be nearly zero under conditions where large and nearly empty pores with small hydraulic conductivities dominate. Such a condition exists where a moving wetting-front encounters a layer of coarse sand or gravel. The hydraulic potential in soil near the wetting-front may be of the order of -100 cm of water and that in the sand below, say, -10^3 or -10^4 cm; the potential gradient would be large. Despite this, flow is nearly zero because so few contacts between sand grains exist and the cross section for liquid flow is so small. Before any appreciable flow can occur, the hydraulic potential in the finer soil must rise to a value near zero at which time some of the pores or channels in the sand will begin to fill to create an appreciable hydraulic conductivity. This is illustrated in Figure 9-11 where a layer of coarse sand in a silt loam soil restricts downward penetration of water. Coarse materials, such as straw or other organic matter, or holes in the

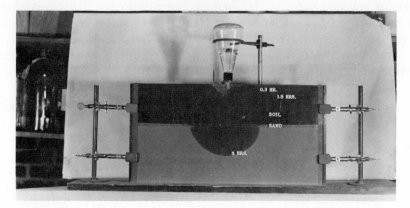

Fig. 9-11. Water retention in soil above a sand layer. (Only after water potential in soil above rises to nearly zero potential does water begin to penetrate the layer, after which movement into soil below no longer is restricted.)

soil created by burrowing insects and animals, restrict rather than aid flow as long as the hydraulic potentials surrounding them are too low for them to fill with water. Dry soil often persists through the wet season beneath straw turned under by plowing. Only when large pores and channels connect with the surface where free water can get to them, or are beneath the water table, do such channels contribute appreciably to liquid flow.

Fine pores in hard pans and clay pans also seriously restrict flow. Such materials wet up rapidly for short distances when first contacted by water because of the high absorptive capacity of fine pores. However, as the distance through which water must move in fine pores increases, the rate of flow decreases; in tight clay it becomes extremely slow. Flow in such materials often is so slow that water tables build up above them. Rapid initial wetting followed by gross restriction of flow is illustrated in Figure 9-12.

Water retention following wetting, or redistribution of water, is greatly affected by stratification. Claypans and hard pans often create serious waterlogging because retention is so pronounced above such layers. Coarse

FIG. 9-12. Water retention in soil above a clay layer. (Although water enters the fine pores of the clay layer easily, the low permeability of the clay prevents rapid movement through. Water tables can be built up over such layers.)

layers act much the same as a check valve. Water tables cannot be maintained above a coarse layer; however, since a coarse layer restricts flow at relatively high hydraulic potentials, retention of water above such layers often is appreciably more than would be the case in the absence of such a layer. In a particular situation where a coarse gravel layer exists at a depth of about 45 cm in a fine sandy loam, water flow ceases at about —10-cm hydraulic potential. At least three times as much water is retained and available for plant use in the soil above this layer as would be retained if the gravel layer were not present.

Vapor Flow in Soil

Vapor flow in soil takes place continuously whenever a vapor path and vapor pressure gradients exist. Vapor flow usually is neglected under conditions where liquid flow is appreciable. However, a point eventually is reached as liquid flow decreases where vapor flow dominates and may have considerable practical importance under some conditions. Vapor flow of a practical magnitude often occurs where sizable temperature gradients exist in soil.

Equations for Vapor and Liquid Flow

Vapor flow is a diffusion process; it depends upon the size and tortuosity of the diffusion path and upon the gradient of vapor concentration or vapor pressure. No completely satisfactory theory exists to describe vapor flow or vapor-liquid flow in soil systems. The simple diffusion equation, which says that flux is proportional to the gradient of the concentration, is inadequate in describing flow of vapor in soil. However, by special treatment of the diffusivity term Jackson (1964a, 1964b, 1964c; 1965) has expressed the flux of liquid and vapor as

$$q = -D_{\Theta_v} \frac{\partial \Theta}{\partial x} \qquad (9\text{-}53)$$

where $D_{\Theta v}$ (Jackson, 1965) is the sum of two temperature-dependent terms:

$$D_{\Theta_v} = D_{\Theta}^{\circ} \exp\left[-E_{\Theta} \frac{(1/T - 1/T_0)}{R} \right] + D_{\text{vap}}^{\circ} \frac{P_0}{P} \exp\left[-E_v \frac{(1/T - 1/T_0)}{R} \right]$$
$$(9\text{-}54)$$

where D° is the diffusion coefficient at the reference temperature T_0, T is the absolute temperature, E_{Θ} and E_v are the activation energies for liquid diffusion and for vapor diffusion, P_0 and P are the ambient pressure and reference pressure, and R is the ideal gas content. It has been possible to separate out liquid and vapor flow with this equation.

Vapor concentration gradients may exist in soil because of two different phenomena. As indicated in Chapter 8, vapor pressure and the matric plus osmotic potentials are so related that water content and vapor pressure are also related, although not uniquely so. Thus at constant temperature there is vapor-pressure gradient if there is a water-potential gradient. The second, and often the more important gradient, is that which accompanies temperature gradients. Gradients due to gravitation are sometimes important but are neglected in the present discussion. Philip and deVries (1957) have recognized both moisture- and temperature-induced gradients in their equation for water flux, J_w,

$$J_w = -D_T \, \text{grad} \, T - D_{\Theta} \, \text{grad} \, \Theta \qquad (9\text{-}55)$$

where the diffusivities, D_T and D_w, are defined as

$$D_T = D_{T(\text{liq})} + D_{T(\text{vap})} \qquad (9\text{-}56)$$
$$D_{\Theta} = D_{\Theta(\text{liq})} + D_{\Theta(\text{vap})} \qquad (9\text{-}57)$$

Liquid flow dominates at high water contents and the liquid diffusivities in equations 9-56 and 9-57 dominate, although $D_{T(\text{liq})}$ often is small. Vapor flow dominates at low water contents, particularly where temperature gradients are large, and the diffusivities for vapor in equations 9-56 and 9-57 are the most important. In the application of equations 9-55, 9-56, and 9-57 to water flux where vapor flow is an important part, vapor diffusivities are larger than can be calculated from ordinary diffusion theory. It is necessary to consider, as pointed out by Philip and deVries (1957), that what is described as vapor flow actually may be a combination of vapor condensation at one end and rapid liquid flow followed by evaporation at the other end of a liquid channel. The net liquid-vapor step flow is more rapid than if the flow were entirely in the vapor state.

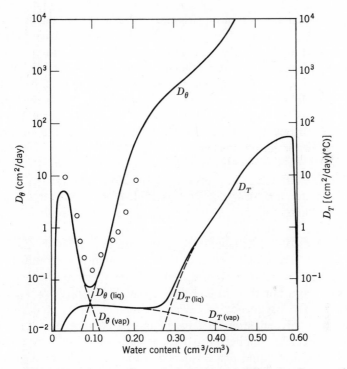

FIG. 9-13. Isothermal diffusivity D_θ and nonisothermal diffusivity D_T as a function of water content. Dashed lines represent the vapor and liquid component diffusivities. Open circles represent D_θ values as measured by an evaporation experiment. (Gee, 1966.)

Gee (1966), using the equations of Philip and deVries on a silt loam soil in a closed column where nondestructive neutron water-content measurements were made, has obtained values for the component diffusivities given in equations 9-56 and 9-57. These are shown in Figure 9-13.

Another approach to water flow under the influence of temperature gradients involves the theory of nonequilibrium thermodynamics. Taylor and Cary (1964) have expressed water flow, both liquid and vapor, as the sum of two terms. Each consists of a driving force expressed as a gradient and a phenomenological coefficient:

$$
\begin{aligned}
J_w &= -L_{ww} \frac{d\Phi}{dx} - L_{wq} \frac{d}{dx} (\ln T) \\
&= -L_{ww} \frac{d\Phi}{dx} - \frac{L_{wq}}{T} \frac{dT}{dx}
\end{aligned}
\tag{9-58}
$$

However, when two or more driving forces occur in the same system simultaneously, they interfere with each other so that strict proportionality between flux and the individual driving forces no longer exists. Thus determination of the phenomenological coefficients is complicated. For steady flow at high hydraulic potential, Cary (1965) has separated out the contributions to liquid and to vapor flow by using this approach. A major utility of this approach to flow is that the general idea applies to flux of various components, such as heat and soluble materials, as well as to water and water vapor. Relationships among coefficients in various flows given in the theory facilitate their evaluation.

Despite the difficulties in expressing vapor flow analytically, the contribution to water flux of vapor flow is evident in a number of situations in nature. One of the most striking and important is circulation of water in a soil that is subjected to a temperature gradient. Gurr, Marshall, and Hutton (1952), in a laboratory demonstration, established a temperature gradient across a soil column that was initially at a uniform water content and contained a chloride salt. Water flow in the vapor phase took place from hot to cold and flow in the liquid phase was from cold to hot. As a consequence, the water content was appreciably higher toward the cold end after 18 days. The salt content near the hot end became appreciably higher than that at the cold end because salt was transported in the liquid phase but not in the vapor phase (Figure 9-14). A similar phenomenon often occurs in nature in arid irrigated areas. Liquid flux to the surface carries soluble salts where they are deposited as water evaporates near or at the surface. Evaporation during the daytime when the surface is warm removes water to the atmosphere and also causes vapor flow downward into the soil where it condenses. This increases the water content gradient toward the surface, which promotes return liquid flow. Day-night temperature reversals contribute to the process and the net effect is an accumulation of salt. Sometimes powdery or crusted salt is built up in layers several centimeters thick on the tops of irrigation furrows where, in addition to the foregoing processes, irrigation in the furrow and capillary rise of water in the hill further adds to salt accumulation in the surface.

The solar still (Jackson and van Bavel, 1965) provides additional evidence of the extent of vapor flow in the presence of temperature gradients. A solar still is made by digging a bowl-shaped hole about 1 meter in diameter and ½ meter deep. The hole is held in place by a plastic sheet with soil around its edges. A rock is placed in the middle to form an inverted cone and a container for catching water from the drip point is placed beneath at the apex of the cone. Vapor flow from the soil

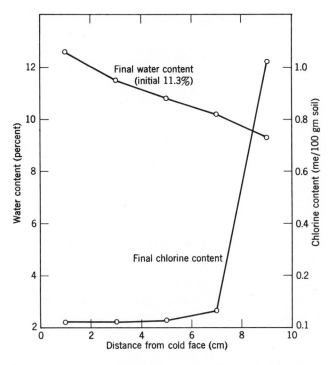

Fig. 9-14. Water and chloride distribution in a soil column warmed at one end after a period of 18 days. Water and chloride content were uniform at the beginning. (Gurr et al., 1952.)

which is warmed by the sun condenses on the bottom of the plastic and drips from the cone. Succulent vegetation placed at the bottom of the hole will increase the water yield. From ½ liter to several liters of water can be produced daily from appropriate sites in desert areas.

Whether or not vapor flow associated with day-night temperature cycles is of practical importance to plant growth depends upon local conditions. However, there is little doubt that such flow often has considerable practical importance, particularly in supplying small amounts of water needed for germination of seeds in dry-land areas.

References

Buckingham, E. (1907). Studies on the movement of soil moisture. *U.S. D. A. Bur. Soils, Bull.* 38.

Cary, J. W. (1965). Water flux in moist soil: Thermal versus suction gradients. *Soil Sci.*, **100**:168–175.

Childs, E. C. (1969). *The Physical Basis of Soil Water Phenomena.* Wiley-Interscience, New York.

Crank, J. (1956). *The Mathematics of Diffusion.* Clarendon Press, Oxford.

Fatt, I., and H. Dykstra (1951). Relative permeability studies. *Petrol. Trans. AIME*, **192**:249–255.

Gardner, W. R. (1956). Calculation of capillary conductivity from pressure plate outflow data. *Soil Sci. Soc. Am. Proc.*, **20**:317–320.

Gee, Glendon W. (1966). Water movement in soils as influenced by temperature gradients. Ph.D. Thesis, Washington State Univ.; Diss. Abst. No. 1671B (Vol. 27).

Gurr, C. G., T. J. Marshall, and J. T. Hutton (1952). Movement of water in soil due to a temperature gradient. *Soil Sci.*, **74**:335–345.

Jackson, R. D. (1964). Water vapor diffusion in relatively dry soil: I. Theoretical considerations and sorption experiments. *Soil Sci. Soc. Am. Proc.*, **28**:172–176.

Jackson, R. D. (1964). Water vapor diffusion in relatively dry soil: II. Desorption experiments. *Soil Sci. Soc. Am. Proc.*, **28**:464–466.

Jackson, R. D. (1964). Water vapor diffusion in relatively dry soil: III. Steady-state experiments. *Soil Sci. Soc. Am. Proc.*, **28**:466–470.

Jackson, R. D. (1965). Water vapor diffusion in relatively dry soil: IV. Temperature effects on sorption diffusion coefficients. *Soil Sci. Soc. Am. Proc.*, **29**:144–148.

Jackson, R. D., and C. H. M. van Bavel (1965). A solar still . . . for survival. In anonymous article *Agr. Res.*, October, 1965.

Kirkham, Don (1945). Proposed method for field measurement of permeability of soil below the water table. *Soil Sci. Soc. Am. Proc.*, **10**:58–68.

Kirkham, Don, and W. L. Powers (1971). *Advanced Soil Physics,* Wiley-Interscience, New York.

Luthin, J. N., and Don Kirkham (1949). A piezometer method for measuring permeability of soil *in situ* below a water table. *Soil Sci.*, **68**:349–358.

Morel-Seytoux, H. J. (1969). Introduction to flow of immiscible liquids in porous media, in *Flow Through Porous Media*, J. M. DeWiest, Ed. Academic Press, New York, Chapter 11.

Philip, J. R. (1957). The theory of infiltration: 4. Sorptivity and algebraic infiltration equations. *Soil Sci.*, **84**:257–264.

Philip, J. R., and D. A. deVries (1957). Moisture movement in porous materials under temperature gradients. *Trans. Am. Geophys. Union*, **38**:222–232.

Rawlins, Stephen L., and Walter H. Gardner (1963). A test of the validity of the diffusion equation for unsaturated flow of soil water. *Soil Sci. Soc. Am. Proc.*, **27**:507–511.

Stewart, Gordon L. (1962). Water content measurement by neutron attenuation and applications to unsaturated flow of water in soil. Ph.D. Thesis, Washington State Univ.; Diss. Abst. No. 63-3034.

Taylor, S. A., and J. W. Cary (1964). Linear equations for the simultaneous flow of matter and energy in a continuous soil system. *Soil Sci. Soc. Am. Proc.*, **28**:167–172.

CHAPTER 10

Soil Water— The Field Moisture Regime

The soil profile in the field is a very dynamic and complex system. The vitality of this system is due in no small part to its place in the hydrologic cycle. A schematic diagram of the hydrologic cycle is shown in Figure 10-1. It is the purpose of this chapter to consider that part of the cycle which is modulated and regulated by the soil profile through its effect upon the fluxes of water into and through the soil.

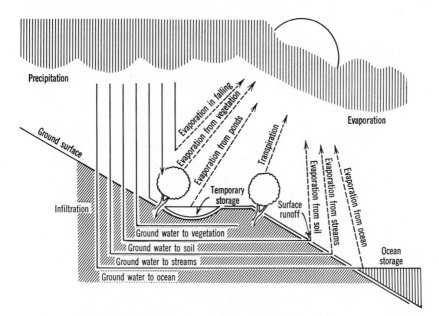

Precipitation

Evaporation in falling

Evaporation from vegetation

Evaporation from ponds

Transpiration

Evaporation

Ground surface

Infiltration

Temporary storage

Surface runoff

Evaporation from soil

Evaporation from streams

Evaporation from ocean

Ground water to vegetation

Ground water to soil

Ground water to streams

Ground water to ocean

Ocean storage

FIG. 10-1. The hydrologic cycle.

353

THE FIELD WATER BALANCE

It is useful to consider first the water balance of the entire soil profile in terms of individual processes which may be conveniently defined and characterized. For the moment the depth of the profile is left unspecified.

$$P + I - R = E + D + \Delta W \qquad (10\text{-}1)$$

The terms on the left-hand side of equation 10-1 represent, respectively, the precipitation (including dew and frost), applied irrigation water, and surface runoff. The sum of these is the net addition of water to the soil profile in some yet to be specified time period. The precipitation is a climatological factor and discussion of its probable frequency and amount will not be considered here. Irrigation will be discussed in some detail in Chapter 11 and runoff is taken up in Chapter 12. On the right-hand side of equation 10-1 are evaporation (which is taken to include plant transpiration unless otherwise specified), drainage or deep percolation, and, finally, the change in the water content of the soil profile.

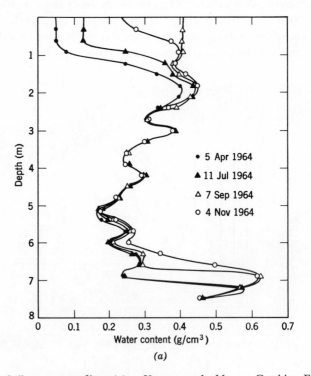

Fig. 10-2. Soil water profiles. (a) = Young sand, Mount Gambier Forest; (b) Kalangadoo sand, Penola Forest. (Holmes and Colville, 1970.)

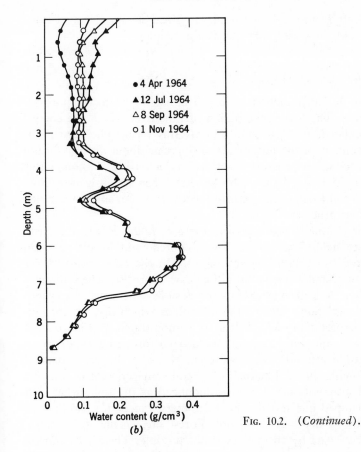

FIG. 10.2. (*Continued*).

The last two quantities may be either positive or negative. All other quantities in the equation are positive. A negative drainage would occur in the case of capillary rise of water from below; a situation that is not uncommon in the presence of a shallow water table. The several components of the water balance equation are of considerable importance to others than soil scientists. Hydrologists must know how much of the precipitation will result in direct runoff and in deep percolation to ground water. Meteorologists are very concerned with the evaporation component since it plays a significant role in the energy budget of the earth.

In this chapter we consider each of the terms of the water balance equation, with the exceptions previously noted, and discuss them quantitatively on the basis of the equations that describe water movement in soil. Many of the conclusions presented will be semiquantitative and preliminary in nature since it is only recently that analysis of field data has begun to shift from a reliance upon empirical statistical correlations to consideration of the underlying physical principles.

Steady-State and Equilibrium Cases

The last term in Equation 10-1 is limited in magnitude to the porosity of the soil profile. The longer the period of time over which the energy balance is calculated, the less significant is this term. The actual soil profile water content in the field tends to fluctuate about an equilibrium or steady-state condition that is described by $\Delta W = 0$. Although this steady-state condition may be more hypothetical than real, it is instructive to consider it in some detail since it often serves as a convenient reference state for transient processes.

In practice it is usually possible to measure P, I, and R with adequate precision in the field. The profile water content and its changes can also be measured by straightforward means. On the other hand, direct measurements of E and D are often difficult or imprecise. These difficulties can and often do lead to erroneous conclusions concerning the evaporation and drainage. Some idea of the difficulties which may be encountered in the interpretation of field data are demonstrated by Figure 10-2. In this figure, the soil water contents under two forested sites are given as a function of depth for several different times of the year (data of Holmes and Colville, 1970). Fluctuations in soil water content in response to rainfall and evaporation are very pronounced at the soil surface but are virtually damped out at a depth of 4 meters, except for the fluctuations at the 7-meter depth at the Penola Forest site. Unambiguous interpretation of these data by themselves is not possible. The water content measurements present a reasonable picture when they are combined with water potential measurements in the soil profile. Figure 10-3 shows the matric potential distribution with depth for the two sites for the wettest and driest sampling dates. At the Mount Gambier Forest it is obvious that there is very little difference in matric potential between the 1-meter and the 6-meter depths. Hence the hydraulic head gradient is almost identical with the gravitational gradient and the direction of water movement must be downward. There is an upward gradient at the soil surface due to evaporation. If the minimum in matric potential at the 7-meter depth is real (as seems probable), then the higher water potential at the 8-meter depth represents a charge of water which entered the profile prior to and is moving downward ahead of the one represented by the higher potential in the 1- to 6-meter range.

The situation in the Penola Forest is significantly different. The hydraulic head gradient is downward between the surface and the 3-meter

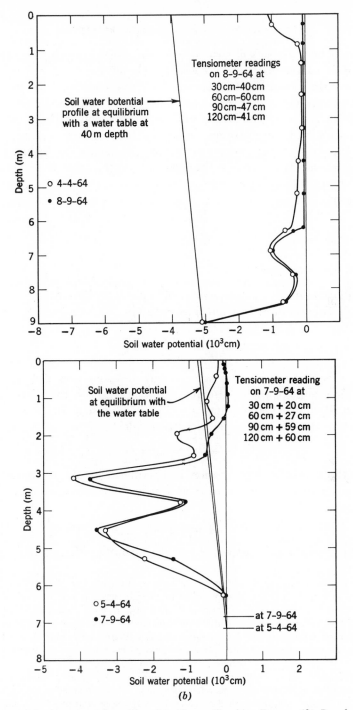

FIG. 10-3. Matric potential profiles. (a) Mount Gambier Forest; (b) Penola Forest. (Holmes and Colville, 1970.)

depth and upward between the 6- and the 5-meter depth. Furthermore, there is a maximum in the potential at about the 4-meter depth. The minima at the 3- and 5-meter depth indicate two successive drying periods with recharge between them. This situation in which water is moving from both above and below cannot continue indefinitely; it obviously represents a transient situation. Whether the long-term average results in net upward or net downward movement cannot be deduced from the data given. At the Mount Gambier site, on the other hand, the situation appears to be more nearly a steady state with a net downward movement.

Water Movement

Downward Movement

Steady-state downward movement of water when flow is in one dimension only follows directly from Darcy's law:

$$q = -k\left(\frac{d\phi}{dz} + 1\right) \qquad (10\text{-}2)$$

where q is the volume of water flowing per unit area and unit time in the direction, k is the unsaturated permeability,* ϕ is the matric potential in head units. At equilibrium $q = 0$ in which case $d\phi/dz = 1$. Integration of equation 10-2 gives the expected result:

$$\phi = -z + \text{constant} \qquad (10\text{-}3)$$

where the constant of integration is determined by the reference level. If a water table exists at depth d, at which point ϕ is set equal to zero, then

$$\phi = -z - d \qquad (10\text{-}4)$$

Since the term depth denotes a positive quantity, the negative sign in

---

* In Chapter 9 the conductivity factor in flux equations for unsaturated flow is given as $k\lambda$ with λ being a dimensionless number between 0 and 1 which depends upon the size and properties of the liquid-filled flow channel, and with k being the conductivity factor for saturated flow. At saturation λ vanishes and the unsaturated flow equation reduces to the Darcy equation for saturated flow. Such a concept has considerable theoretical and some practical value. However, k and λ usually are not separated in practice and a single symbol for unsaturated conductivity has merit. For this reason, and in keeping with contemporary usage, unsaturated conductivity is designated as k in this and subsequent chapters. The context of the discussion or explicit statements reveal whether k is meant to represent saturated or unsaturated conductivity.

front of d is introduced to keep the sign convention consistent. A positive depth is interpreted as a negative elevation.

When $q \neq 0$, it is necessary to know the relation between the permeability k and the matric potential ϕ in order to integrate equation 10-2. We can indicate this integration by

$$z = - \int \frac{d\phi}{(q/k) + 1} \qquad (10\text{-}5)$$

where the limits on the integral are chosen in accordance with the choice of reference for the gravitational potential. Since q may take on positive or negative values, equation 10-5 can be used to represent upward as readily as downward flow by suitable specification of the limits of the integral. Analytical solutions for equation 10-5 are given for a limited number of functional relations for $k(\phi)$ (Gardner, 1958).

Two points concerning equations 10-4 and 10-5 are worthy of emphasis. The first is that the absolute value of ϕ must be less than the depth to an existing water table if the movement is to be in the downward direction. For example, if a water table is within 2 meters of the soil cm. The second point is that as the water-table depth increases the matric surface, the matric potential, on the average, must be higher than -200 potential gradient $d\phi/dz$ becomes increasingly smaller than the gravitational gradient, which remains equal to unity. For a water table sufficiently deep, the total potential gradient may be set equal to the gravitational gradient throughout much of the soil profile and we obtain the simple result

$$q = k \qquad (10\text{-}6)$$

The flux equals the unsaturated conductivity for a sufficiently deep water table. The soil water content adjusts to that value necessary to give the appropriate conductivity, the flux is independent of the water-table depth. This is the situation under the Mount Gambier Forest as shown in Figure 10-2. It is sometimes erroneously assumed that there is no water movement if the water content of the soil is not changing. However, this is generally not the case, particularly below the plant root zone where steady-state flow is more nearly the rule than the exception.

For equation 10-2 to be applicable it is necessary that the flux density q be less than the saturated permeability of the soil. Otherwise, one has saturated flow. It is possible to apply equation 10-2 to layered soils if the $k - \phi$ relation is known for each layer and if the integration is performed layer by layer by matching the potentials at each boundary. It may sometimes be the case that the conductivity of a given layer is less than q even though other layers have a higher permeability. This results in a perched water table and a positive pressure in the soil water just above the impeding layer. The head gradient across the impeding layer will then be greater than unity and that in the layer above will be less than unity. However, the higher water content tends to offset the reduction in head gradient so that the effect of an impeding layer may not be appreciable over more than a short distance above it, unless it is substantially lower in permeability. For an example of the layered case see Srinilta, Nielsen, and Kirkham (1969).

A situation that occurs frequently in the field is one in which the surface layer of the soil is less permeable than the subsurface horizons. In this case, if a thin layer of water is ponded on the soil surface, the flux (steady state) through the soil is less than the saturated permeability of the lower horizons but greater than the permeability of the surface crust. This is due to the matric potential in the lower horizons, which tends to pull the water into the soil by virtue of an increased potential gradient or hydraulic head gradient. If the relation between the permeability and the matric potential is known, it is possible to calculate the steady-state flux into the soil for a given surface crust conductance. An analysis of this problem is given by Hillel and Gardner (1969).

This same situation frequently occurs beneath ponds and canals where there may exist an unsaturated region between the pond or canal bottom and a saturated region below the water table at some depth with steady percolation of water to the water table. Such a situation is illustrated in Figure 10-4, which is taken from a paper by Jeppson and Nelson (1970). The concept sometimes expressed by hydrologists that such a situation represents a lack of hydraulic continuity is at the very least misleading since the seepage rate is as great or greater than if the water table were shallow enough to result in complete saturation between.

Upward Movement or Capillary Rise

Equations 10-2 and 10-5 apply to upward as well as downward steady-state movement of water, as has already been pointed out. If the total potential at the soil surface ($z = 0$) is less than the potential at the water table ($z = d$, $\phi = -d$), the hydraulic gradient is upward and this will be the direction of water movement. A major difference exhibited by the upward flow regime is that the matric potential gradient is opposite in sign to the gravitational gradient and must exceed it in absolute value in order for there to be upward movement. This requires that the matric potential become increasingly negative with increasing height above the water table. A more negative matric potential means a lower water content and a lower capillary or unsaturated permeability. It so happens

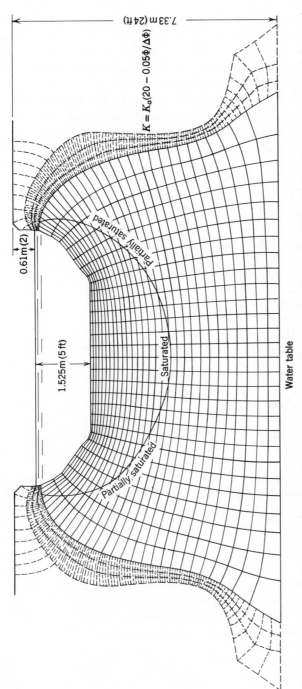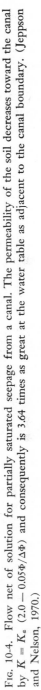

FIG. 10-4. Flow net of solution for partially saturated seepage from a canal. The permeability of the soil decreases toward the canal by $K = K_a (2.0 - 0.05\Phi/\Delta\Phi)$ and consequently is 3.64 times as great at the water table as adjacent to the canal boundary. (Jeppson and Nelson, 1970.)

that the reduced permeability offsets any increase in gradient brought about by reducing the matric potential at the soil surface. The result is a very steep water potential and water content gradient at the soil surface and a limit on the rate at which water can move upward from a water table under steady-state conditions.

Figure 10-5 shows the matric potential plotted as a function of height above a water table for Yolo light clay (Gardner and Fireman, 1958). The evaporation rate during the experiment is reported by Moore (1939).

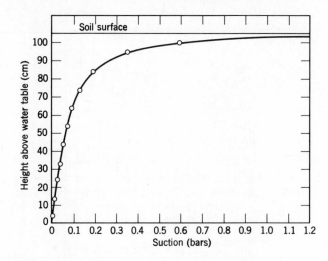

FIG. 10-5. Soil suction (negative matric potential) as a function of height above a water table for Yolo light clay. (Gardner and Fireman, 1958.)

The evaporation rate during the experiment was 0.08 cm/day and the unsaturated permeability-matric potential relation is described reasonably well by the empirical relation

$$k = \frac{1}{((\phi/20)^2 + 1)}$$

where the matric potential ϕ is given in centimeters.

The maximum flux that can be transmitted upward from a water table by a given soil can be calculated readily if the relation between the permeability and the matric potential is of the form

$$k = \frac{K}{(\phi/\phi_0)^n + 1} \tag{10-7}$$

where n and ϕ_0 are constants with $n > 1$. The result is of the form

$$z = \frac{\phi_0}{n \sin \pi/n} \frac{K}{q_m} \left(\frac{K}{q_m} + 1\right)^{(1-n)/n}$$ (10-8)

where q_m is the maximum flux possible. When $q_m \ll K$ equation 10-8 can be simplified and solved explicitly for q_m:

$$q_m = K \left(\frac{\phi_0 n}{\pi \phi_0} \sin \frac{\pi z}{n}\right)^{-n}$$ (10-9)

Since $n > 6$ for coarse-textured soils, this reduces to a good degree of approximation in the very simple formula

$$q_m = K \left(\frac{\phi_0}{z}\right)^n$$ (10-10)

Equation 10-9 gives the correct relative effect of z upon q_m even for small values of n, as long as n is greater than one; the absolute value, however, may be in error by 10 percent or more.

It should be pointed out that equation 10-2 and the solutions which follow for both downward and upward steady-state water flux in the soil neglect any effect of thermal gradients. These effects may be very important at or near the soil surface, as will be discussed later. The effect of temperature at greater depths tends to be small relative to isothermal forces. Since the thermal effects are most pronounced in the vapor phase, the soil must be sufficiently dry to permit substantial vapor movement; as has been pointed out, this does not require continuity of the vapor phase. Philip's calculations (1957a) show that for a light clay with a water table below 2 meters a downward heat flux inhibits evaporation more than an upward flux of the same magnitude increases the flux. In the case considered, this inhibition amounts to a factor of two for the water table below 5 meters for a downward heat flux density of 0.002 cal/(cm²)(sec). The upward flux in the absence of the thermal gradient is less than 0.01 cm/day; thus the absolute magnitude of the effect is relatively small.

Transient Processes in the Field

True equilibrium exists fleetingly, if at all, in the field. The steady state is also the exception rather than the rule. The complexity of the field moisture regime is so great as to render it almost incapable of exact quantitative description. As in all branches of science, certain processes are simplified and idealized in order to obtain their approximate analytical or numerical characterization. Since linear combinations of solutions of the equation describing the flow of water in soil are also solutions of the equation, it is the frequent practice to combine solutions for simple cases in order to describe the more complex situations.

Two types of mathematical expression are found in the literature that describe the movement of water in soils. One is derived empirically from experimental observations in the field with only minor concern for the physical principles involved. The other is derived from flow equations and gives insight into the manner in which various soil physical properties influence water movement. It is difficult at the present time to apply the more rigorous mathematical expressions to large, heterogeneous areas. Emphasis here is on the more rigorous approach, but some of the more useful empirical expressions are also considered.

The Response Time of a Soil Profile

Before treating the individual transient flow processes in specific detail, it is worthwhile to consider briefly some general conclusions that can be deduced from the flow equation. It can be shown for a linear diffusion system that the characteristic time for propagation of a disturbance is related to the characteristic length and the diffusivity:

$$t = \frac{L^2}{D} \tag{10-11}$$

In a nonlinear system, such as that represented by the unsaturated flow equation, some average value for the diffusivity must be used. This must be a weighted average for many purposes since water movement in the region of highest diffusivity often predominates. A convenient weighted average diffusivity is given by the expression

$$\bar{D} = \int_{\Theta_{min}}^{\Theta_{max}} \Theta^2 D(\Theta) \, d\Theta \tag{10-12}$$

where the integration is performed over the range of water contents of interest.

A more nearly exact expression for equation 10-12 is discussed by Crank (1956). A plot of the weighted diffusivity for an exponential relation between diffusivity and water content is given by Gardner (1959). A reasonable average diffusivity for the drying of a soil might be 10 cm²/day. For a diurnal drying cycle $t = 1$ day and $L = 10^{1/2} = 3.16$ cm; this represents the limiting depth to which we would expect to see daily fluctuations in soil water content. It can be shown for periodic processes which repeat themselves with a frequency of n cycles per day that the amplitude of any disturbance that is applied at the soil surface will diminish with depth according to the exponential relation

$$A = A_0 \exp - x \left(\frac{\pi n}{D}\right)^{1/2} \qquad (10\text{-}13)$$

where A is the amplitude of the variation in water content (or water potential) at depth x and A_0 is the amplitude at the soil surface. Equation 10-13 is of questionable value near the soil surface where the diffusivity may vary over a wide range; but the diffusivity becomes more nearly constant as the amplitude damps out with depth and the degree of approximation becomes quite good.

By the way of example, consider the hypothetical case in which it rains every 10 days and where the average diffusivity for both the infiltration and drying processes might be of the order of 100^2 cm^2/day. The depth of penetration of this impulse will be about 25 cm. For the same diffusivity over a period of 365 days, which might represent an alternating rainy and dry season, the depth of penetration will be about 150 cm. The time required always increases as the square of the distance involved. It can be seen therefore that high frequency events will tend to merge with depth so that individual rainfall events become indistinguishable well below the root zone and only annual events or those with even lower frequency are identifiable.

The value of this sort of approximate analysis lies in part in its indication of whether it is reasonable to consider that a given system is represented by a steady state or whether one must consider in detail the dynamics of the transient process.

INFILTRATION

Infiltration, or the downward entry of water into soil, is one of the more important processes in the soil phase of the hydrological cycle. Inasmuch as water entry into soil is caused by matric as well as gravitational forces, this entry may occur in the lateral and upward directions as well as downward. Infiltration usually refers to the downward movement and this aspect is emphasized in this section. Two- and three-dimensional movement are considered briefly since they are also of interest in the field. The matric forces usually predominate over the gravitational force during the early stages of water entry into soil so that many conclusions that are reached concerning the early stages of infiltration are valid in the absence of gravity.

The mathematical treatment of infiltration that is based upon solution of the flow equation has been reviewed in detail by Philip (1969). There are many infiltration equations that have originated in the analysis of field data. The more exact solution of the equation gives greater insight into the physics of infiltration but, as of this writing, the empirical equations still have considerable currency because they contain parameters that can be adjusted to account for complexities which have been eliminated in the mathematical analysis to render the problem soluble.

The significance of the infiltration process in the hydrological cycle is readily apparent. It is this process that divides precipitation or other surface water between overland flow and subsurface flow. It is this subsurface flow which is of particular interest in soil physics. It has long been recognized that the ability of a soil to absorb water by infiltration is not constant; it is very significant that the infiltration rate tends to decrease with time during an infiltration process. The approach in the development of field infiltration equations has usually been to attempt first to drive a mathematical expression for this decreasing infiltration rate and then to attempt a physical explanation of the process. The basic mathematical approach starts with physics and shows that the decreasing rate is an inevitable consequence of the decreasing matric potential gradient that is associated with the ever deeper penetration of the wetting front during infiltration.

Flow Equations

The Kostiakov Equation

Kostiakov (1932) proposed the equation

$$I = Kt^a \tag{10-14}$$

where I is the cumulative infiltration after time t and K and a are constants. K is not to be identified with the hydraulic conductivity of the soil other than indirectly. The parameters in equation 10-14 have no particular physical meaning and are evaluated from experimental data.

The Horton Equation

Horton (1933, 1939) devoted a great deal of attention to the investigation of infiltration and developed an equation which he felt was consistent with his physical concept of the process. The infiltration rate is given in Horton's model by the equation

$$i = i_f + (i_0 - i_f) \exp(-\beta t) \tag{10-15}$$

which, for purposes of comparison with equation 10-14, integrates to give a cumulative infiltration

$$I = i_f t + \frac{(i_0 - i_f)}{\beta} [1 - \exp(-\beta t)] \qquad (10\text{-}16)$$

In equations 10-14 and 10-15, i_0 is the initial infiltration rate at $t = 0$, i_f is the final constant infiltration rate which is achieved at large times; β is a soil parameter which describes the rate of decrease of infiltration.

Horton (1940) felt that the reduction in infiltration rate with time after the initiation of infiltration was largely controlled by factors operating at the soil surface. They included swelling of soil colloids and the closing of small cracks which progressively sealed the soil surface. Compaction of the soil surface by raindrop action was also considered important where it was not mitigated by crop cover. Horton's field data, similar to those of many other workers, indicated a decreasing infiltration rate for 2 or 3 hours after the initiation of storm runoff. The infiltration rate eventually approached a constant value which was often somewhat smaller than the saturated permeability of the soil. Air entrapment and incomplete saturation of the soil were assumed to be responsible for this latter finding. Horton used an exponential function to describe the decreasing infiltration rate since it fit the data reasonably well; many "exhaustion" processes in nature can be described by this mathematical expression.

The Green and Ampt Equation

Green and Ampt (1911) earlier derived an infiltration equation that was based upon a very simple physical model of the soil. It has the advantage that the parameters in the equation can be related to physical properties of the soil (Philip, 1954). Their approach is worth further study by those interested in other flow processes since it lends itself to the solution of a number of other problems. Physically, Green and Ampt assumed that the soil was saturated back of the wetting front and that one could define some "effective" matric potential at the wetting front. These assumptions, when combined with Darcy's law, lead to the expression

$$L - \phi_L \ln\left(\frac{1 + L}{\phi_L}\right) = \frac{Kt}{f} \qquad (10\text{-}17)$$

where L is the depth to the wetting front, ϕ_L is the potential (matric) at the wetting front, K is the permeability of the soil, f is the soil porosity, and t is the time. The cumulative infiltration is just

$$I = fL \tag{10-18}$$

which, unfortunately, cannot be solved explicitly from Equation 10-17. However, the instantaneous infiltration rate $I = dI/dt$ can be expressed as a function of the depth of wetting, which is a very convenient circumstance from the standpoint of mathematical modeling of the infiltration process for a computer. It is interesting to note that the Horton equation (10-15) can be derived by using the Green and Ampt approach. This was shown by Gardner and Widstoe (1921) but requires that there exist an inverse proportionality between the water potential and the total water content of the soil, as well as a constant permeability.

Exact Solutions of the Flow Equation

Exact solution of the flow equations for infiltration requires numerical procedures too complex for anything other than computer solution. A procedure which lends itself to convenient computation for infiltration into a homogeneous soil at a uniform initial water content has been developed by Philip (1957b). The solution appears in the form of an infinite series of terms containing powers of $t^{1/2}$:

$$i = St^{1/2} + (A_2 + K_0)t + A_3t^{3/2} + A_4t^2 + \cdots \tag{10-19}$$

where S is a parameter which Philip calls the sorptivity, A_2, A_3, and A_4 are constants characteristic of the soil, and K is the hydraulic conductivity of the soil at the final water content (usually assumed to be saturation, but not always so). The first term in equation 10-19 is exactly the expression one would obtain for horizontal entry of water into soil under the same initial and boundary conditions, but with gravity not a factor. The other terms are a consequence of the gravitational field. Terms beyond the fourth in the infinite series are generally negligably small for computational purposes. These terms become significant as t gets larger and equation 10-19 eventually fails for large values of t. For large t values

$$i = Kt \tag{10-20}$$

It has been erroneously assumed by some workers that the second term in equation 10-19 gives the final infiltration rate at large values of t but it can be seen that this assumption is in error by the value of A_2. Evaluation of the parameters in equation 10-19 requires the aforementioned numerical procedures (details discussed in Philip, 1969).

One conclusion from the analysis by Philip is that the wetting front assumes a constant shape when the infiltration rate approaches a constant value and that this front moves down through the soil with a constant velocity. Figure 10-6 is from an infiltration experiment reported by Davidson et al. (1963). A zone of almost constant water content extends immediately down from the soil surface. This is often referred to as the

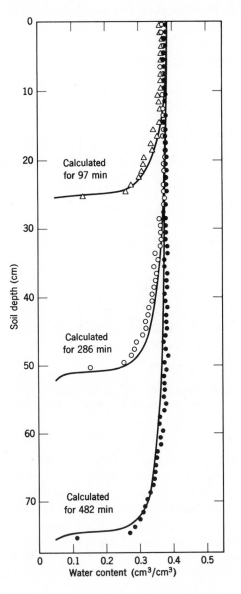

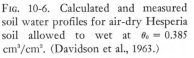

FIG. 10-6. Calculated and measured soil water profiles for air-dry Hesperia soil allowed to wet at $\theta_0 = 0.385$ cm³/cm³. (Davidson et al., 1963.)

"transmission zone." The wetting front is seen to be extremely steep below the transmission zone, and is constant in shape as nearly as one can discern from the data. Such a steep wetting front is characteristic of infiltration into relatively dry soil and is easily discernable by eye. It is obvious why the assumption of a uniform water content and a constant permeability back of a pronounced wetting front has been the basis for the mathematical models of infiltration of Green and Ampt and others.

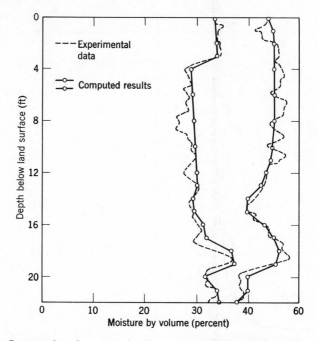

FIG. 10-7. Computed and measured soil moisture profiles before and after 51½ hr infiltration using two wetting capillary pressure curves, and variable porosity and permeability. (Green et al., 1970.)

The rigorous analysis which leads to the equation 10-19 neglects two important factors which cannot always be neglected in the field. These are the possible entrapment of air in the soil profile during infiltration and the existence of a surface crust similar to that to which Horton attributed the falling infiltration rate. The analysis does show that, even in the absence of a crust, the infiltration rate must decrease with time due to the decreasing matric potential gradient. However, Philip's analysis tends to predict a much slower rate of decay that acts over a much longer time than is usually found with field data. Green et al. (1970) took air entrapment into account in their theoretical analysis and found that this could explain discrepancies between theory and experiment. Figure 10-7 shows experimentally determined water content during infiltration in the field compared with the numerical calculations that are based upon the flow equation.

Infiltration Cases

In Nonhomogeneous Soil Profiles

One reason that field infiltration data do not necessarily agree with theoretical calculations is that soil profiles are seldom uniform with depth nor is the water content distribution uniform at the initiation of infiltration. These two effects usually tend to reduce the infiltration rate more rapidly than would otherwise be estimated. In the presence of a shallow water table, the approaches just discussed are likely to be less than satisfactory. This particular problem has been investigated in some detail by Freeze (1969). Figure 10-8 gives the pressure head (matric potential), total head, and water content as a function of depth for successive times during infiltration into three different soils in the presence of shallow water tables. These curves were obtained by numerical solution of the flow equation.

In dealing with nonhomogeneous profiles, it is usually most convenient to divide the profile up into layers or horizons in which each horizon is assumed to be homogeneous. Childs and Bybordi (1969) have extended the Green and Ampt approach to stratified soils in which the conductivity decreases with depth. Hillel and Gardner (1969, 1970) have taken the same approach with soils with an impeding crust on the soil surface. It is of interest to note that in the presence of a crust the infiltration increases during the early stages of infiltration as the square root of time. This is also true where a crust is not present. At sufficiently large values of time, the infiltration is described by the (approximate) expression

$$I = K_f t + E \ln (1 + Ft) \qquad (10\text{-}21)$$

where I is the cumulative infiltration, K_f is the final steady-state infiltration rate, and E and F are parameters whose values depend upon the properties of the soil and of the surface crust.

A layer at some depth below the soil surface tends to reduce the infiltration rate whether it is of coarser or finer texture than the surface layer. If the texture is finer, the reduction in infiltration is due directly to the lower saturated permeability. A coarse layer below a fine layer

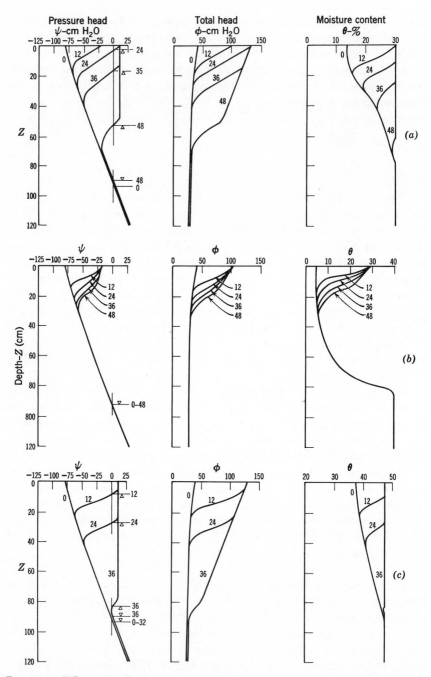

Fig. 10-8. Effect of soil type on ground-water recharge and discharge. (a) Del Monte sand; $R = 0.13.5$ cm/min; $O = 0.001315$ cm/min, (b) Rehovot sand; $R = 0.13.5$ cm/min, $O = 0.001315$ cm/min, (c) Grenville silt loam; $R = 0.13.5$ cm/min; $O = 0.001315$ cm/min.

tends to reduce the infiltration rate for reasons that can be explained from an examination of Figure 10-8. It can be seen that in every case the pressure head (matric potential) is negative at the wetting front. This means that the larger pores in the coarser-textured lower layer do not fill with water and the partial unsaturation of this region results in a very low unsaturated permeability. Both situations are illustrated very dramatically in Figure 10-9 in which the infiltration rate is plotted

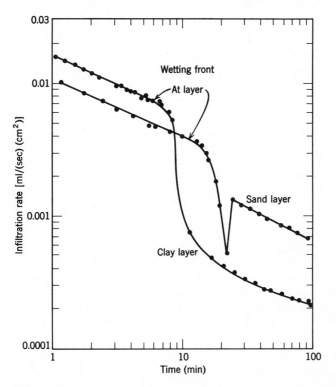

Fɪɢ. 10-9 Effect of sand and clay layers on infiltration rate into Palouse silt loam as a function of time. (Miller and Gardner, 1962).

as a function of time. When the wetting front reaches the clay layer, the infiltration rate is immediately reduced and continues to decrease. In the case of the sand layer, there is an immediate reduction of infiltration rate; however, when the matric potential at the wetting front increases somewhat with increasing water content of the soil above the sand, more pores fill, the permeability increases, and there is an actual increase in infiltration rate. The rate does not increase to that which would exist in the absence of the sand layer.

When Rainfall Is Limiting

The infiltration case described thus far has assumed that the rainfall rate was sufficient to maintain a thin layer of water on the soil surface. If the rainfall rate never exceeds the final infiltration rate that is given by any of the equations, there will be no runoff. The water content of the soil will adjust in such a way that the average potential gradient above the wetting front will be unity and the permeability (unsaturated) will be equal the rainfall rate.

If the rainfall rate is greater than the final steady infiltration rate and continues for a sufficiently long time, then the infiltration equations that have been given do not apply exactly. There will be an initial period

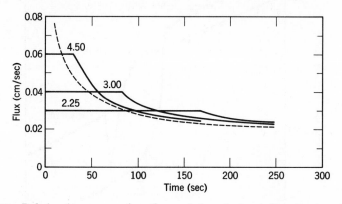

Fig. 10-10. Relation between surface flux and time during infiltration into Rehovot sand due to rainfall (solid lines) and flooding (dashed line). The numbers labeling the curves indicate the magnitude of the relative intensity. (Rubin, 1968.)

of time after rainfall begins during which the soil can absorb water as fast as it falls. The absorption capacity of the soil is decreasing during this period of time and there will eventually come a time when it falls below the rainfall rate. The surface depressions will fill at this point and runoff will begin only after they are filled.

No simple expression has been developed for the prediction of the length of time that a given rainfall rate can be accommodated by a soil. Numerical calculations by Rubin (1968) suggest that as a first approximation the rainfall rate will exceed the capacity of the soil to absorb water at about the time the cumulative rainfall approaches the cumulative infiltration that is predicted for the case of continuous surface ponding of water. This is illustrated by Figure 10-10, adapted from the paper by Rubin. The surface flux or infiltration rate is plotted as a function of time in this figure. The lower the rainfall rate, the longer that rate is maintained constant. If one takes the area under each curve, which is a measure of the cumulative infiltration, it turns out that it is approximately the same for each curve and is not greatly different from the dashed line representing continuous flooding. This finding suggests a way of handling intermittent rainfall. It is not unusual in many regions to find that the rainfall rate varies markedly during any one storm. It is possible for the rate to exceed the infiltration capacity of the soil for a while and then decrease to a value below this. If this latter period of time is a matter of minutes, it can probably be disregarded. If rainfall ceases for a period of hours, then there is at least a partial restoration of infiltration capacity due to drainage of water out of the soil profile or evaporation and transpiration if the period of time is long enough.

In Two and Three Dimensions

The foregoing discussion dealt with infiltration in one dimension and, more specifically, with downward entry of water into soil. Several problems of two- and three-dimensional infiltration are of interest since they are related to the problem of measuring infiltration rates in the field and application of irrigation water.

Infiltration from a semicircular furrow and from a hemispherical cavity are considered in some detail by Philip (1968). In both cases, as was also the case in one-dimensional infiltration where water was continuously available at the absorbing surface, the infiltration rate decreases until it approaches a constant rate. This constant rate can be estimated if the dimensions of the cavity or furrow are specified and the diffusivity of the soil is known. Talsma (1969) found good agreement between field measurements with Philip's approximate treatment of the problem. The shape of the wetting front depends very much upon the relative importance of matric and gravitational forces during infiltration. If the matric forces predominate, the wetting front is almost spherical. If gravity is more important, as in the case of very coarse-textured soils, then the wetting front is elongated and is more nearly ellipsoidal in shape. Figure 10-11 shows lines of equal water content for large times during infiltration from a cylindrical cavity. The constants shown are dimensionless. The abscissa is the ratio of the distance from the center of the cavity, x, to the radius of the cavity, r_0, and the ordinate is the ratio of the depth, Z, to r_0. R_0 is a parameter which is proportional to the ratio of the permeability of the soil and its average diffusivity and is a measure of the relative importance of gravity and matric forces. $R_0 = 0$ would correspond to negligible gravity. Gravity is quite important at $R_0 = 0.25$.

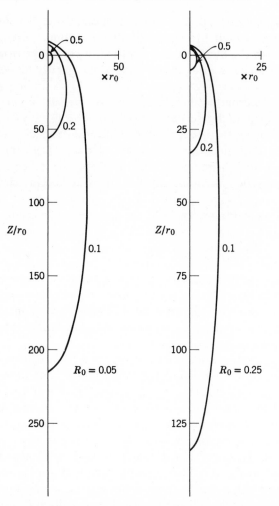

FIG. 10-11. The limiting moisture distribution as $t \to \infty$ during infiltration from a cylindrical cavity. Numerals on the curves represent values of $(\Theta_* - \Theta_0)/(\Theta_1 - \Theta_0)$. Distributions are for the cases $R_0 = 0.05$ and 0.25. (Philip, 1968.)

Philip (1968) has given expressions for the final infiltration rate for all three dimensions which are worth examining:

One dimension: $\qquad i_f = K$ $\hfill$ (10-21a)

Two dimensions: $\qquad i_f = K\left[\dfrac{1}{2} + \dfrac{1}{\pi} + \dfrac{K_1(2R_0)}{2K_0(2R_0)}\right]$ $\hfill$ (10-21b)

Three dimensions: $\qquad i_f = K\left(\dfrac{3}{4} + \dfrac{1}{4R_0}\right)$ $\hfill$ (10-21c) where i_f is the infiltration rate per unit area of soil surface at steady state, K is the saturated permeability of the soil, K_1 and K_0 are modified Bessel functions, which can be evaluated from tables of functions, and R_0 for a soil which is initially dry is given by

$$R_0 = \frac{Kr_0}{4D^*\Theta_1} \tag{10-22}$$

In equation 10-22 r_0 is the radius of the source of water, Θ_1 is the water content of the soil at the point of entry of water into the soil, and D^* is the average or weighted-mean diffusivity which gives the correct infiltration rate at small times. D^* can be evaluated numerically and is virtually identical with D as given in equation 10-12.

Wooding (1968), using the same simplifying approximations, has developed an expression analogous to equation 10-21c that is for a shallow-ponded source of water rather than a hemispherical cavity. He makes the additional assumption that the unsaturated permeability is related to the matric potential by the expression

$$k = (\text{constant})(\exp \alpha\phi) \tag{10-23}$$

where α is a constant. Wooding then suggests the following equation for the steady infiltration rate per unit area:

$$i_f = K\left(1 + \frac{4}{\pi\alpha r_0^2}\right) \tag{10-24}$$

Just as there is a limit to the rate at which water can move upward from a water table, there is also a limit to the lateral movement of water during infiltration in two and three dimensions in the presence of gravity.

The significance of the lateral flow of water in multidimensional systems can be illustrated by two examples. The percolation test is a common procedure for the evaluation of the suitability of a soil for a septic tank drainage field. A hole is dug into the soil in this test and water is introduced for some specified period of time. It is usually implicitly assumed that this time will be long enough to achieve the final steady infiltration rate. This rate will be given approximately by equation 10-21c. However, the tiled field for a septic tank drainage system more nearly represents two-dimensional flow, and equation 10-21b is more applicable. This represents a smaller infiltration rate for a given R_0. An impeding layer tends to build up in time at the point of entry of water into the soil profile. This additional resistance tends to reduce the effective value of K and hence the rate of sorption of water.

A ring infiltrometer is commonly employed to evaluate the infiltration capacity of surface soil. This is usually a metal ring, 1 to 2 ft in diameter, which is pressed a short distance into the soil and filled with water. The rate of water loss from the ring is taken as an estimate of the infiltration rate. Equation 10-24 gives an estimate of the error involved in this procedure. The constant α will seldom exceed 0.05 cm, and a value of 0.01 cm is more common. For the latter value and a ring diameter $r_0 = 30$ cm, the infiltration rate will be about 40 percent higher than the infiltration rate for ponded infiltration over a large area. A large value for α represents a coarse-textured soil and a small value a fine texture.

REDISTRIBUTION AND DRAINAGE

Redistribution of Water in Soil Profiles

After Complete Wetting

The nature of redistribution of water in a soil profile after addition at the surface ceases depends upon the amount of water added and upon whether or not it is sufficient so that the advancing wetted front reaches wet soil below. In humid areas with plentiful rainfall or in arid areas where heavy irrigation sometimes is practiced, the wetting front often will advance into already wet soil during infiltration or shortly afterward. Drainage from such a profile at cessation of irrigation proceeds at a decreasing rate until the hydraulic gradient becomes zero, at which time the downward flux becomes zero. The pull of gravity is responsible for such flow. This force is opposed by resisting matric forces which increase as the water content decreases. Flow ceases when the gravitational and matric potentials are equal but opposite in sign. Should infiltration continue without ponding at the soil surface the hydraulic gradient becomes unity. With ponding at the surface, gradients exceeding unity are possible.

After Incomplete Wetting

Outward movement from a water source into dry soil occurs under wetting conditions as long as water is applied continuously and the water content-energy relationship is given by a wetting or sorption curve. However, when water application ceases, further movement of water involves both wetting and drying, or hysteresis; water moves out of the region of the source and into the drier region beyond. Drying which occurs in the wetter part of the soil follows a scanning curve on the interior of the hysteresis envelope (see Figure 8-4) where, for any given water potential, the water content is greater than it was at the same potential during wetting. Equilibrium is achieved in a horizontal soil column wetted from one end and with evaporation prevented when all of the soil comes to the same matric potential. At uniform potential the soil which has dried down to the equilibrium value will be significantly wetter than that which has wet up to this same potential. The practical difference in water content at two ends of such a soil column is further accentuated by the extremely slow rate at which equilibrium is approached. Rate of flow in such a system becomes progressively slower with time because of the marked decrease in the hydraulic gradient and overall unsaturated conductivity which occurs as equilibrium is approached. Approach to equilibrium often is so slow that equilibrium water conditions rarely exist in nature.

If the wetting soil column is vertical, as in many field conditions, equilibrium, or a zero hydraulic gradient, requires that the matric and gravity potentials be everywhere equal and of opposite sign so that their sum is zero. Under such conditions, ignoring hysteresis, a soil column at equilibrium should be wetter at the bottom than at the top. However, because of hysteresis such a water content distribution never would exist except where the entire column is wetted and allowed to equilibrate. Distributions more commonly are like those shown in Figure 10-12 (Gardner et al., 1970) where sandy loam soil has been wet with different quantities of water and water content distributions have been followed for several weeks. Here the effect of hysteresis and low rate of flow gives rise to significant water storage in the upper part of the soil profile with drier soil below, despite the downward pull of gravity.

Empirically, it has been observed by a number of workers that the decrease in water content in the initially wetted zone obeys the equation

$$w = a(t + c)^{-b} \qquad (10\text{-}25a)$$

where a, b, and c are constants. Gardner et al. (1970) have shown that these constants can be related to the soil water diffusivity and permeability if certain simplifying assumptions are made. When the time t in equation 10-25a is more than a day or so the constant c can usually be neglected.

Equation 10-25a with $c = 0$, written logarithmically, is

$$\log w = \log a - b \log t \qquad (10\text{-}25b)$$

and it may be seen that values of the constants a and b can be obtained directly from a logarithmic graph if empirical data plot as a straight line. Water content in various layers of a sandy loam soil profile during drainage after irrigation (Ogata and Richards, 1957) are shown logarithmically in Figure 10-13. These obey the relation $w = 0.256\ dt^{-0.128}$

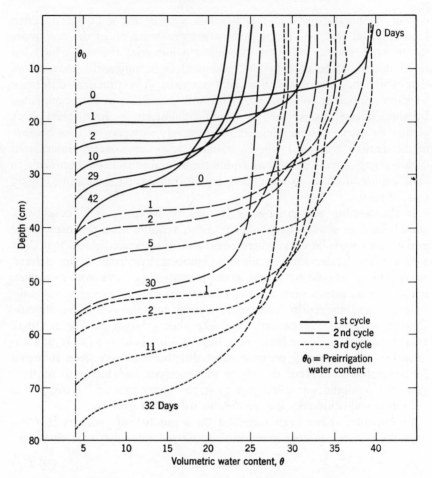

FIG. 10-12. Successive water content profiles during redistribution cycles following one, two, and three irrigations of 5 cm each. (Gardner et al., 1970.)

where w is the water content in $cm^3 cm^2$ ($=$ cm) of water in profile depth d for time in days after cessation of irrigation.

Rate of water loss from a wetted profile, or the drainage rate, may be obtained by differentiation of equation 10-25a with time

$$dw/dt = -ba(t + c)^{-(b+1)} \qquad (10\text{-}26a)$$

Since parameters a and b may be obtained from a logarithmic plot where c may be neglected (see equation 10-25a), drainage rate may be directly computed. Drainage rate from the soil profile is of considerable interest

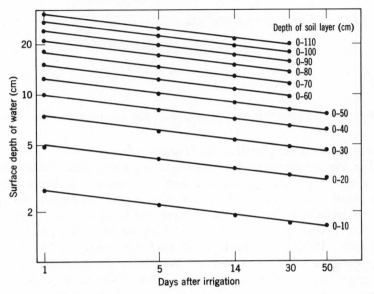

Fig. 10-13. Surface depth of water in various layers as related to time (logarithmic scale). (Ogata and Richards, 1957.)

in many practical problems relating to water use, fertilization, pollution, and particularly in connection with evaluation of the field capacity.

The Field Capacity Concept

Field capacity has been defined as the water content of a soil profile, usually the rooting zone, which has been thoroughly wetted by irrigation or rainfall and after the subsequent rate of drainage out of the profile has become negligibly small. Often the period of time required to reach a negligible drainage rate has been arbitrarily taken as 2 days. The rate of drainage from some soil profiles quickly drops to a sufficiently low value over about a 2-day period so that the measurement of profile water content is reasonably reproducible. However, for soils where water retention obeys equation 10-25a or where drainage is in accord with its differential, equation 10-26a, there is no unique point in time where an unequivocal negligible rate of drainage could be chosen. And, indeed, significant drainage may occur for many days. A simple definition of field capacity under such conditions is not possible. However, it is possible to assign arbitrarily a drainage rate to be considered negligible and to determine the water content of a profile at which such a drainage rate is achieved. This is best achieved by combining equations 10-25a and

10-26a to eliminate t, again neglecting c, to yield

$$w = a^{1/(b+1)} b^{-b/(b+1)} \left(\frac{dw}{dt}\right)^{b/(b+1)} \tag{10-26b}$$

and

$$\log w = \log [a^{1/(b+1)} b^{-b/(b+1)}] + \frac{b}{(b+1)} \log \ (-\text{drainage rate}) \quad (10\text{-}26c)$$

where dw/dt is designated as the drainage rate.

The logarithmic plot of profile water content w against drainage rate from the profile (rate is negative since w is decreasing, hence need for the negative sign in equation 10-26c so that a logarithm may be taken) permits assignment of a profile drainage rate to be regarded as negligible for any desired purpose and the selection of the corresponding "field capacity." Putting this value for w back into equation 10-25a permits determination of the time required to achieve the selected "negligible" drainage rate. Such times vary from a few hours to weeks, depending upon choice of a drainage rate which may be regarded as negligible and upon the water transmission properties of the profile. What constitutes negligible drainage depends upon the particular problem under consideration. A rate which may be considered unimportant relative to the daily transpiration rate may be significant when summed over an entire year. No single criterion exists for the neglect of downward drainage.

Many attempts have been made to relate field capacity to water retention at a particular matric potential, often to the $-\frac{1}{3}$ bar moisture percentage, or in older literature, to the moisture content of a soil sample wetted and subjected to a force of 1000 g in a centrifuge (the moisture equivalent), which corresponds to about $-\frac{1}{3}$ bar. Such attempts ignore the fact that water retention in a profile depends on transmission properties of the entire profile and the hydraulic gradients which exist rather than on only the energy state of water at a particular point in the profile. At profile drainage rates of a few millimeters per day, water potentials ranging from as much as -0.005 in highly stratified soils to as little as -0.6 bar in deep dry-land soils have been observed in rooting zone soil. Considering the extremely flat nature of most desorption curves in this wet range, huge errors in water retention estimates may be expected from the arbitrary association of field capacity with a particular water potential like, say $-\frac{1}{3}$ bar. Additionally, a profile containing a tight claypan or hardpan, if wet excessively, may retain near-saturation conditions for a long period of time quite independently of the texture or structural state of the rooting-zone soil; such factors would be involved in determining the water content at a particular matric potential.

Use of the term field capacity often is misleading; it has been used to mean practically everything from "wet but not saturated" to a specific water content such as the $-\frac{1}{3}$ bar percentage. The problem is associated with the impossibility of finding any unique state of water in highly dynamic and variable systems which can have a useful and consistent meaning under all practical circumstances. There is need for a term to be applied somewhat nonspecifically to a rather wet but unpuddled soil in the field, or to a sample in a state in which it might exist in the field after drainage has slowed appreciably. Although the word "capacity" implies something more specific, perhaps the term "field capacity," by virtue of its long usage in this sense, should retain this meaning. But when the term is intended to mean the water content of a profile after drainage has become "negligible," some indication must be given as to precisely what constitutes negligible drainage. Perhaps the terminology should include designation of the drainage rate, say, "field capacity for 2 mm/day drainage." However, increasing recognition of the fact that most soils do not really have a unique water retention capacity could lead to the abandonment of the term for use in this sense.

When a particular energy state of water in soil is to be specified, quite apart from the dynamic problem of water content changes in a field profile, the term field capacity should be avoided entirely. Rather, the matric potential should be given.

Experimental observations in both laboratory and field reveal that homogeneous soil profiles wet to sufficient depth tend to drain at an almost uniform water content. If there is no evaporation at the soil surface, then the downward flux of water at any depth L is given by the equation

$$q = \frac{dW}{dt} = -k\frac{dH}{dz} \qquad (10\text{-}27)$$

where W is the water content of the soil between $z = 0$ and $z = L$, k is the unsaturated conductivity at $z = L$, and dH/dz is the hydraulic gradient at $z = L$. To the extent that the water content gradient can be assumed negligible, $dH/dz = 1$ and equation 10-27 reduces to

$$\frac{dW}{dt} = -k \qquad (10\text{-}28)$$

Equation 10-27 lends itself readily to the determination of the unsaturated conductivity in the field (Ogata and Richards, 1957; and Rose, Stern, and Drummond, 1965). In this procedure, a soil profile is wetted to a depth of 2 or 3 ft, covered to prevent evaporation, and allowed to drain. The water content is measured as a function of time and the hydraulic head is determined, usually by the use of tensiometers. The tensiometer data can also be used to calculate the diffusivity directly from the same experiment (Gardner, 1970). If the water content above $z = L$ is uniform, then $dW/dt = L \, d\Theta/dt$. Substituting this expression into equation 10-28 and remembering the definition of the diffusivity gives

$$D = -L \left(\frac{d\phi}{dt}\right) \left(\frac{dH}{dz}\right) \qquad (10\text{-}29)$$

Drainage and Deep Percolation

Equation 10-28 applies particularly well to drainage out of a root zone, when the gravitational potential is predominant. An important exception to this is the situation in which there is a water table that is not too far below the draining region. In that case, the hydraulic gradient is less than unity and the rate of drainage is slower than the unsaturated permeability. This problem can be solved exactly by numerical procedures that use computer techniques. A number of approximate solutions have been developed. One of these (Youngs, 1960) uses the equivalent of the assumptions made by Green and Ampt (1911) for infiltration. Youngs obtains the following equation for the cumulative drainage of water out of a profile that drains to a water table:

$$\frac{Q}{Q_\infty} = 1 - \exp\left(\frac{Kt}{Q_\infty}\right) \qquad (10\text{-}30)$$

where Q is the quantity of water per unit area which has drained at time t, Q_∞ is the total drainage at infinite time, and K is the hydraulic conductivity. The reduction in the drainage rate with time in equation 10-30 is due solely to the decrease in hydraulic gradient with time. In practice, this equation overestimates the drainage rate at large times because of the decrease in the conductivity as the soil drains. This decrease is usually neglected in the solution of most drainage problems, such as those to be discussed in the following chapter. It is customary to assume that a soil drains most of its pore space at a relatively high matric potential with a conductivity that is equal to the saturated hydraulic conductivity. This simplifying assumption is not always justified.

It can be shown that for a fixed water-table depth even very nonhomogeneous soils tend to drain in such a way that the rate of drainage out of the profile can be related in a simple fashion to the water content of the soil. This provides a very useful simplification for the calculation of the field water balance where the assumption of a unique relation between the soil profile water content and the drainage rate is justified. Figure 10-14 shows the drainage rate plotted as a function of the soil water content for Plainfield sand (Black et al., 1969). The capillary conductivity is also shown in the same graph and can be seen to be very nearly equal to the drainage rate; this is in accordance with equation

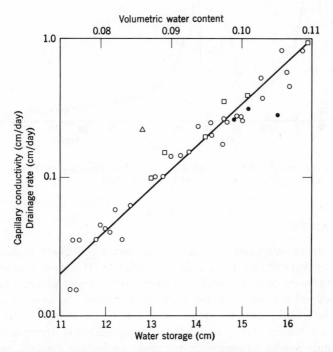

Fig. 10-14. Lysimeter drainage rate as a function of water storage (circles) and capillary conductivity of Plainfield sand as a function of soil water content (squares). (Black et al., 1969.)

10-28. Figure 10-15 depicts the drainage rate for the summer season for the soil profile as predicted from the line in Figure 10-14 and compared with the actual drainage rate as measured with a suction lysimeter. The data of Davidson et al. (1969) and the widespread use of the empirical equation 10-26 suggest that this approach may be applicable to a wide range of soil profiles and that it may be possible to substitute a characteristic drainage curve for a profile instead of the field capacity concept.

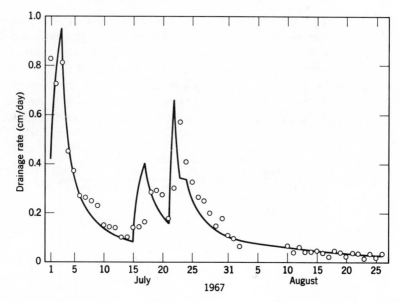

Fig. 10-15.   Predicted drainage (solid lines) from bare Plainfield sand compared with that measured by lysimeter (circles). (Black et al., 1969.)

EVAPORATION

The rate of evaporation from a wet, bare soil surface is determined by meteorological factors, and in particular by the energy available to change the state of water from the liquid to the vapor phase. This problem is beyond the scope of this book. It was shown in an earlier section that under steady-state conditions the soil could supply water to the surface at a limiting rate and no faster. Evaporation is determined under these conditions by atmospheric processes or by the rate of movement of water to the soil surface, whichever is the less dominant.

The same circumstance exists in the transient drying of a soil. The soil surface is initially wet and the rate of evaporation is governed by factors external to the soil. The soil surface becomes progressively drier during the initial or "constant-rate" phase of drying. Eventually, as the surface approaches that water content where the vapor pressure is very nearly that in the atmosphere at the soil boundary, the rate becomes limited by the rate of water movement to the soil surface. This is a problem then of simultaneous movement of water and heat and its exact solution is exceedingly difficult.

It turns out, rather fortunately, that if one is concerned primarily with average daily rather than minute by minute evaporation, the solution of the flow equation under isothermal conditions provides a reasonable first approximation to the evaporation rate. The evaporation rate decreases continuously with time during the second or "falling-rate" stage of drying.

If the constant-rate phase is very brief, due to a high potential evaporation rate, the flow equation predicts that the evaporation rate should decline as the square root of time and that the cumulative evaporation should increase proportionately to the square root of time. Figure 10-16 (Black et al., 1969) illustrates the cumulative evaporation from Plainfield sand plotted against $t^{1/2}$. The circles are experimental data and the dashed

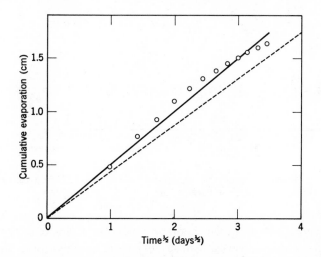

Fɪɢ. 10-16. Cumulative evaporation from bare Plainfield sand as a function of $t^{\frac{1}{2}}$. Circles = lysimeter date; solid line = best fit by eye of data; dashed line = $\bar{D} =$ 10 cm²/day from laboratory data. (Black et al. 1969.)

line is the evaporation which would have been predicted from an independent measurement of the soil water diffusivity. Black et al. (1969) assumed that each time it rained or the soil was irrigated the process started over again from time $t = 0$. They then predicted the evaporation from a lysimeter for an entire season by using only rainfall and irrigation inputs. The calculation results are shown in Figure 10-17. The smooth curve represents the theoretical calculation and the circles the actual evaporation. Two curves are shown for August because the lysimeter was covered for calibration for 13 days. The upper curve assumes that evaporation was not hindered by this process and the lower curve assumes that no flow occurred in the profile during this period. The actual evaporation should lie between these two bounds.

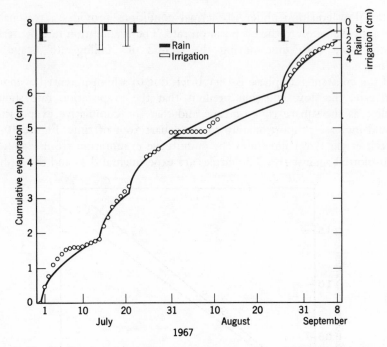

Fig. 10-17. Predicted accumulative evaporation from bare Plainfield sand compared with that measured by lysimeter (data = circles). (Black et al., 1969.)

The square root of time-evaporation relation is for a soil which is wet infinitely deep. The rate drops off more rapidly for a finite soil sample or a finite depth of wetting. It is still convenient to use the ratio (D_0tL^2) as a dimensionless variable against which to plot the evaporation. D_0 is the soil water diffusivity at the soil surface in this expression. Figure 10-18 shows the cumulative evaporation as a fraction of the water applied to a soil sample as plotted against the square root of this variable. The initial stage follows a straight line, which indicates the square root of time behavior as in the case of an infinite depth of wetting. The later stages exhibit a more rapidly decreasing evaporation rate. Most of the water has evaporated by the time $D_0tL^2 = 1$. The data in Figure 10-18 were obtained by Gardner and Gardner (1969).

Figure 10-19 shows data from the same study in which the initial evaporation rate was relatively low so that the constant-rate phase persisted for about 10 days. The smooth curve for this set of data points is identical with the curve for the data for a high potential evaporation rate except that it is displaced along the X-axis until it intersects the dashed line. The dashed line is a theoretical calculation of the limit to

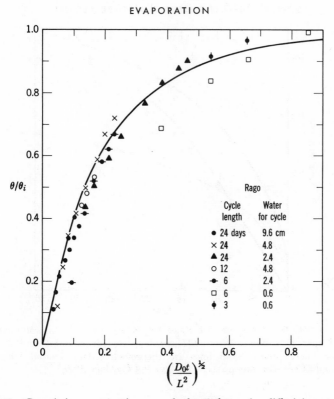

Fig. 10-18. Cumulative evaporation as calculated from the diffusivity equation for Rago soil. Variables are dimensionless with $D_o = 1.0$, and $D_i/D_o = 200$. (Gardner and Gardner, 1969.)

the constant-rate phase, as derived by Gardner and Hillel (1962). Figure 10-19 illustrates an important point concerning the conservation of water by surface mulches. A surface mulch tends to reduce the initial rate of evaporation and prolong the constant-rate phase. The total evaporation during this period is less than that which would exist in the absence of a mulch. However, if the evaporation process continues for a sufficiently long time, the two rates converge and little water is saved by the mulch. The effectiveness of a mulch depends very much upon the frequency of rainfall or irrigation.

In the Presence of Plant Cover

The predication of evaporation from soil under plant canopies or between rows in the case of row crops is very difficult. One cannot simply

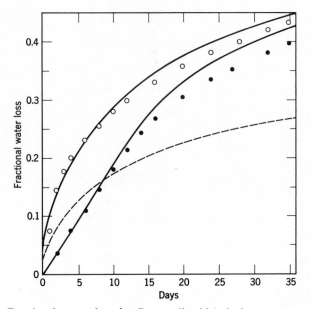

Fig. 10-19. Fractional water loss for Rago soil which had 9.6 cm water added to dry soil. Upper data points show 0.72 cm/day loss for constant rate with 0.18 cm/day for lower points. Solid lines are theoretical curves. Dotted lines shows calculated end of constant rate phase. (Gardner and Gardner, 1969.)

cover the soil surface and measure the evaporation or transpiration from the plant surfaces, subtract this from the sum of evaporation and transpiration, and obtain the evaporation by difference. This is because much of the energy which strikes the covered soil surface remains in the plant canopy and results in increased transpiration. Moreover, it is very difficult to measure the evaporation or transpiration independently in the presence of the other.

One relatively successful attempt was reported by Black, Tanner, and Gardner (1970). They were able to measure the energy that is available at the soil surface for evaporation and to estimate the ability of the soil to supply water from the surface water content. They then assumed that the evaporation was limited by the energy or by the soil-water flux, whichever was smaller. Independent calculations of the transpiration were made on the basis of leaf stomatal-resistance values and appropriate meteorological measurements in the canopy. The ability to separate the two processes in this experiment may very well have been fortuitous and the general applicability of the approach remains to be shown.

The need for a solution to this problem is made urgent by the fact that most models for the prediction of evaporation and transpiration from plant canopies have failed except where the soil surface was assumed to be at either 100 or 0 percent relative humidity. Neither assumption is valid for more than a part of the time.

SUMMARY

An understanding of each component in the field water balance is essential for the efficient management of water resources. It has often been necessary in the past to estimate one or more of these components by the difference between the others. This has often led to significant errors. It is to be expected that a combination of soil profile measurements and meteorological measurements will reduce greatly the uncertainty in many field water balance calculations. Some of the important reasons for requiring knowledge of these components will be discussed in Chapter 11.

References

Black, T. A., W. R. Gardner, and G. W. Thurtell (1969). The prediction of evaporation, drainage, and soil-water storage for a bare soil. *Soil Sci. Soc. Am. Proc.*, **33**:655–660.

Black, T. A., C. B. Tanner, and W. R. Gardner (1970). Evapotranspiration from a snap bean crop. *Agron. J.*, **62**:66–69.

Childs, E. C., and M. Bybord (1969). The vertical movement of water in stratified porous material. I. Infiltration. *Water Resources Res.*, **5**:446–459.

Crank, J. (1956). *The Mathematics of Diffusion.* Oxford Press, London.

Davidson, J. M., D. R. Nielsen, and J. W. Biggar (1963). The measurement and description of water flow through Columbia silt loam and Hesperia sandy loam. *Hilgardia*, **34**:601–617.

Davidson, J. M., L. R. Stone, D. R. Nielsen, and M. E. Larue (1969). Field measurement and use of soil properties. *Water Resources Res.*, **5**:1312–1321.

Freeze, R. Allan (1969). The mechanism of natural ground-water recharge and discharge. I. One-dimensional, vertical, unsteady, unsaturated flow above a recharging and discharging ground-water flow system. *Water Resources Res.*, **5**:153–171.

Gardner, H. R., and W. R. Gardner (1969). Relation of water application to evaporation and storage of soil water. *Soil Sci. Soc. Am. Proc.*, **33**:192–196.

Gardner, W. R. (1958). Some steady-state solutions of the unsaturated flow equation with application to evaporation from a water table. *Soil Sci.*, **85**:228–232.

Gardner, W. R. (1959). Solutions of the flow equation for the drying of soils and other porous media. *Soil Sci. Soc. Am. Proc.*, **23**:183–187.

Gardner, W. R. (1970). Field measurement of soil water diffusivity. *Soil Sci. Soc. Am. Proc.*, **34**:382.

Gardner, W. R., and M. Fireman (1958). Laboratory studies of evaporation from soil columns in the presence of a water table. *Soil Sci.*, **85**:244–249.

Gardner, W. R., and D. Hillel (1962). The relation of external evaporative conditions to the drying of soils. *J. Geophys. Res.*, **67**:4319–4325.

Gardner, W. R., D. Hillel, and Y. Benyamini (1970). Post-irrigation movement of soil water. I. Redistribution. *Water Resources Res.*, **6**:851–861.

Gardner, Willard, and John A. Widstoe (1921). The movement of soil moisture. *Soil Sci.*, **11**:215–232.

Green, Don W., Hassam Dabiri, Charles F. Weinaug, and Robert Prill (1970). Numerical modeling of unsaturated ground water flow and comparison of the model to field experiment. *Water Resources Res.*, **6**:862–874.

Green, W. Heber, and G. A. Ampt (1911). Studies in soil physics. I. The flow of air and water through soils. *J. Agr. Sci.*, **4**:1–24.

Hillel, D. I., and W. R. Gardner (1969). Steady infiltration into crust-topped profiles. *Soil Sci.*, **108**:137–142.

Hillel, D., and W. R. Gardner (1969). Steady infiltration into crust-topped profiles. *Soil Sci.*, **108**:137–142.

Hillel, D., and W. R. Gardner (1970). Transient infiltration into crust-topped profiles. *Soil Sci.*, **109**:69–76.

Holmes, J. W., and J. S. Colville (1970). Forest hydrology in a Karstic region of Southern Australia. *J. Hydrology*, **10**:59–74.

Horton, Robert E. (1933). The role of infiltration in the hydrologic cycle. *Trans. 14th Ann. Meeting Am. Geophys. Union*, pp. 446–460.

Horton, Robert E. (1939). Analysis of runoff-plot experiments with varying infiltration capacity. *Trans. Am. Geophys. Union, Part IV*, pp. 693–694.

Horton, Robert E. (1940). An approach toward a physical interpretation of infiltration capacity. *Soil Sci. Soc. Am. Proc.*, **5**:399–417.

Jeppson, R. W., and W. R. Nelson (1970). Inverse formulation and finite difference solution to a partially saturated seepage from canals. *Soil Sci. Soc. Am. Proc.*, **34**:9–14.

Kostiakov, A. N. (1932). On the dynamics of the coefficient of water percolation in soils and the necessity for studying it from a dynamic view for purposes of amelsoration. *Trans. 6th Com. Int. Soc. Soil Sci., Russian Part* A:17–21.

Miller, D. E., and W. R. Gardner (1962). Water infiltration into stratified soil. *Soil Sci. Soc. Am. Proc.*, **26**:115–119.

Moore, R. E. (1939). Water conduction from shallow water tables. *Hilgardia*, **12**:383–346.

Ogata, G., and L. A. Richards (1957). Water content changes following irrigation of bare-field soil that is protected from evaporation. *Soil Sci. Soc. Am. Proc.*, **21**:355–356.

Philip, J. R. (1954). An infiltration equation with physical significance. *Soil Sci.*, **77**:153–157.

Philip, J. R., (1957a). Evaporation, and moisture and heat fields in the soil. *J. Met.*, **14**:354–366.

Philip, J. R. (1957b). The theory of infiltration: 4. Sorptivity and algebraic infiltration equations. *Soil Sci.*, **84**:257–264.

Philip, J. R. (1968). Absorption and infiltration in two- and three-dimensional systems. Water in the unsaturated zone. *UNESCO Symposium*, Wageningen, Vol. I, pp. 503–525.

Rose, C. W., W. R. Stern, and J. E. Drummond (1965). Determination of hydraulic conductivity as a function of depth and water content for soil *in situ*. *Aust. J. Soil Res.*, **3**:1–9.

Rubin, J. (1968). Numerical analysis of ponded rainfall infiltration. Water in the unsaturated zone. *UNESCO Symposium*, Vol. I, pp. 440–450.

Srinilta, Sam-arng, D. R. Nielsen, and Don Kirkham (1969). Steady flow through a two-layer soil. *Water Resources Res.*, **5**:1053–1063.

Talsma, T. (1969). Infiltration from semi-circular furrows in the field. *Aust. J. Soil Res.*, **7**:277–284.

Wooding, R. A. (1968). Steady infiltration from a shallow circular pond. *Water Resources Res.*, **4**:1259–1273.

Youngs, E. G. (1960). The drainage of liquids from porous materials. *J. Geophys. Res.*, **65**:4025–4030.

CHAPTER 11

Soil Water—
Plant Relations

One of the major pathways for water transfer from the soil to the atmosphere is through the vegetative cover. Since all life processes take place in an aqueous environment, soil water plays a vital role in the growth of plants and other living organisms. The importance of the effect of soil water upon plant response justifies special consideration of this subject.

In Chapter 10 it was pointed out that the soil profile modulates the hydrological cycle through its effect upon infiltration and drainage and by means of its storage capacity. The vegetative cover of the soil plays an equally profound role. The plant root system usually provides a path of lower resistance to water movement than does a drying soil surface. Plants have little storage capacity for water compared with the amounts which pass through them each day but they provide a hydraulic connection between the soil and the atmosphere which, collectively, they regulate to suit their evolutionary purposes. In order to understand this role and the influence of the soil water regime upon plant response, it is necessary to discuss the internal water relations of plants in some detail.

INTERNAL WATER RELATIONS OF PLANTS

Characterization of the state of water in the plant must start with a consideration of the water potential. Just as in the soil, water in the plant tends to move along water potential gradients, from regions of higher energy to those of lower energy. In plants the picture is not as simple as in soil because of the presence of biological membranes which tend to permit the passage of water but are selective in the transmission of solutes. The total water potential in a plant leaf cell is usually partitioned into three components:

$$\psi = \psi_m + \psi_r + \psi_p \tag{11-1}$$

where the first term is the matric component, the second the osmotic, and the third the pressure or turgor. Water held by matric forces corresponds roughly to "bound" water, which is regarded as held by adsorption on colloidal particles. Solutes tend to be excluded from regions of bound water so that where the matric component is significant the osmotic is less important. Conversely, in regions where the osmotic component is predominant, the matric forces are presumed to be small. If we omit the matric term from equation 11-1, we have the equivalent of the widely used expression

$$DPD = OP - TP \qquad (11\text{-}2)$$

In equation 11-2 the diffusion pressure deficit (DPD) is the same as the total potential except that the negative sign is implied by the word deficit in the title. The OP is the osmotic pressure or negative of the osmotic potential. The TP is the turgor pressure or turgor potential. In living cells the total potential and its components are not completely

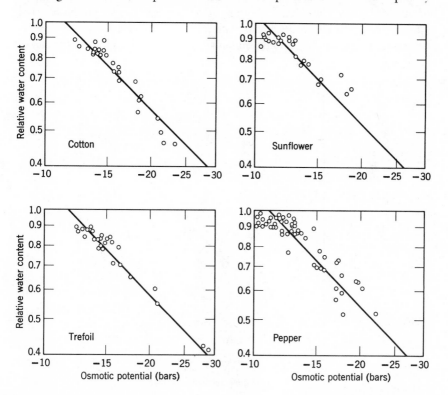

Fig. 11-1. Leaf relative water content as a function of osmotic potential. (Gardner and Ehlig, 1965.)

independent. This is due to the elastic and semipermeable properties of the cell walls. An increase in the turgor pressure results in an expansion of the cell walls. This is accomplished by the uptake of water. Unless this uptake is also accompanied by a proportional uptake of solutes, the solute concentration decreases with a consequent increase in the osmotic potential. Solute transport across the membranes can and does occur but at a rate that is generally slower than the rate of water movement so that the immediate response of a cell to any change in water potential is a change in its water content or degree of hydration.

The components of the water potential in a leaf cell are not completely independent. On a short-term basis, if the total potential is specified, this determines both the osmotic and the turgor potential, as well as the degree of hydration. A comprehensive discussion of the hydration aspects of plant tissues and methods of measurement of plant water potential is given by Barrs (1968). A convenient measure of the hydration of plant tissue is the ratio of the water content at any given degree of hydration to the water content at zero total water potential. This ratio is known as the *relative water content*. As the relative water content decreases, if the solute content remains constant, the solute concentration

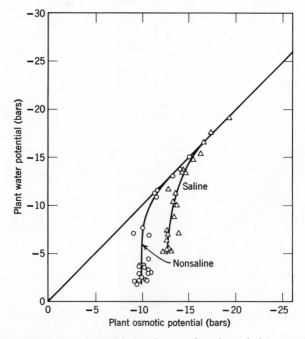

FIG. 11-2. Plant leaf water potential (total) as a function of the osmotic potential component. (Gardner and Ehlig, 1968.)

must increase proportionately. This results in a decrease in the osmotic potential.

Figure 11-1 shows the relation between the osmotic potential and the relative water content for four different plant species, as determined by Gardner and Ehlig (1965). The data are plotted on a logarithmic scale and the straight line has a slope of 45°, as would be predicted if the solute content were to remain constant and the amount of bound water were negligible. If a plant is growing in a saline soil solution, then over a period of time the solute content of the cells tends to adjust accordingly. The rate of adjustment varies from species to species. Figure 11-2 shows the relation between the total water potential and the osmotic potential for bell pepper on both saline and nonsaline substrates (Ehlig and Gardner, 1968). If we neglect the matric potential, we can use equation 11-1 or Figure 11-2 to obtain the turgor potential by subtracting

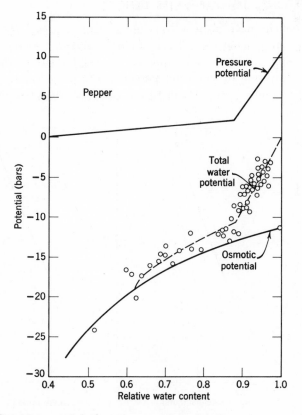

FIG. 11-3. Components of leaf water potential of bell pepper as a function of relative water content. (Gardner and Ehlig, 1965.)

the osmotic potential from the total potential. Since both of these potentials are negative, with the osmotic the more negative, we get a positive turgor potential.

All three potentials are plotted as a function of relative water content for a nonsaline pepper plant in Figure 11-3. Of particular interest is the abrupt change in slope of the pressure potential relation at a leaf water content of about 0.85. This corresponds to a total water potential of about −11 bars for pepper and coincides with the appearances of marked symptoms of visible wilting. The change in slope corresponds to a change in the elastic modulus of the leaf tissue and explains the wilting symptoms. As will be expanded upon later, this also corresponds roughly with the point at which the stomates are virtually completely closed.

PLANT RESPONSE TO LEAF-WATER DEFICITS

Water deficits have many effects upon plant response and these have been reviewed at length by Slatyer (1967, 1969) and Salter and Goode (1967) among others. It now appears reasonably certain that the first effect of a lowering of the water potential of a plant leaf is the stomatal closure that results from a loss of turgor pressure. Many factors influence

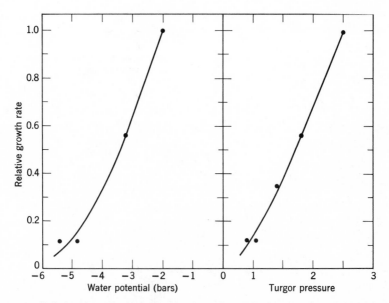

Fig. 11-4. Relative growth rate as a function of total leaf water potential and turgor potential for onions (Millar et al., 1971.)

stomatal behavior but loss of leaf turgor almost invariably tends to result in stomatal closure. Since the stomates are the pathway for the carbon dioxide essential to photosynthesis, closure increases the resistance of CO_2 uptake. A reduction in the rate of net photosynthesis is the result. Thus there is a direct relation between leaf-water potential and growth rate, other factors being constant. Such a relation is shown for onions in Figure 11-4 by Millar et al. (1971). In the onion plants in their investigation the turgor pressure did not go above about 3 bars. Kanemasu and Tanner (1969) plotted the leaf-water potential, stomatal resistance, and net assimi-

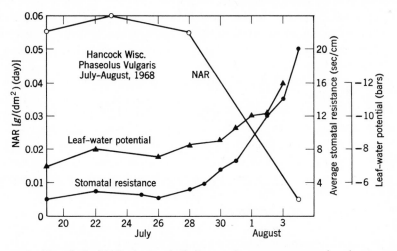

Fig. 11-5. Trends in NAR (net assimilation rate), average stomatal resistance, and leaf-water potential during the first drying cycle. (Kanemasu and Tanner, 1969.)

lation rate (NAR) as a function of time during the season (Figure 11-5). Similar results have been reported by a number of workers, including Troughton (1969).

RELATION OF LEAF-WATER POTENTIAL TO SOIL WATER

One of the major problems impeding research on soil-plant-water relations in the past has been the inference of plant response from soil water behavior. Veihmeyer and his colleagues (Veihmeyer and Hendrickson, 1933) felt that their experiments showed that soil water content had little influence upon plant response until the soil water content went below the "permanent wilting point," a concept which Hendrickson and Veihmeyer (1945) developed out of the "wilting coefficient" of Briggs and Shantz (1911). The permanent wilting point was the water content of a soil below which a plant could not extract sufficient water to regain turgor when transpiration was prevented overnight. It can be seen from the relation between turgor pressure and leaf elasticity in Figure 11-3 why this point became associated with the water content at 15 atm or −15 bars (Richards and Weaver, 1943). When the water potential of the leaf is less than about −12 bars, depending upon the species, the leaf appears wilted. In order to extract water from the soil, the plant must be at a lower water potential than that of the soil. When the soil-water potential is lower than −15 bars the plant cannot recover sufficient turgor to alleviate the wilting symptoms.

In a classical review, Richards and Wadleigh (1952) dissented from the concept of equal water availability and argued that water became decreasingly available as the water potential of the soil declined. Although this view has been increasingly accepted, it was difficult for some time to prove on the basis of existing field data. Stanhill (1957) reviewed the pertinent literature to that time and concluded that the evidence was largely on the side of plant response over a wide range of soil water content. More recently this view has become well established, as will be indicated in the following sections. A major cause of the difficulty in the interpretation of earlier field data was the variability of both the climatic factors and the soil-water potential over a growing season. In addition, most growth data were obtained from yields for an entire season, thus averaging overall variations.

In order to relate the leaf-water potential to the soil water status and to other environmental factors, one must consider the dynamics of the transport of water through the soil-plant-atmosphere system. It has been assumed for some time that transport through the entire system could be expressed by a transport equation analogous to Ohm's law or Darcy's law:

$$E = \frac{\psi_1 - \psi_r}{R} \tag{11-3}$$

where E is the transpiration flux through the plant, ψ_1 is the potential in the leaves at the evaporating surface, and ψ_r is the potential outside the root surface.

van den Honert (1948) and others who followed considered the very low water potential values corresponding to the vapor pressure outside the plant leaf and concluded that most of the resistance to flow was in the vapor phase. The inclusion of the vapor phase in the pathway resistance presents conceptual difficulties and can be very misleading. There is a phase change from liquid to vapor at the evaporating surface and one cannot treat this process as a resistance. It is also a serious over-simplification to treat flow through the root membranes in terms of a simple resistance since these membranes restrict the flow of solutes. It is therefore best to consider separately the water loss from the plant leaves, the flow through the plant, and the flow to the plant roots.

Transpiration from Plant Leaves

The relation between the water potential and relative humidity in Chapter 8 (equation 8-2) shows that in the range of leaf-water potential of usual interest the vapor pressure at the internal surfaces of the substomatal cavities must be very nearly that of pure water. The internal surfaces are acting essentially as wet surfaces. The rate of water loss through the stomates is inversely proportional to the resistance of the diffusion path and directly proportional to the vapor pressure difference between the inside of the cavity and some specified point outside. The diffusion resistance can be divided into an external diffusion resistance in the air r_a and an internal resistance r_i, which depends upon the stomatal aperture. The evaporation from the leaf is then related to the potential evaporation (for which see Chapter 7 by the expression

$$E = \frac{E_p}{1 + [\gamma/(s + \gamma)](r_i/r_a)} \tag{11-4}$$

where E_p is the potential evaporation, γ is the psychrometer constant, and s is the slope of the curve relating temperature to saturation vapor pressure. Equation 11-4 has been put forth by Penman, McIlroy, and Monteith, and is reviewed by Tanner (1968). The plant exerts its influence on transpiration through variation of the stomatal resistance r_i. The extent of this influence depends upon the air resistance and the energy budget. The air resistance depends in turn upon the wind speed, leaf geometry, and density of the vegetative canopy.

It is frequently assumed that plant transpiration is proportional to the potential evapotranspiration. This is a reasonable assumption only as long as r_i is small compared to r_a. As the leaf loses turgor, r_i becomes increasingly large and the proportionality eventually fails. There is generally a very abrupt decrease in rate of transpiration as the leaf dehydrates beyond the wilting point.

Transport Through the Plant

Water transport through the plant involves a complex set of pathways and it cannot be said that it is thoroughly understood in any quantitative sense. The relation between the flux through the plant and the water potential difference between the root surfaces and the leaves is nonlinear. It is known that the killing or cutting of the tips of plant roots tends to reduce the resistance to flow. Root resistance is very sensitive to poor aeration, low temperatures, and metabolic poisons which inhibit root respiration. Figure 11-6 shows the relation between the flux through an onion plant and the potential difference across the plant, as measured by Millar, Gardner, and Goltz (1971). Similar curves are obtained for plants growing in nutrient solutions or soils at different water potentials. It is probable that the explanation for the nonlinearity lies in the coupled transport of ions and water across the root membranes. Whatever the

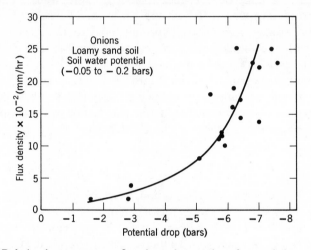

Fɪɢ. 11-6.   Relation between water flux through an onion plant and the total potential between leaves and roots. (Millar, Gardner, and Goltz, 1971.)

explanation, the effect is to cause the water potential of the plant during the day to follow that of the soil, with only a modest effect due to transpiration rate. Even very low transpiration rates, as might be found at night, result in a rather significant difference between the leaf-water potential and that around the roots.

Figure 11-7 shows a continuous record of both plant leaf- and soil water potential for bell pepper under greenhouse conditions (Rawlins et al., 1968). The soil water potential measurements were obtained with an *in situ* thermocouple psychrometer at the 25-cm depth and tensiometers at 25-, 55-, 85-, and 115-cm depths. Leaf-water potential was obtained from leaf-water content measurements obtained by beta-ray absorption and calibrated by means of the psychrometer. Leaf-water potential measurements, as seen in Figures 11-6 and 11-7, were very difficult

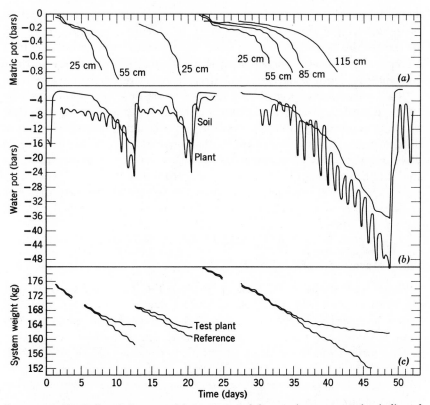

FIG. 11-7. (*a*) Soil matric potential as measured by tensiometers at the indicated soil depths. (*b*) Total soil water potential as measured by a thermocouple psychrometer at the 25-cm depth, and plant water potential as inferred from a calibrated beta gage. (*c*) Actual weight of the soil-plant system (upper curve) and hypothetical weight that would have occurred had transpiration been equal to the potential transpiration (lower curve). The lower curve was obtained by matching the transpiration rates of the test plant and an evaporimeter plant during the period of maximum transpiration following each irrigation. All parameters are functions of time elapsed from the beginning of the experiment. (Rawlins et al., 1968.)

to obtain until the development of the thermocouple psychrometer. Such measurements are now made routinely on detached samples and *in situ* measurement methods are improving rapidly. The lower set of curves in Figure 11-7 shows the weight of the entire soil column (30 cm in diameter and 120 cm deep) as a function of time compared with that of a test plant adequately supplied with water and used as a measure of the potential evaporation in the greenhouse. Divergence of the two curves indicates a decreasing rate of transpiration from the test plant.

This occurred on days 7 and 35 when the soil-water potential fell below −12 bars. After the first drying cycle, the test plant failed to return to the original transpiration rate for 2 or 3 days after irrigation. It is a commonly observed phenomenon that the stomates do not regain their full aperture until a day or two after stress is relieved. This may be related to the production of a growth hormone in the roots. (Kirkham, 1971). Figure 11-7 illustrates another interesting feature of water uptake by root systems. There was very little water uptake from the lower third of the soil column until the water potential in the upper two-thirds was quite low. This was despite the fact that the root system had proliferated quite extensively throughout the entire column. This appears to be a consequence of a high resistance to flow due to the small diameter of the lower roots amplified by the flux-potential difference relation shown in Figure 11-6.

Water Movement to Plant Roots

Figure 11-7 shows the difference between the leaf-water potential and the *average* soil-water potential at several depths in the soil. Included in this potential difference therefore is the potential gradient in the soil itself, if any, in the vicinity of the plant roots. An estimate of this gradient can be made from the solution of the flow equation. If we assume that the root can be simulated by a cylinder of radius a and that adjacent roots are a distance $2b$ apart, then the approximate steady-state solution for this geometry is

$$\psi_b - \psi_a = \left(\frac{q}{4\pi k}\right) \ln \left(\frac{b^2}{a^2}\right) \qquad (11\text{-}5)$$

where ψ_b is the potential midway between roots and is that which would be measured by any finite sized measuring device, ψ_a is the potential at the plant root-soil boundary, q is the volume of water taken up per unit length of root per unit time, and k is the unsaturated conductivity. We have neglected any contact resistance at the root-soil interface.

Figure 11-8 (Gardner, 1960) shows the potential at the plant root plotted as a function of the soil water potential for Pachappa sandy loam for three different rates of water uptake. Except for unusual circumstances, the lowest rate indicated is probably more nearly that which occurs in nature and may, in fact, be high for a fully developed root system. The main point is that over short distances there is little water potential or water content gradient in the vicinity of the roots until the soil becomes very dry. Even then, the flux rate tends to decrease markedly due to plant wilting, so in many cases one can probably assume

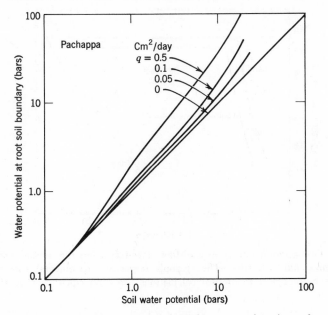

FIG. 11-8. Water potential at root-soil boundary as a function of average soil water potential for different rates of water uptake q. (Gardner, 1960.)

that the roots are at very nearly the same potential as the surrounding soil.

It can be shown (Lang and Gardner, 1970) that in a sufficiently dry soil, the soil can limit the rate of water movement to plant roots just as it does to an evaporating surface. Transpiration from wilted plants appears to be limited largely by the soil properties and by the root extent. The stomates open just far enough and often enough to permit water loss only as fast as it is obtained. The transpiration rate decreases with decreasing soil water content and/or soil water potential. Figure 11-9 (Gardner and Ehlig, 1963) shows the transpiration rate for birdsfoot trefoil as a function of water content for three soil textures. This very nearly linear relation between transpiration rate and soil water content is found for both uniform and nonuniform root systems and for a wide variety of conditions. As yet no really adequate model of water uptake by nonuniform plant root systems exists. Those which have been developed thus far fail to take into account the nonlinear nature of the flux-potential curve and do not predict correctly the uptake from the lower part of the root system. On a larger scale, water content gradients of considerable magnitude can and do exist and account for significant movement of water into the root zone. The equations developed in Chapter

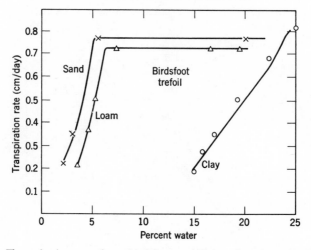

Fig. 11-9. Transpiration rate from birdsfoot trefoil as a function of soil water content for three different soils. The symbols represent experimental data and the smooth curves represent theoretical calculations based on equation 11-5. (Gardner and Ehlig, 1963.)

9 for capillary rise can be used to estimate the contribution of a ground-water table to transpiration.

An additional word about root uptake patterns is in order here. Root uptake patterns are often inferred from water content changes in the various parts of the soil profile. Most of these inferred patterns neglect downward drainage, which continues longer than the presumed one or two days after irrigation. Water lost from the top foot of soil represents not only water taken up by the plant but water lost to the second foot. The result is to overestimate uptake in the upper foot and underestimate it in the second foot. When allowance is made for this redistribution, the uptake pattern tends to be more uniform. The opposite is the case when a shallow water table is present.

Relation of Plant Growth to Soil Water

Since the average daytime leaf-water potential tends to be a constant amount which is lower than the soil water potential, the relation between growth rate and soil water potential is similar to Figure 11-4. This is illustrated for onions in Figure 11-10 for three different soil textures (Millar et al., 1971) where the stomatal conductance, which is proportional to the growth rate, is plotted against water potential. The same data plotted as a function of soil-water content are found in Figure 11-11.

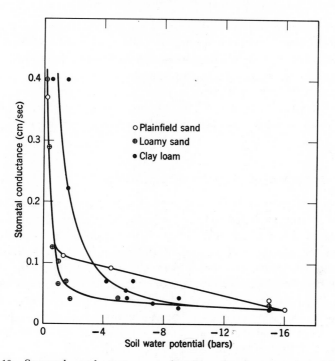

FIG. 11-10. Stomatal conductance as a function of soil water potential for seed onions, three different soil textures. (Millar et al., 1971.)

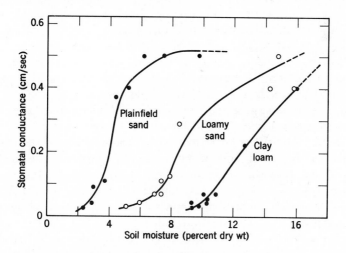

FIG. 11-11. Stomatal conductance as a function of soil water content for seed onions. (Millar, Gardner, and Goltz, 1971.)

In the coarse-textured sand there is a range of water contents over which the water potential does not decrease appreciably so there is little influence upon growth until it drops markedly. The curve for the clay is more gradual, again because of the shape of the curve that relates matric potential to layer content of the soil. There is little question about the very definite decrease in growth rate at the high water potential. Even a small decrease in matric potential results in a decrease in growth rate.

WATER RELATIONS OF SOIL MICROORGANISMS

Soil microorganisms bear some similarity to higher plants in that one can define components of the total potential according to equation 11-1. A significant difference is that lower organisms do not transpire appreciably and the time rate of change of hydration is relatively less. In the case of higher plants, the turgor pressure tends to decrease as the total potential decreases. This is not necessarily true for lower forms. Adebayo et al. (1971) found that the turgor pressure for the fungi *Mucor hiemalis* and *Aspergillus wentii* tended to remain constant or increase somewhat as the total potential decreased. Total potentials as low as 40 bars were achieved and the turgor pressure ranged from about 4 to 11 bars for the *Mucor hiemalis* and around 15 bars for the *Aspergillus wentii*.

Soil microorganisms are very nearly in equilibrium with the soil around them at all times because of their small size. (The same is true of plant seeds imbedded in the soil.) The effect of soil water potential on microorganism response varies from species to species and cannot be generalized. For example, some species of fungi, such as those just described, can tolerate extremely low water potentials relative to those which inhibit growth of higher plants. Figure 11-12 shows the rate of growth of the diameter of a colony of *Phytophthora cinnamoni* as a function of water potential. The matric potential was varied by varying the soil water content. The osmotic potential was varied with KCl solutions at a constant soil water content (Adebayo and Harris, 1971).

Soil water potential is correlated with various stages in the life cycle of other soil-borne organisms. For example, Collis-George and Blake (1959) found that the expulsion of the larval mass of the nematode *Anguina Argostis* was completely inhibited at matric potentials below 10 to −100 mbars, depending upon the texture of the porous media. They suggested that the textural influence was due to differences in the capillary conductivity of the different textures due to different water contents. In view of the small sizes of the masses, it would seem that the influence of soil water content as operating through some other mech-

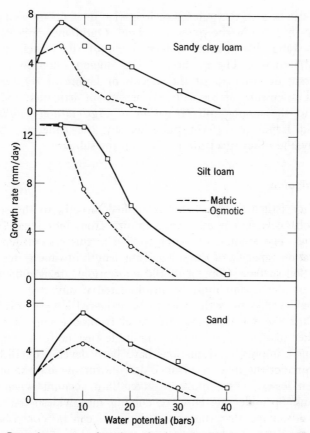

FIG. 11-12. Growth rate as a function of substrate water potential for *P. cinnamoni* for three different soil textures. (Adebayo and Harris, 1971.)

anisms. Osmotic potentials that extended well below the matric potential values had no effect upon expulsion.

IRRIGATION

Except perhaps in the rain forests of the world, nature does not normally supply adequate water with the frequency and in the amount for optimum plant growth; the only recourse is for man to do so. Large irrigation projects now exist in almost every part of the world and ambitious engineering works have been constructed in order to realize them. It is easily recognized that evapotranspiration places a daily demand upon the soil reservoir. From consideration of the relative magnitude of the potential transpiration in arid regions and the storage capacity of a soil profile, it can be seen that rainfall or irrigation must occur with reasonable frequency throughout the growing season. Optimum application of this irrigation water demands an understanding of the principles of water movement into soil. The method of introducing water into the soil is an important determinant of the success or failure of the venture. For a detailed discussion of virtually all aspects of irrigation the reader is referred to the monograph *Irrigation of Agricultural Lands* (Hagan, Haise, and Edminster, 1967) published by the American Society of Agronomy and other specialized works on the subject.

Surface Irrigation

In surface irrigation the water is applied directly to the soil surface through channels which may vary in size from individual furrows to large basins. The amount of water applied in this system depends upon the infiltration capacity of the soil and the length of time water is ponded upon the soil surface. Control of the infiltration rate is difficult so that the time of application must be manipulated if any degree of control is to be achieved. Unless the area to be irrigated is very small the time required for the surface water to travel from one end of the system to the other may be of sufficient magnitude that uniform irrigation becomes almost impossible. Equations have been developed (Bishop et al., 1968) to predict the time of advance down a furrow and the unavoidable percolation losses. These must be taken into account when designing the systems. Sprinkler systems are coming into widespread use. They have the advantage that the application rate can be controlled by the design of the system, as long as it does not exceed the infiltration capacity of the soil. This reduces the requirements for land leveling and generally involves lower labor costs.

Subirrigation

Subirrigation is practiced widely in certain humid areas. In the Netherlands, drainage is essential during the spring months when rainfall exceeds evaporation. In the summer, when the opposite is the case, the drain lines are used to supply water to the root system. The water table then becomes concave between the tiles. One can use the same formulas developed for drainage to estimate the spacing needed for irrigation, provided the appropriate rate of evapotranspiration is employed. Only the spacing of the tiles is generally subject to control so the feasibility of such systems depends upon the transpiration demand, the permeability of the soil, and the nature of the root system.

Subirrigation in arid regions poses additional problems over and beyond those encountered in humid regions. The most serious is the hazard of accumulation of soluble salts. Some means must be provided in such systems for the periodic leaching of the profile by surface application of water. Where high-quality water, such as that obtained by desalination, is to be used in arid regions, subirrigation by means of a constant, steady supply of water through buried plastic tubes is considered promising. Burying the tubes is intended to reduce evaporation losses from the soil surface and a steady rate of application just sufficient to meet the plant requirements minimizes percolation losses. The winter rains in Mediterranean-type climates may be sufficient to provide whatever leaching is needed.

Irrigation Scheduling

Physical and economic constraints usually place limitations upon the timing and amount of irrigation water applied to field crops. The plant response to the soil water and the storage capacity of the root zone provide further limitations. Actual practice usually represents a compromise which may provide less than optimum growth. Control of the amount of irrigation water applied is feasible under sprinkler irrigation since the application rate is independent of the soil properties and the initial soil water content. Smaller, more frequent applications can generally be managed more efficiently from the standpoint of percolation losses.

If one attempts to minimize the number of irrigations, then it is important to know how much water can be stored in the soil. Referring back to Chapter 10, we see that the drainage rate decreases as the soil water content decreases. The frequent assumption that the profile drains to some field capacity within a few days may be an acceptable approximation, if the drainage curve characteristic of the soil has been established. The lowest water potential or water content acceptable depends upon the requirements of the crop. The growth response curve is such that little benefit is gained by allowing the soil to go beyond about -1 bar. The reduction in transpiration is not commensurate with the reduction in growth.

The depth at which the water potential is measured and the potential at which irrigation are indicated is somewhat flexible. Fluctuations in the soil-water content damp out with depth, as was explained in Chapter 10.

Two general approaches to the scheduling of irrigation have been developed. The first uses some measure of the actual soil-water content or soil-water potential to determine when irrigation is needed. Tensiometers are particularly suited to fruit orchards and citrus groves. Electrical resistance blocks are also used. They are more simple to operate and have a wider range but are less precise. A variation in this procedure is to measure the plant-water status. The relative water content of the leaves gives an indication of their water potential; when it begins to decrease below some predetermined value, irrigation is indicated. The procedure is more adaptable to experimental work than routine irrigation.

In the second approach the water budget of the soil profile is estimated by using meteorological data to calculate the evapotranspiration. This method lends itself particularly to the management of large regional projects where irrigation requirements must be anticipated in advance, if possible. Local adjustments are normally required since evapotranspiration estimates for individual fields are not feasible. As computer modeling of the entire energy and water budgets of large areas is improved, irrigation forecasts should improve irrigation scheduling considerably, since it may permit irrigation district managers to give more flexibility to individual farmers.

References

Adebayo, A. A., and R. F. Harris (1971). Fungal growth response to osmotic as compared to matric water potential. *Soil Sci. Soc. Am. Proc.* in press.

Adebayo, A. A., R. F. Harris, and W. R. Gardner (1971). Turgor pressure of fungal mycelia. *Trans. Br. Mycol. Soc.* In press.

Barrs, H. D. (1968). Determination of water deficits in plant tissues, in *Water Deficits and Plant Growth*, T. T. Kozlowski, Ed. Academic Press, New York, pp. 236–347.

Bishop, A. A., M. E. Jensen, and W. A. Hall (1968). *Irrigation of Agricultural Lands*, R. M. Hagan, H. R. Haise and T. W. Edminster, Eds. Am. Soc. Agron. Monograph 11, Chapter 43.

Briggs, L. J., and H. L. Shantz (1911). Application of wilting coefficient determinations in agronomic investigations. *J. Am. Soc. Agron.*, 3:250–260.

Collis-George, N., and C. D. Blake (1959). The influence of the soil moisture regime on the expulsion of the larval mass of the nematode *Anguina Agrostis* from Galls. *Aust. J. Biol. Sci.*, 12:247–256.

Ehlig, C. F., and W. R. Gardner (1968). Effect of salinity on water potential and transpiration in pepper (Capsicum frutescens). *Agron. J.*, 60:249–253.

Gardner, W. R. (1960). Dynamic aspects of water availability to plants. *Soil Sci.*, 89:63–73.

Gardner, W. R., and C. F. Ehlig (1963). The influence of soil water on transpiration of plants. *J. Geophys. Res.*, 68:5719–5724.

Gardner, W. R., and C. F. Ehlig (1965). Physical aspects of the internal water relations of plant leaves. *Plant Physiol.*, 40:705–710.

Hagan, R. M., H. R. Haise and T. W. Edminster, Eds. (1968). *Irrigation of Agricultural Lands*. Am. Soc. Agron. Monograph 11.

Hendrickson, A. H., and F. G. Veihmeyer (1945). Permanent wilting percentage of soils obtained from field and laboratory trials. *Plant Physiol.*, 20:517–539.

Kanesmasu, E. T., and C. B. Tanner (1969). Stomatal diffusion resistance of snap beans. I. Influence of leaf-water potential. *Plant Physiol.*, 44:1547–1552.

Lang, A. R. G., and W. R. Gardner (1970). Limitation to water flux from soils to plants. *Agron J.*, 62:693–695.

Millar, A. A., W. R. Gardner, and S. M. Goltz (1971). Internal water status and water transport in seed onion plants. *Agron. J.* In press.

Rawlins, S. L., W. R. Gardner, and F. N. Dalton (1968). *In situ* measurement of soil and plant leaf water potential. *Soil Sci. Soc. Am. Proc.*, 32:468–470.

Richards, L. A., and C. H. Wadleigh (1952). Soil water and plant growth, In *Soil Physical Condition and Plant Growth*, B. T. Shaw, Ed. Am. Soc. Agron. Monograph Vol. 2, Academic Press, New York, pp. 73–251.

Richards, L. A., and L. R. Weaver (1943). Fifteen atmosphere percentage as related to the permanent wilting percentage. *Soil Sci.*, 56:331–339.

Salter, P. J., and J. E. Goode (1967). *Crop Responses to Water at Different Stages of Growth*. Commonwealth Agr. Bur., Farnham Royal, Bucks, England.

Slatyer, R. O. (1967). *Plant-Water Relationships*, Academic Press, London.

Slatyer, R. O. (1969). Physiological significance of internal water relations to crop yield. *Physiological aspects of crop yield*. Am. Soc. Agron. Symposium (Lincoln, Nebraska), Chapter 4.

Stanhill, G. (1957). The effect of differences in soil moisture status on plant growth: A review and analysis of soil moisture regime experiments. *Soil Sci.*, 84:205–214.

Tanner, C. B. (1968). Evaporation of water from plants and soils, in *Water Deficits and Plant Growth*, T. T. Kozlowski, Ed. Academic Press, New York, Chapter 3.

Troughton, J. H. (1969). Plant water status and carbon dioxide exchange of cotton leaves. *Aust. J. Biol. Sci.*, 22:289–302.

van den Honert, T. H. (1948). Water transport in plants as a catenary process. *Disc. Faraday Soc.*, 3:146–153.

Veihmeyer, F. J., and A. H. Hendrickson (1933). Some plant and soil-moisture relations. *Am. Soil Survey Assoc. Bull.*, 15:76–80.

CHAPTER 12

Soil Water Management

As man has intensified his use of the land he has found it increasingly important to give attention to the management of soil water. Whether he is concerned with increased production of food and fiber or preservation of an ecosystem, man finds that his manipulation of the hydrological cycle has a profound influence on many processes in the biosphere. This chapter will deal with the more conscious efforts to manage the soil water regime.

SURFACE WATER MANAGEMENT

The annual precipitation in humid regions exceeds the potential evaporation. Inasmuch as the soil profile and the ground water aquifers have a limited capacity for transmission of water, surface drainage must accommodate the excess. For any given region, a natural drainage network has evolved through geomorphological processes which is characteristic of the transmissibility of the soils and the quantity of water to be drained. It is frequently, though not invariably, the case that man-made alterations in the vegetative cover of a large region tend to increase the ground water runoff. The plowing under of grasslands and the clearing of forests result in a decrease in evapotranspiration and thus an increase in deep percolation and ground water runoff. The result of having large areas is even more obvious.

The seasonal distribution of rainfall in many humid areas results in excess water in the spring. Low-lying areas subject to runoff from adjacent areas and areas of relatively shallow soils with flat topography often exhibit lower infiltration and higher direct runoff. An obvious and simple practice to enhance surface runoff from soil areas is that of land forming. In this practice the surface of the soil is smoothed so that depressions are removed. Artificial drainage channels are provided where natural channels do not exist or are unsatisfactory. Even small depressions in the soil surface can result in a highly nonuniform distribution of water in the soil profile.

In areas of less rainfall, land forming has been practiced for exactly the opposite reason—increased infiltration. Terraces may be constructed to store surface water until it can infiltrate rather than to permit it to run off. Smoothing the soil surface can be extremely important where flood irrigation is practiced since the amount of water that infiltrates the soil depends upon the length of time it is ponded upon the surface. An uneven topography with many ridges and depressions can result in extremely poor water distribution.

Surface Mulching and Fallowing

In continental areas such as the Great Plains of the United States, there is sufficient rainfall in most years to produce a crop. Except for very wet years, the rainfall is far from optimum. Any reduction in unwanted water loss results in almost proportional increases in crop production. Stubble mulch tillage is often practiced in some areas. As much dead vegetative cover is left on the soil surface as possible. This cover tends to reduce surface runoff or snow loss and, to a lesser extent, evaporation losses from the soil surface.

The long-term beneficial effect of a mulch depends very much upon the frequency and amount of rainfall. Over a long rainless period the water loss from a mulched soil tends to approach that from the same soil in the absence of a mulch. Surface mulches of almost any kind tend to have a greater effect upon the thermal regime of the soil than upon evaporation losses. A mulch has its greatest inhibiting effect upon evaporation the first few days after a rainfall (Gardner and Gardner, 1969). A highly reflective cover will result in a decrease in evaporation through a decrease in the net radiation. Complete mulches such as impermeable plastic films or asphalt sprays (Waggoner et al., 1960) alter the evaporation markedly. When placed between rows of transpiring plants, the effect of a mulch depends a great deal upon the thermal mixing in the canopy. Some of the heat that would otherwise go into evaporation now goes into transpiration. However, in some cases there is a sufficient reduction in evapotranspiration through such mulching to be of benefit.

Although a surface mulch may have only a modest effect upon evaporative losses from the soil, it has a profound effect upon the temperature and moisture regime of the surface horizon. Figure 12-1 shows the soil water content in the surface 6 in. in a citrus grove. Cahoon et al. (1961) found little moisture conservation under nine woodshaving mulches and

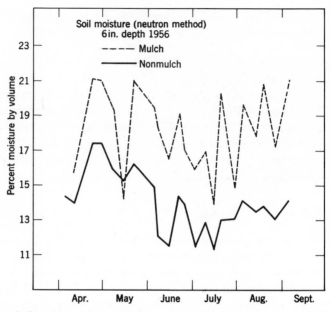

F IG. 12-1. Soil moisture determinations made with the neutron moderation method in orchard I at a depth of 6 in., under woodshavings mulch and no-mulch treatments. (Cahoon et al., 1961.)

little or no increase in yield due to the mulch. The surface soil water content was higher at almost all times under the mulch. A mulch therefore significantly alters the environment of soil microorganisms.

Most soils tend to limit evaporation from their surface within a day or two after a rainfall. When vegetative cover is present, water loss continues at the potential rate for a much longer time. Where the rainfall supply is marginal at best, the practice of summer fallowing may be adopted. In this practice, vegetative cover is prevented over an entire season and the land is allowed to lie fallow. A surface mulch may be added in an attempt to further reduce surface water loss. It is the hope that a sufficient quantity of the rainfall which is thereby conserved from evaporation will remain within the soil profile so that this water will be available for plant use at the beginning of the next growing season. The conserved water must not all be lost through drainage if this system is to be beneficial; there must be adequate storage in the soil profile to take advantage of it. No benefit would accrue in regions where the soil profile storage is replenished each season. It can be seen that if summer fallowing were practiced over a very large area, one would expect a significant increase in percolation to ground water, since some increase in percolation is inevitable

Surface Shading and Cover

Rather extreme measures for limitation of evaporation through surface cover may be justified under certain circumstances. Mechanical shading by means of wood or metal strips intercepts the radiant energy and causes a local decrease in the net radiation on the crop or soil. If the light levels are not reduced below those required by the plants, the reduction in transpiration and leaf temperatures may be beneficial. In the future, desalination of sea water may provide a substantial fraction of available fresh water in many arid regions. The high cost of purification of this water makes mandatory its most efficient use. In such cases, large areas may be covered with transparent films which transmit light in the photosynthetic region but prevent the loss of water from the system. The water is then recycled many times before it is eventually lost through percolation. Such a system entraps a large amount of heat, which must be removed by ventilation. It also requires highly efficient management.

Induced Runoff

Another practice, which has ancient origins (Tadmor et al., 1961) and which is subject to renewed interest as the population of arid regions increases, is the modification of the nature and topography of the soil surface in order to produce runoff from a certain fraction of the surface. This runoff is applied to the remainder of the surface. By adjusting the ratio of the area of runoff to the area of runon, the infiltration in the runon area can be any amount desired. A similar result is achieved without actual surface manipulation in some dryland areas through the adjustment of plant-row spacing, as with sorghum production in the Great Plains of the United States. A row spacing which provides completed vegetative cover of the soil surface at maturity requires more water than is generally available. Decreasing the plant density reduces the actual transpiration below the potential evapotranspiration, since the plant and its root system transmit water from the soil to the atmosphere more readily than bare soil. This is nothing more than a variation of the natural plant distribution which occurs in arid ecosystems where the plants are distributed to use all of the available water.

Surface water management, at best, is capable of only moderate influence on the soil water regime. A very important aspect of such management, particularly from the standpoint of agricultural lands is erosion control. Discussion of this aspect of soil water management is presented in Chapter 13. Where major manipulation of the soil water is required, subsurface means are required.

DRAINAGE

The importance of soil aeration to plant growth was discussed in Chapter 6. There must be an exchange of gases between the atmospheres of the soil and of the air to provide a suitable environment for root growth and microbiological activities. Oxygen enters the soil atmosphere and carbon dioxide leaves it primarily through the process of gaseous diffusion. If the soil pore space is filled with water, this exchange cannot take place. The discussions in Chapter 5 clearly showed the role of soil oxygen in root proliferation. Increased CO_2 concentrations in the root zone and decreased O_2 uptake inhibit water and nutrient uptake by the plant.

Lack of aeration impedes biological activity. Organic matter decomposition is retarded and the rate of mineralization of organic nitrogen compounds is decreased. Some pathogenic organisms become more active under anaerobic conditions, which accentuate problems with soil-borne diseases. The rate of mineralization of nitrogen is also affected by soil temperature. Consequently, poorly drained soils in the temperate region, with their high heat capacities, are slow to warm up in the spring. These cold soils retard the release of nitrogen to growing plants. Not only do low soil temperatures affect nutrification processes, but they also delay germination of spring-planted seeds and the subsequent establishment of the crop.

In addition to the impacts of excess water on aeration and soil temperatures, wet soils are easily compacted by both machine and animal traffic. These compaction effects increase the bulk density of the soil, which results in reduced root development due to greater mechanical impedance and lower oxygen amounts. As in the case with surface water, excess ground water in the humid regions may be the result of infiltration of rainfall into the soil profile or may be due to ground-water inflow from adjacent areas. The problem is somewhat different in arid regions. Irrigation on a wide scale readily overtaxes the natural drainage system, whose capacity is adapted to much smaller rates of percolation to ground water. Waterlogging of the soil profile is the frequent result. However, even in areas where internal drainage is adequate and waterlogging does not occur, soil salination is almost inevitable unless care is taken to ensure adequate drainage. History is replete with the decline of ancient civilizations that resulted from salination of soils. All irrigation water carries a certain content of dissolved salts. These are left behind in the soil profile when the water is evapotranspired. If these salts are not leached out, they accumulate to levels that are toxic to crop plants.

Specification of the Drainage System

In order to design an adequate drainage system several factors must be taken into account. Whether it consists of lines of tile, tube wells, or open ditches, the role of such a system is to shorten the distance which water must flow through the porous medium, by increasing the extent of surface flow or flow through pipes. It is axiomatic that water will not leave the soil profile and enter a tile drain or ditch unless the potential energy of the water in the drain is lower than that in the soil. This means that a drain must be located below the water table, which is defined as the locus of points at which the matric potential is zero. For the same reason, the water table cannot be lowered below the level of the drain. Thus the desired depth of water table must be known in order to specify the drainage system. This requires a knowledge of the aeration requirements of the plants, their rooting habits, and the soil properties. Unfortunately, it is still much easier to design a drainage system with a given capability than to specify what that capability ought to be. In regions such as the Netherlands, drainage is necessary in order to provide adequate aeration. Lowering the water table too much results in a deficiency of water during the summer when evapotranspiration exceeds rainfall. Thus providing too much drainage is not only costly in terms of installation of excess capacity but in terms of plant yields. The optimum water-table depth depends upon the crop, the soil, and the rainfall distribution; it is usually determined from experience.

While steady-state calculations are often made in order to determine drainage requirements, the periodic nature of rainfall often results in fluctuating water tables. The rate of fall of the water table may then be as important, or more important, than the steady-state level to which it returns.

Leaching Requirement

In arid regions where salinity is a hazard, the drainage requirement carries an additional constraint. The leaching requirement is a concept developed by L. A. Richards based on the conservation of mass. In order to maintain a steady-state concentration of salts in the soil profile, the amount of salt that enters the soil must equal, on the average, to that which leaves it. The amount entering is the quantity of irrigation water times the concentration of salts in the irrigation water. If I is the quantity of irrigation water and C_i is its concentration, then the quantity of salts introduced into the profile in a given time is IC_i. At steady state

$$IC_i = DC_d \tag{12-1}$$

where D is the quantity of drainage water and C_d is its concentration. It is further assumed that for any given crop there is a limiting concentration of the drainage water which must not be exceeded if acceptable yields are to be maintained. If one denotes this by C_{max}, then the leaching requirement is defined as

$$\text{LR} = \frac{D}{I} = \frac{C_i}{C_{max}} \tag{12-2}$$

The leaching requirement represents that fraction of the irrigation water which must pass through the root zone as drainage in order to maintain the salt concentration of the soil solution below the maximum allowable level. This level depends upon the salt tolerance of the crop.

Drainage Problems

Drainage problems are formulated in terms of the equation for flow of water in saturated soil known as Laplace's equation, as discussed in Chapter 9. Such problems are known as boundary-value problems. In order to specify the problem uniquely, it is necessary to delineate mathematical boundaries corresponding to surfaces in the real physical system and to specify the fluxes across these boundaries, the hydraulic head on the boundaries, or a combination of the two. It is sometimes though not always convenient to take the soil surface as the upper boundary. In many cases it would be desirable to take the water table as a boundary, but often the position of the water table is not known until the problem has been solved. If an impermeable layer exists within a few meters of the soil surface, this is usually significant and is taken as the lower boundary. If the impermeable layer is very deep, the boundary is taken at infinity.

The number of conceivable geometrical arrangements of boundary conditions which one might meet in the field is almost limitless. A large number of solutions of Laplace's equation have been published to date. The availability of high-speed digital computers now makes it possible to solve virtually any problem one can specify. It will be possible to consider a few simple approximate solutions here. Rigor will be sacrificed for simplicity.

Rectilinear Flow

Many flow problems can be reduced to a fair degree of approximation to problems in one-dimensional flow. The flow equation for the steady state in this case reduces simply to Darcy's law. An example of this is the ponded water situation illustrated in Figure 12-2. A permeable loam soil of thickness d and hydraulic conductivity k overlies a layer of gravel that rests on an impervious clay. Water is ponded to a height t above the soil. Two ditches at a distance a apart penetrate the soil to the gravel. The walls are vertical and considered impermeable to the

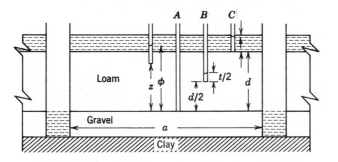

Fig. 12-2. Geometry for rectilinear vertical seepage of ponded water into gravel penetrated by drainage ditches. (Kirkham, 1957.)

water except where they enter the gravel; then their permeability becomes infinite. Water stands in the ditches to the upper layer of the gravel.

According to Darcy's law (Chapter 9), the volume of water draining through the soil in unit time is equal to

$$Q = k \left(\frac{d + t}{d} \right) a \qquad (12\text{-}3)$$

where Q is the volume of water in cubic feet per day per foot of ditch, k is the hydraulic conductivity in feet per day, d is the depth of the soil in feet, t is the depth of the ponded water in feet, and a is the distance between ditches in feet. This simply states that the volume of flow is directly proportional to the hydraulic conductivity, the hydraulic head, and the length of flow surface; it is inversely proportional to the depth of the soil. If t approaches zero, the velocity of flow per unit area will be equal to the hydraulic conductivity.

An examination of piezometers A, B, and C in Figure 12-2 enables one to distinguish between the two important terms in flow problems, *pressure* and *hydraulic head.* Recall that pressure head refers to the height of a water column which can be supported at a given point in the soil by the soil-water pressure at that point. For example, the pressure head at the base of piezometers *A*, *B*, and *C* is zero, $t/2$, and t, respectively. Hydraulic head at the same point signifies the distance from a given reference level to the top of the water column that can be supported by the pressure at this point. The hydraulic head at the base of piezometers *A*, *B*, and *C* is zero, $(d/2 + t/2)$, and $(t + d)$, respectively.

As the thickness of the ponded water approaches zero, the pressures at the bases of the piezometers and at all other points in the soil also

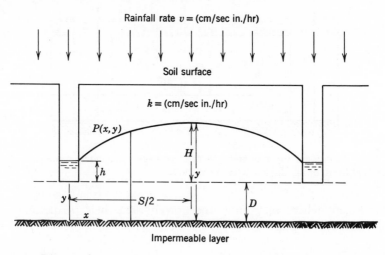

FIG. 12-3. Diagram for Hooghoudt's drain-spacing formula (Luthin, 1966, used with permission of John Wiley and Sons.)

become zero. This means that there will be no pressure along the ditch walls which will prevent the seeping of water outward. It is important to note this point since seepage at the face of a ditch has sometimes been overlooked.

Another example of one-dimensional flow is illustrated in Figure 12-3. Flow in this system is not actually one dimensional, but a rather good estimate of the relation between the drainage rate and the height of the water table can be obtained by assuming that it is. These assumptions are due initially to Dupuit (1863). It is assumed that all streamlines in such a system are horizontal and that the velocity of flow along these streamlines is proportional to the gradient of the hydraulic head but independent of the depth. For these assumptions to be acceptable, *H* in Figure

12-3 must be small relative to D. If these two assumptions are applied, one obtains the relation

$$q = -ky\frac{dy}{dx} \qquad (12\text{-}4)$$

where q is the horizontal flux towards the drain per unit length perpendicular to the plane of the figure. But the flux q must increase with increasing x in proportion to the rainfall or percolation rate v. Thus

$$q = -v\left(\frac{S}{2-x}\right) \qquad (12\text{-}5)$$

If one combines equations 12-4 and 12-5 and integrates, one obtains for $y = (D+H)$ when $x = 0$

$$y^2 - (D+h)^2 = \frac{v(Sx - x^2)}{k} \qquad (12\text{-}6)$$

At the midpoint between drains $x = S/2$, $y = (D+H)$:

$$(D+H)^2 - (D+h)^2 = \frac{v}{k}\frac{S^2}{2} \qquad (12\text{-}7)$$

The total flow to any drain is given by the product $vS = Q$ so that

$$S = \frac{2k}{Q}[(D+H)^2 - (D+h)^2] \qquad (12\text{-}8)$$

If one assumes that h is negligibly small compared with D, this reduces to the equation given by Hooghoudt (1940), which is similar to the formula derived by Aronovici and Donnan (1946). This equation has been widely used to calculate the drain spacing from the permeability, flux, and desired water table level. It can be seen from equation 12-7 that the spacing does not increase linearly with the permeability but as the square root of permeability and inversely as the square root of the flux.

If H is small compared to D, instead of equation 12-4 one can write

$$q = -kD\frac{dy}{dx} \qquad (12\text{-}9)$$

which results in a more simple expression but one with limited applicability. One application is illustrated in Figure 12-4. We are interested here in calculating the rise in the water-table elevation y due to an increase in infiltration rate v. This might be due to introduction of irrigation, increased precipitation due to climatic changes, or increased percolation due to a change in vegetative cover with a consequent decrease in evapotranspiration. The flux across any plane x must increase with increasing

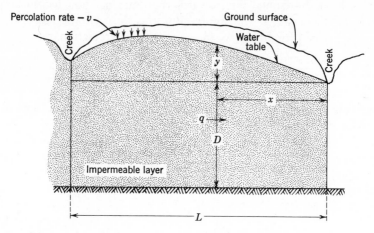

Fig. 12-4. Diagram for calculating water table rise due to irrigation.

distance from the creek in proportion to the percolation rate v. As before, one can write

$$q = -v\left(\frac{S}{2} - x\right) \tag{12-10}$$

where S is some arbitrary distance from the creek where $x = 0$. If equations 12-9 and 12-10 are equated and integrated, one obtains the result

$$y = \frac{v}{kD}\left(\frac{Sx}{2} - \frac{x^2}{2}\right) + y_0 \tag{12-11}$$

where y_0 is the value of y at $x = 0$. At $x = S/2$ this becomes

$$y = \left(\frac{v}{kD}\right)\frac{S^2}{8} \tag{12-12}$$

Equation 12-12 and Figure 12-3 illustrate how the topography and the hydrological cycle are related. If y exceeds the elevation of the ground surface at any point, surface runoff will occur. Equation 12-12 will be used in the discussion of a falling water table.

Transient Flow-Falling Water Table

The equations derived here assumed a constant rate of supply of water per unit area of soil. This permits one to calculate the steady-state elevation of the water table. Since equations 12-8 and 12-11 are equations describing ellipses, the shape of the water table predicted by these equations is elliptical. It is also important to know how rapidly the water table will drop if the infiltration were to suddenly stop. Here a very important assumption is almost always made. It is assumed that a certain constant fraction of the pore space will drain as soon as the matric potential becomes negligible. This fraction is known as the *specific yield*. It is sometimes assumed that there is an undrained region just above the water table known as the *capillary fringe*. This might be only a few centimeters in a sandy soil, but greater and less well-defined in a finer textured soil. In practice, the water content distribution above a falling water table does not decrease with elevation so abruptly as is assumed for purposes of calculation. This leads to errors in predicting the water content of the drained soil but still gives good estimates of the rate of flow to drains. Let us assume that f cm of water will drain for each centimeter of drop in the water table, and that the downward percolation in Figure 12-3 is coming from the draining pores rather than from the soil surface. Then one can write

$$v = -f \frac{dy}{dt} \tag{12-13}$$

where dy/dt is the rate of fall of the water table and t is the time. But from equation 12-12,

$$v = 8k \frac{Dy}{S^2} \tag{12-14}$$

Equating these two relations for v gives a differential equation whose solution is

$$y = y_0 \exp\left(-\frac{8kDt}{fS^2}\right) \tag{12-15}$$

where $y = y_0$ when $t = 0$.

This type of exponential decay is to be expected when the falling water table results in a decreasing hydraulic head difference and when there is little change in the flow geometry. Equation 12-15 differs from the one derived by Glover (see Dumm, 1954) only in that Glover's equation has a factor π^2 instead of the factor 8 in the exponent. This is because Glover assumed a flat water table at time $t = 0$, whereas the foregoing derivation starts with the elliptical water table.

If the maximum allowable time for y to decrease to some desired value is specified, equation 12-15 can be used to calculate the necessary drain spacing, provided that k and f are known.

Two- and Three-Dimensional Flow

In actual drainage problems flow is seldom truly one dimensional. Much of the effort devoted to the mathematical solution of drainage problems in the past has been concentrated upon solving Laplace's equation in two dimensions:

$$\frac{d^2H}{dx^2} + \frac{d^2H}{dy^2} = 0 \qquad (12\text{-}16)$$

An infinite number of solutions of this equation exists and the problem is to find that solution which matches the specified boundary conditions. Some problems can be solved by analytical procedures. The analogy between Darcy's law and Ohm's law has been used in the construction of electric analogs. These may consist of electrical conducting paper, tanks of electrolyte solution, or networks of resistors. The advent of high-speed digital computers has made it possible to obtain solutions to almost any conceivable problem and the number now extant is much too large to reproduce a representative sample here.

The solution of Laplace's equation is most readily understood when it is represented graphically as a flow net. An example of such a net is shown in Figure 12-5, which is the solution to the ditch drainage problem illustrated in Figure 12-3. This solution was obtained analytically by Kirkham (1950). The lines that intersect the soil surface and the face of the ditch are streamlines which show the actual direction of flow. At right angle to the streamlines are the equipotential lines or lines of equal hydraulic head. It can be seen that a drop of water which infiltrates midway between two ditches must travel a much greater distance before it reaches the ditch than one that starts near the ditch. Since the potential difference between the ends of all the streamlines is the same, the average gradient decreases with increasing length of streamline. The average velocity of flow also decreases. Since the distance increases with length, the time of travel from the surface to the ditch increases approximately as the square of the length of the streamline. This is important to the movement of dissolved substances in the soil solution.

Figure 12-6 shows the flow net for a similar problem except that the flow is down a sloping, impermeable layer. The problem of hillside seeps often occurs in areas with heavy subsoils. Water flows down the slope at the juncture between the surface soil and the hardpan until it seeps through to the immediate surface. The correct drainage procedure under these conditions is to place the drains above the seepage area at right angles to the plane of water flow. The seepage water is intercepted and removed from the slope. Maximum interception of the seepage water is achieved when the drains or ditches are placed on the impervious layer (Childs, 1946). Open ditches or buried drain tile are equally effective with land slopes of 1 ft in 30 or less. The ditch is a more effective interceptor on steeper slopes; buried tile permit as much as 25 percent

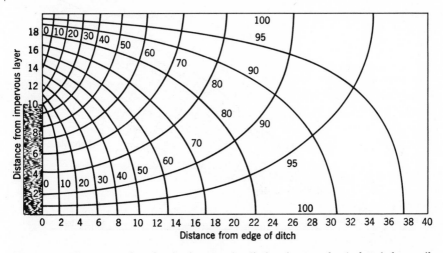

FIG. 12-5. Flow net when level of water in ditches is at a depth $h/2$ below soil surface and distance $2s = 4h$; only net between $x = 0$ and $x = s$ is shown. (From Kirkham, 1950.)

of the water to pass by. In Figure 12-6 the equipotentials and streamlines are shown for a tile drain on an impervious bed on a slope of 1 ft in 3. It should be noted that the equipotentials in the more pervious surface layer are almost perpendicular to the impervious bed, except in the immediate vicinity of the tile drain. Streamlines are parallel to the bed until they approach the drain tile.

Radial Flow

If the soil profile is very deep, the depth of any impermeable layer becomes increasingly less important. Most of the resistance to flow in such a drainage system occurs in the vicinity of the drain. The solution of Laplace's equation (Chapter 8) in two dimensions for radial flow gives

$$H = \left(\frac{Q}{2\pi k}\right) \ln r \qquad (12\text{-}17)$$

If in Figure 12-3 we had an auger hole rather than a long ditch, our problem would become a problem in radial flow with cylindrical geometry. If we make the equivalent of the Dupuit assumption, equation 12-17

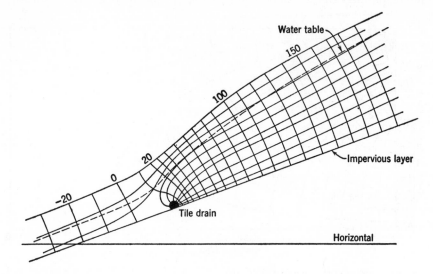

F_IG. 12-6. Interception by drain laid on impermeable bed with bed gradient of 1 in 3. (Childs, 1946.)

is the appropriate solution. Let a be the radius of the hole, H_a the head of water in the hole, and H_r the head at distance r from the axis of the drain. Then

$$Q = (H_r - H_a)\,\frac{2\pi k}{\ln\,(r/a)} \qquad (12\text{-}18)$$

One obtains a similar expression for a horizontal tube drain in a uniform deep soil. If the spacing between drains is less than the depth of the drain below the soil surface, then H_r is the head at the distance at $r = S/2$, where S is the distance between drains. The maximum head that can occur will be midway between adjacent drains. Instead of equation 12-18 we have, referring to Figure 12-3,

$$Q = (H - h) \frac{2\pi k}{\ln (S/2a)} \tag{12-19}$$

This equation differs by a small constant from that given by Hooghoudt (1940).

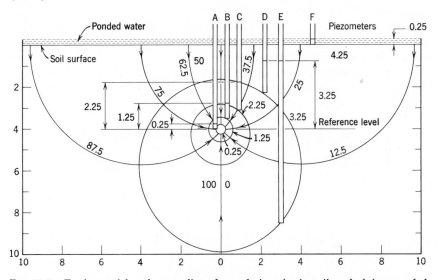

FIG. 12-7. Equipotential and streamlines for a drain tube in soil underlying ponded water. In this model, $d = 4$, $r = 0.25$, $s = 0.25$, and $t = 0.25$. The reference level is the axis of the drain. Six piezometers have been installed to show the different pressures at the various equipotentials. The bottom of piezometer A is situated on the axis of the drain. Therefore its height is 0.25. The bottoms of piezometers D and E touch the 3.25-foot equipotential. The water pressure in D is 1.55 ft of water; that in E is 7.7 ft. The streamlines are at right angles to the equipotentials. They are designated as 0, 12.5, 25, 37.5, 50, 62.3, 75, 87.5. and 100. They are expressed as percentages of the volume of flow. The difference between the values of adjacent lines represent the percentage of total flow that takes place between these lines. Although the value of Q between streamlines is the same $(0.125Q)$, the amount of water entering the soil per unit area over the tile drain is 16.6 percent as compared with 1.9 between streamlines 25 and 12.5. This difference is nearly ninefold. (Kirkham, 1957.)

If, on the other hand, the distance between drains is greater than the depth of the drain, the problem is more difficult. Let water be ponded on the surface to a depth t and the drain is a distance d below the soil surface as is illustrated in Figure 12-7. This problem was also solved by Kirkham (1957). If one imagines a source of water a distance d above the soil surface with symmetrical flow above and below the soil surface, it turns out that one can obtain an almost exact solution:

$$Q = \frac{2k(d + t - h)}{\ln\,(2d/r)} \qquad (12\text{-}20)$$

The numbers given in Figure 12-7 are dimensionless so any unit of length can be used, as long as the permeability is given in the same units. As the depth of the drain is increased, the numerator in equation 12-20 increases linearly; the denominator increases much more slowly since it increases logarithmically with d. Thus the flux to the drain increases with the depth to the drain in a manner which is illustrated in Figure 12-8. An analysis of the various geometric factors that affect the rate of seepage of water into tile drains indicates that the depth of tile placement is the most important single factor.

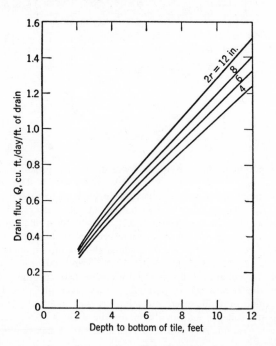

FIG. 12-8. Drain tube inflow, Q, as a function of depth, $d + r$, to bottom of drain. Hydraulic conductivity = 1 in./da, no ponded water ($t = 0$, r = radius of drain). (Kirkham, 1949.)

Intermediate Cases

In general, the geometry of the system does not permit such a simple approach. It is sometimes possible to combine the linear and radial flow-solutions in such a way to obtain a reasonable approximation to some problems. More often, more sophisticated methods are necessary. The reader is referred to the specialized literature for details.

The presence of an impervious layer has one of the greatest influences upon the water-flow pattern in the soil since it limits the flow region. Figure 12-9 shows the flow nets for three different depths to the impermeable layer. It is noted that the depth to the impervious layer not only affects the streamline pattern but also influences the height of the water table. When the depth of the impervious stratum is 5 meters below the tile, only one streamline is found to penetrate below the 2-meter depth (Figure 12-9a). One would expect, therefore, that a hardpan at 2 meters would have little effect upon the water table. This is verified in Figure 12-9b. The water table at the midpoint between the drains is raised from 0.9 to 1.05 meters above the axis of the tile. Placing the tile on the impervious layer raises the water table at the midpoint to 1.8 meters above the plane of the tile. The streamlines in Figure 12-9c show that the water entering the tile horizontally from the side tends to flow along the juncture between the pervious and impervious layers. Figure 12-10 shows the effect of drain depth upon the flow net when the soil depth is constant.

It is observed that the distance of a given streamline from the drain increases with the depth of the drain. The water uptake per unit area of soil, however, decreases with increasing drain depth. The 50 percent streamline is 4.1 ft from the drain when the tile is placed 4 ft deep; it increases to 7.3 ft when the drain is 9 ft deep.

When the homogeneous soil overlies an impervious layer that is relatively close to the surface, the relationship between drain depth and drain flux varies from that shown in Figure 12-8. Calculations show that the drain flux increases with tile depth to a maximum at a distance of 1 to 2 ft above the impervious layer (Figure 12-11); this depends upon the depth of the hardpan. The flux then decreases as the impervious layer is approached. This maximum is about 10 to 12 percent greater than the drain flux when the tile is semiembedded in the impervious layer. It is seen that a drain placed at a depth of 4 ft in a soil with an impervious layer at 8 ft will discharge as much water as one embedded in the claypan. The decreased gravitational head is compensated by flow that takes place through the lower part of the drain. The data in Figure 12-11 indicate that a 6-in. drain will have the maximum flux if placed about 3 ft deep, if the impervious layer occurs at a depth of 4 ft; maximum flux will take place if the tile is placed at the 6-ft depth when the impervious layer is at 8 ft. In the case of an impervious layer at 4 ft, just as rapid seepage will take place if the tile is placed 2¼ ft

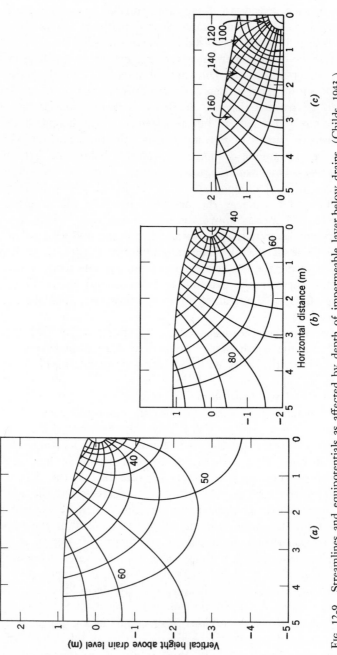

Fig. 12-9. Streamlines and equipotentials as affected by depth of impermeable layer below drains. (Childs, 1943.)

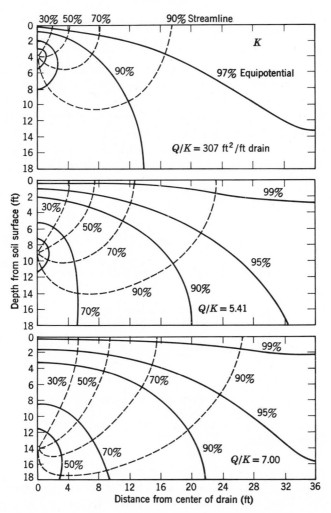

FIG. 12-10. Examples of flow nets for drain lines at three different depths beneath the soil surface. (After Luthin, 1966, used with permission of John Wiley and Sons.)

above this layer as compared with a drain placed upon the hardpan. Placing tile on an impervious layer will reduce flow by 32 percent compared with the flow that would occur if this layer were deep enough so as not to influence the streamlines that enter the drain from below. There would only be a 10 percent reduction if the impervious layer were 2 ft below the drain.

A similar effect is found in stratified soils in which the hydraulic conductivity of two distinct layers are quite different, though not zero.

Kirkham (1951) studied the problem of stratified soils by setting up
a model in which the surface layer was a sand with 10 times the hydraulic
conductivity of a clay subsurface layer. Then he reversed the layers
and placed the clay on top. Analyses were made in one instance by
assuming that the tile is placed in the upper stratum; in the second instance the tile is buried in the lower stratum. When the sand and the
drains are in the upper layer, 95 percent of the flow into the drains
took place in the sand; only 70 percent of the flow occurred in the
upper layer when it consisted of clay. Kirkham found little flow along
streamlines that extended deeper than twice the depth of the upper stra-

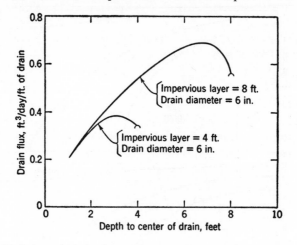

Fɪɢ. 12-11. Variation of drain flux with depth of drain in relation to impervious
layer. (Kirkham, 1947.)

tum. Consequently, it is concluded that the layers below twice the thickness of the surface layer will have little effect upon soil drainage, irrespective of their hydraulic conductivities. The drainage rate changes very
little with variations in the hydraulic conductivity of the subsurface stratum unless the tile is placed at the juncture of the two layers.

A different picture is obtained when the tile is placed in the lower
stratum. Here the lower layer controls the rate of drainage. With sand
as the surface layer, the streamlines diverge considerably as they pass
from the sand into the clay. There is a large head loss in the clay.

Flow Factors in Relation to Drainage System Design

Studies of flow systems as well as experience lead to several general
conclusions with respect to the design of a drainage system.

Open Ditch versus Tile Drains

It is obvious that an open ditch permits water to seep through a relatively larger surface than the buried tile. Consequently, the water table is lower under open-ditch drainage. As would be expected, there is a greater reduction in height of the water table immediately adjacent to a ditch than to a tile.

Two practical considerations must be weighed before deciding the relative efficiency of the open ditch versus the tile for draining land. First, the soil should not be so impervious from the surface downward as to make an efficient functioning of a tile system impossible. Open ditches to remove surface water are about the only practical means of removing excess water from such lands. Second, the presence of open ditches in the field may seriously affect economical field operations. Subsurface tile will prove to be the practical solution to such a problem.

Spacing of Drains

The flow rate into a drain is independent of the drain spacing if they are spaced more than 20 ft apart (Kirkham, 1949). When the drain inflow rate is plotted as a function of spacing, the curves flatten out at spacings greater than 20 ft, irrespective of the depth to the impervious layer. That is, the flow rate into a given drain does not change with spacing. Therefore just twice as much water will be removed by drains spaced 50 ft apart as those with a 100-ft spacing. There will be just twice as many drains in the field.

Diameter of Tile Drains

The drain diameter does not play a major role in an efficient drainage system, even though the drain flux is affected somewhat (Kirkham, 1949). The curves in Figure 12-8 point out that the size of the tile has a much smaller effect than the depth to which they are placed. At the 2-ft depth, the 6-, 8-, and 12-in. tile increased the drain flux over the 4-in. drain by 6.6, 10.5, and 13.3 percent, respectively. These percentage increases at the 12-ft depth were 8, 14, and 23, respectively. Figure 12-8 does show that a 12-in. tile placed 9 ft deep will discharge as much water as a 6-in. tile at 10.5 ft. It should be remembered that the increase in drain diameter is only a matter of inches, whereas the length of the streamlines is in many feet. Therefore the total frictional resistance to ground water flow throughout the entire system is decreased by a very small percentage.

Spacings between Individual Tiles

The rate of flow into a 6-in. tile (4 ft deep) spaced $\frac{1}{16}$, $\frac{1}{8}$, and $\frac{1}{4}$ in. apart increases approximately 10, 22, and 36 percent, respectively, over the normal $\frac{1}{32}$-in. spacing. If the tiles are embedded in gravel, the rate of flow increases about 180 percent. Embedding in gravel gives an effect of an infinite spacing between tile, if all theoretical conditions are met.

Perforated Pipes

Perforated pipes have been recommended as a substitute for ordinary drainage tile for subsurface drainage. An electrical analog analysis has shown that flow into perforated tubes varies with the number and size of the holes (Kirkham and Schwab, 1951). Flow increases approximately proportionally to the number of holes per foot for four holes or less per foot. Ten holes per foot represents the upper limit of increased flow. In comparing perforated drain tubes with the tile drains, the data indicate that for 6-in. tile placed at a depth of 4 ft, 12 half-in. holes per foot are required to equal the seepage rate of tile placed $\frac{1}{8}$ in. apart; 17 holes are needed to equal the $\frac{1}{4}$-in. spacing; and 31 holes are necessary to balance the $\frac{1}{2}$-in. spacing. There are tremendous increases in the rate of flow if the perforated drain tubes are embedded in gravel.

Design of Drainage Systems in Relation to Soil Properties

Summarization of the previous discussions on soil drainage leads to the following conclusions:

1. The hydraulic conductivity of the upper soil layers is the most important soil parameter in drainage problems.
2. The need for drainage is a function of the aeration porosity of the soils as it affects the hydraulic conductivity.
3. A system of surface drains is required for those soils where the hydraulic conductivity of the soil from the surface downward is so low as to prevent the functioning of subsurface drains.
4. Impervious strata affect the pattern of flow into drains.
5. The pattern of equipotentials and streamlines in homogeneous soils is not changed by differences in hydraulic conductivity; only the rate of water removal is influenced.

It is apparent therefore that the design of drainage systems requires an evaluation of the hydraulic conductivity of the soil profile.

Hydraulic Conductivity Measurements in Situ

There are several techniques that can be used to determine hydraulic conductivity *in situ* in the presence of a water table (Luthin, 1966). The simplest is the single auger-hole method as illustrated in Figure 12-12. The rate of rise of water in the hole, dy/dt, is directly proportional

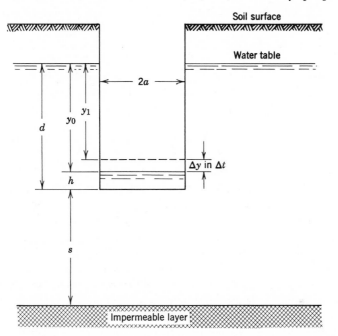

Fig. 12-12. Geometrical quantities for auger-hold method for a homogeneous soil. (Luthin, 1966, used with permission of John Wiley and Sons.)

to (1) the hydraulic conductivity of the soil, (2) the surface of seepage, which includes the circumferential flow plus flow through the bottom, and (3) the hydraulic head, which is the difference between the height of the water table and the water in the bottom of the hole. It is inversely proportional to the cross-sectional area of the hole and a constant, which depends on the radius of the hole, the depth of the water table from the bottom of the hole, and the height of water in the hole at the time of measurement.

The rate of rise of water in the hole due to seepage through the walls of the hole is expressed by equation 12-21:

$$\frac{dy}{dt} = -k\left(\frac{2ad}{a}\right)\frac{y}{S} \tag{12-21}$$

where k is the hydraulic conductivity, a is the radius of the hole, d is the depth of the hole below the water table, y is the difference between the height of the water table and the height of the water in the hole, and S is a constant. S may be determined with the aid of sand-tank experiments; its value has been established as equal to $ad/0.19$.

The rate of rise of water due to seepage through the bottom of the hole is expressed by equation 12-22:

$$\frac{dy}{dt} = -k\frac{y}{S} \qquad (12\text{-}22)$$

When equations 12-21 and 12-22 are added and the resulting equation is integrated between the limits of y_o and y_1 and t_0 and t_1, the hydraulic conductivity can be calculated by equation 12-23:

$$k\left(\frac{m}{da}\right) = \frac{2.3aS}{(2d+a)t} \log_{10}\frac{y_o}{y_1} \qquad (12\text{-}23)$$

The problem of measuring hydraulic conductivity in the absence of a water table is more difficult. It is often assumed that pumping water into a shallow well will give a measure of this permeability. Such a procedure will tend to overestimate the conductivity, the extent depending upon the nature of the soil. The reasons for this have already been discussed in the previous chapter.

Bouwer (1961) has developed a procedure for correcting for the contribution of the matric potential to the infiltration process to permit measurements under such circumstances. He devised a system using a pair of concentric rings. He observes the flow from the individual rings under controlled conditions. Correction for the unsaturated flow is by means of graphs which are obtained by electric analog solutions of the flow equation. His method appears to be the best available for hydraulic conductivity measurements in the absence of a water table.

LEACHING AND RECLAMATION

The problem of salt accumulation in irrigated soils was mentioned earlier in this chapter. These salts are introduced into the soil profile in the irrigation water and are largely left behind by the transpired soil water. They consist mainly of the cations sodium, calcium, magnesium, and potassium and the anions chloride, sulfate, carbonate, bicarbonate, borate, and nitrate. Concentrations sufficiently high to inhibit plant growth are easily achieved. Some of these effects are due to specific ion toxicity. For example, a concentration of about 700 to 1000 ppm of chloride will cause leaf burn in fruit crops. On the other hand, in many plants the effect does not appear to be specific and is correlated with the osmotic potential of the soil solution. The tolerance of a large number of economic crop plants has been measured (U.S. Salinity Laboratory Handbook 60, 1954). In this section, we consider the movement of these salts, as well as other dissolved substances, through the soil profile in the soil solution.

Solute Transport Equations

The principle mechanism for the movement of solutes through soils is convection in the moving soil solution. If the flux density of water moving in the x direction is q and the concentration of a solute in the soil solution is C then the rate of transport, J is

$$J = qC \qquad (12\text{-}24)$$

If the solute is excluded by the plant root system, as is essentially the case with chloride, then the concentration will increase with depth as water is taken up by the plant root system. If, by way of example, the rate of uptake is uniform with depth and is given by w, then

$$q = I - wz \qquad (12\text{-}25)$$

where I is the average rate of irrigation applied at $z = 0$. Under steady-state condition the flux J down through the profile is constant. If equation 12-25 is substituted into equation 12-24, one obtains

$$C = \frac{J}{I - wz} = \frac{IC_0}{I - wz} \qquad (12\text{-}26)$$

The concentration does not increase linearly with depth for a uniform water uptake pattern. Although equation 12-25 was derived by using steady-state assumptions, it can be shown that a similar result obtains for periodic, discrete applications of irrigation water.

Transport of Sorbed Substances

Since soil particle surfaces are highly reactive, many substances are physically adsorbed to a greater or lesser extent. If the sorption isotherm is linear, that is, a constant fraction of the substance is adsorbed regardless of concentration, then a simple equation (12-24) still is used for estimating the flux density of the solute. Adsorption has a great influence, however, on the rate of travel. In the absence of adsorption, a discrete solute peak would move at a rate q/θ, where θ is the water content of the soil. When adsorption does occur, the rate of movement must be reduced to account for the fraction of solute that is adsorbed. If b represents the ratio of adsorbed substance to that in solution per unit volume of soil, then the rate of transport J is

$$J = \frac{q}{b\Theta} \tag{12-27}$$

If the sorption isotherm is nonlinear, b becomes a function of the concentration C, since it is derived from the slope of the isotherm.

Miscible Displacement

It is observed that a sharp peak of solute tends to spread out and disperse as it moves with the soil solution. This spreading is due to molecular diffusion and to the variation in flow velocity that exists from place to place throughout the soil, even in individual pores. The assumption of an average velocity v gives an average rate of displacement of the solute but does not describe the exact distribution. Theories for this dispersion process exist in the chemical engineering literature and have been applied to transport through soils.

A simple treatment of the process can be made by choosing a frame of reference which moves through the soil with the average velocity v. If there were no dispersion, a saline solution would be displaced by fresh water with a sharp boundary between the two. This boundary becomes diffuse when dispersion occurs. Because saline and fresh water are miscible, this is known as miscible displacement to distinguish it from displacement when two immiscible fluids such as oil and water are involved. The equation for the solute distribution is then given by

$$\frac{\partial c}{\partial t} = D \frac{\partial^2 c}{\partial X^2} \tag{12-28}$$

where $X = (x - vt)$. The dispersion coefficient D depends upon the flow velocity and the pore geometry of the soil. Figure 12-13 shows the leaching of a calcium chloride solution by relatively pure water in a field experiment (Gardner and Brooks, 1957). The smooth curves represent theoretical predictions that are based upon a somewhat more rigorous equation than 12-28 but which reduces to 12-28 when at the lower soil depths. The soil in the experiment was rather homogeneous and the dispersion was less than might be found in a more heterogeneous soil.

In general, one pore volume replacement will reduce the initial con-

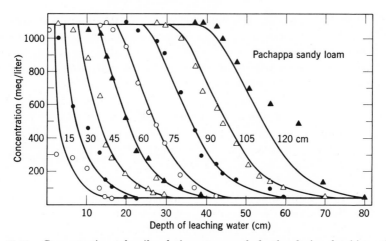

FIG. 12-13. Concentration of soil solution at several depths during leaching of a field plot of Pachappa sandy loam plotted as a function of depth of leaching water. (Gardner and Brooks, 1957.)

centration to 50 percent of its original value; a second volume will reduce it to a small fraction of the original. There tends to be less dispersion and hence more efficient leaching in unsaturated soils, since the largest pores are not filled. Miller et al. (1965) found that intermittent applications of irrigation water in 2-in. amounts were more efficient in terms of water used than continuous application. However, it resulted in sharper peaks than the ponded treatment.

Reclamation of Salt-Affected Soils

Salt-affected soils may suffer from an excess of soluble salts in the soil solution, an excess of exchangeable sodium on the exchange sites, or both. In order to reduce the effect of soil texture and structure upon diagnostic measurements, the U.S. Salinity Laboratory recommends that the salt content of a soil sample be determined by adding sufficient water to bring the soil to a saturated paste. This is a reasonably reproducible water content and is usually somewhat greater than the water content of a saturated, undisturbed soil sample. The solute content is conveniently calculated from the electrical conductivity of the extracted soil solution. Hence salinity levels in the United States are reported, by convention, in terms of the electrical conductivity of the saturation extract. It turns out that the saturation water content for many soils is roughly twice the water content found in the soil two days after irrigation. In that case, the concentration of a saturated extract is about one-half that of the initial irrigation water. A conductivity of the saturation extract of greater than 4 mohms/cm, corresponding to an osmotic potential of about -1.5 bars, is considered saline. This might correspond to a concentration of about three parts per thousand on a weight basis for a typical soil solution.

Removal of the saline water from a soil profile is accomplished by displacing it with nonsaline or higher quality water. The amount of water required is determined by the heterogeneity of the soil and the dispersion process discussed earlier. Where water is ponded on the soil surface, the amount of water passing through any given region in the soil profile may be determined largely by the drainage pattern. For example, where a tile drain system is needed to remove the leached soil solution, flow above the drains can far exceed that between drains; this will result in very nonuniform leaching. Talsma (1966) concluded from his experiments that complete desalinization is achieved under such circumstances with considerably less leaching water under alternate ponding and draining than under continuous ponding. In the case he studied, the difference amounted to a factor of from 1.5 to 2 in. of water required.

When the exchangeable sodium percentage is above about 15, dispersion of the clay minerals in most soils is sufficient to cause serious infiltration problems. The reclamation of these sodic (or alkali) soils is through the replacement of the exchangeable sodium by calcium and magnesium. This is accomplished by infiltrating a solution down through the soil profile that contains the divalent ions. The preferred adsorption of the calcium by the soil particles results in a displacement of the adsorbed sodium. Bower et al. (1956) showed that this exchange can be described very well by ion-exchange column theory. The major difficulty encountered in this reclamation procedure is the initially low permeability of the soil, which makes it difficult to move water through the soil. It was pointed out in Chapters 1 and 2 that the swelling of the soil colloids depends upon the concentration of the soil solution. The higher the electrolyte concentration of the reclamation water, the greater is the flocculation of the colloids and the permeability. Unfortunately, the solubility of gypsum, which is a convenient source of calcium, is so low that its use is often precluded in a soil in which the permeability is initially low (Reeve and Tamaddoni, 1965). In order to shorten the reclamation time, chloride or anions other than sulfate must be present so that higher electrolyte concentrations in solution are possible. In some special instances, the use of very highly concentrated natural waters may be possible by a process involving successive dilution with higher quality water (Reeve and Bower, 1960).

It is important to anticipate the effect of a given irrigation water upon the soil before an irrigation project is initiated. An irrigation water which is high in dissolved sodium relative to the divalent cations may cause deterioration of the soil physical properties. The exchange theories which are used to predict reclamation may also be used to predict the effect of a given water upon the cation exchange status, as well as the potential hazard due to any boron in the irrigation water. In an analysis of this problem, Bower (1961) also reviewed procedures of estimation of precipitation of constituents from the irrigation water, particularly calcium carbonate.

WASTE-WATER DISPOSAL AND GROUND-WATER RECHARGE

Water is applied to the soil surface in many areas for purposes other than crop irrigation. It is not uncommon for the water table to drop as ground water is developed for use by agriculture or industry. Deliberate attempts to recharge the ground water are often considered where surface runoff occurs regularly. Waste-water disposal is becoming an increasing problem with the urbanization of many areas. The soil is man's oldest filter for the purification of waste water and its greater utilization is receiving increasing attention. These two processes present special common management problems which will be considered briefly.

Recharge

Ground-water recharge usually implies a low water table. If the recharge area is chosen properly, the ground-water mound which builds up under the recharge area should not be a problem. A major difficulty is the maintenance of the permeability of the surface of the soil. Microbial activity, combined with any suspended solids in the recharge water, tends to clog the soil surface. The biological effect is reversible and periods during which the soil surface is allowed to dry tend to partially restore the permeability. Since a surface-impeding layer can cause partial desaturation of the underlying soil horizon under ponded conditions, it sometimes turns out that finer textured soils maintain their permeability for a longer time than coarser textures. They do not desaturate to the same matric potential as the coarser soil. In selecting a recharge area, it must be ascertained that the permeability of all soil horizons and substrata are adequate. If the infiltration rate is too slow and a large area is ponded, evaporation losses may prove to be excessive. Where the surface soil is limiting, shafts, pits, or injection wells may be used to bypass the impeding layer.

Waste Disposal

Waste-water disposal poses additional problems to those of ground-water recharge. The soil acts as a very good bacterial filter and is also presumed to reduce virus levels significantly. Waste disposal on the land in humid regions often requires introduction of greater quantities of water into the soil profile than the internal drainage system is capable of disposing. Infiltration tests which are routinely used to evaluate the suitability of soils for septic tank fields generally overestimate the permeability of the soil. This is for two reasons. First, during the test in which the rate of water movement out of an auger hole is determined, soil matric potential is usually more influential in maintaining the infiltration rate than during the operation of the field. Second, poor aeration leads to biological clogging of the soil.

A further problem in the land disposal of urban wastes is the relatively large quantities of dissolved substances, particularly nitrogen and phosphorus. Trace elements such as selenium, cadmium, copper, and boron can also be problems. The degree of renovation of the waste water depends upon the ability of the soil to adsorb the trace elements and the phosphorus. Renovation insofar as nitrogen is concerned depends upon the management of the system. The Flushing Meadows Project described by Bouwer (1968) represents an excellent pilot study in waste-water renovation. This is only one of several projects under long-term investigation.

References

Aronovici, V. S., and W. W. Donnan (1946). Soil-permeability as a criterion for drainage-design. *Trans. Am. Geophys. Union*, 27:95–101.

Bouwer, Herman (1961). A double tube method for measuring hydraulic conductivity of a soil *in situ* above a water table. *Soil Sci. Soc. Am. Proc.* 25:334–339.

Bouwer, Herman (1968). Returning wastes to the land—A new role for agriculture. *J. Soil and Water Conserv.*, 23 (Sept.–Oct.).

Bower, C. A. (1961). Prediction of the effects of irrigation waters on soils. *Salinity Problems in the Arid Zones*, Proceedings of the Teheran Symposium, UNESCO, pp. 1–7.

Bower, C. A., W. R. Gardner, and J. O. Goertzen (1956). Dynamics of cation exchange in soil columns. *Soil Sci. Soc. Am. Proc.*, 21:202–206.

Cahoon, G. A., L. H. Stolzy, and E. S. Morton (1961). Effects of mulching on soil-moisture depletion in citrus orchards. *Soil Sci.*, 92:202–206.

Childs, E. S. (1946). The water table, equipotentials, and streamlines in drained land: IV. Drainage of foreign water. *Soil Sci.*, 62:183–192.

Dumm, L. D. (1954). Drain spacing formula. *Agr. Engr.*, 35:726–730.

Dupuit, Jules (1863). *Etudes theoretiques et pratiques sur le mouvement des eaux*, ed. 2. Dunod, Paris.

Gardner, H. R., and W. R. Gardner (1969). Relation of water application to evaporation and storage of soil water. *Soil Sci. Soc. Am. Proc.*, 33:192–196.

Gardner, W. R., and R. H. Brooks (1957). A descriptive theory of leaching. *Soil Sci.*, 83:295–304.

Hooghoudt, S. B. (1940). Review of the problem of detail drainage and subirrigation by means of parallel drains, trenches, ditches and canals. *Verslag. Landbouwk, Onderzoek*, 46:515–707.

Kirkham, Don (1949). Flow of ponded water into drain tubes in soil overlying an impervious layer. *Trans. Am. Geophys. Union*, 30:369–385.

Kirkham, Don (1950). Seepage into ditches in the case of a plant water table and an impervious substratum. *Trans. Am. Geophys. Union*, 31:425–430.

Kirkham, Don (1951). Seepage into drain tubes in stratified soil. I. Drains in the surface stratum. *Trans. Am. Geophys. Union*, 32:422–432. II. Drains below the surface stratum. *Trans. Am. Geophys. Union*, 32:433–442.

Kirkham, Don, and G. O. Schwab (1951). The effect of circular perforations on flow of water into subsurface drain tubes. Part I. Theory. *Agr. Engr.*, 32:211–214.

Luthin, J. N. (1966). *Drainage Engineering*. John Wiley and Sons, New York.

Miller, R. J., J. W. Biggar, and D. R. Nielsen (1965). Chloride displacement in Panoche clay loam in relation to water movement and distribution. *Water Resources Res.*, 1:63–73.

Reeve, R. C., and C. A. Bower (1960). Use of high-salt waters as a flocculant and source of divalent cations for reclaiming sodic soils. *Soil Sci.*, 90:139–144.

Reeve, R. C., and G. H. Tamaddoni (1965). Effect of electrolyte concentration on laboratory permeability and field intake rate of a sodic soil. *Soil Sci.*, 99:261–266.

Tadmor, N. H., M. Evenari, L. Shanan, and D. Hillel (1961). The ancient desert agriculture of the Negev. I. Gravel mounds and gravel strips near Shivta. Ktavim Vol. 8, No. 1127-151, *Agr. Res. Sta. Rehovot*, Israel.

Talsma, T. (1968). Leaching of tile-drained saline soils. *Aust. J. Soil Res.*, 5:37–46.

U.S. Salinity Laboratory (1954). Diagnosis and improvement of saline and alkali soils. *U.S.D.A. Agr. Handbook* No. 60.

Waggoner, P. E., P. M. Miller, and H. C. De Roo (1960). Plastic mulching. *Conn. Agr. Exp. Sta. Bull.* 634, 44 p.

CHAPTER 13

Soil Erosion— Water Erosion

Rainfall which reaches the soil surface either enters the soil or runs off or both. Runoff water may be impounded in lakes and reservoirs, or it may reach streams and finally the ocean. Runoff in excess of the carrying capacity of streams overflows the banks, inundates fields, and causes floods. Soil erosion takes place during the runoff process. Valuable topsoil is removed from the land and deposited in fields, flood plains, reservoirs, and lakes. The environment of the landscape is changed. Streams, reservoirs, and lakes become polluted with sediments. If these sediments originate from agricultural lands that had received phosphate fertilization, the phosphorus that is adsorbed on the surfaces of the colloidal clay particles will be carried with the sediment to the water sink into which the stream flows. This adsorbed phosphorus may help to produce algal blooms in lakes and reservoirs.

AN ANALYSIS OF THE PROBLEM OF RUNOFF AND EROSION

Water erosion is due to the dispersive action and transporting power of water, water as it descends in the rain and leaves the land in the form of runoff. If there were no runoff, there would be no erosion. On the other hand, if raindrops could not beat the soil into a state of dispersion and if runoff water could be prevented from bringing soil into suspension as it travels across the surface, there would be no erosion. The dispersive action and transporting power of water are determined by the dispersive effects of falling raindrops and the amount and velocity of runoff and by the resistance of the soil to dispersion and movement. Both are dependent upon (1) climate, primarily rainfall characteristics, (2) topography, the slope and area of the watershed, (3) vegetative cover, and (4) soil, the ability of the soil to resist dispersion and to absorb and transmit water through the profile.

There are two forms of erosion resulting from runoff. The first is sheet erosion, which is a combination of raindrop dispersion and the movement of water in shallow layers more or less uniformly across the soil surface. The second is rill or gully erosion resulting from channelized flow of concentrated water into defined water courses. This type of flow begins soon after runoff starts. The small rills or microchannels (Smith and Wischmeier, 1962) are smoothed over by cultivation but the beginning of a future gully has been established. Continued sheet and rill erosion leads to more and more concentrated flow into these established channels until gully erosion becomes the major loss of soil from the field.

Mechanics of the Erosion Process

Raindrops are the causal factor in soil erosion. They can be large or small and consequently have varying velocities. They can fall with low to high intensities. What actually happens when a raindrop hits the soil? If the soil is dry, the raindrop is absorbed, and the soil becomes moist. As more drops fall, they hit the soil water surface, and considerable splashing occurs. The splash is muddy or turbid. This means that the falling raindrop has brought soil particles into suspension, primarily by breaking down soil aggregates or by detaching soil particles from the soil mass. The turbid water enters the soil, resulting in a clogging of the pores. The continued impact of the raindrops compacts and seals the immediate surface. A soil crust is formed and reduces infiltration (see Chapter 5).

Raindrop Energy

Neal and Baver (1937) obtained data on raindrop fall which indicated that the impact of raindrops per unit area was determined by the number and size of the drops plus any increase in velocity due to the driving force of the wind. Laws and Parsons (1943) related the drop-size distribution of raindrops to intensity of fall and found that the relationship could be expressed by the equation

$$D_{50} = 2.23I^{0.182} \tag{13-1}$$

where D_{50} is the median drop size representing the midpoint of the total volume and I is the intensity in inches per hour. This equation shows that the median drop size increases with the intensity. They found that the terminal velocities of raindrops 1 and 6 mm in diameter were 4.0 and 9.3 m/sec, respectively.

Ellison (1944a) placed metal disks on the soil and then applied artificial rain. He observed that the soil eroded under the action of the raindrops except under the disks. Columns of soil were left standing under these disks in an area where the soil had eroded. He concluded that the relative amount of soil detached from the surface was a function of the size and velocity of the raindrop and the intensity of the rain. The erosive potential of a falling mass of water for a given time and velocity is a function of the drop mass per drop cross section and the square of the drop velocity (Eckern, 1950, 1953). The kinetic energy of the falling drop determined the force of impact and the horizontal area regulated the amount of soil that received this impact.

The dead weight of water that falls during a common thunderstorm in the midwest in 30 min may exceed 100 T/A; this is equivalent to about 2,000,000 ft lb/A (Wischmeier and Smith, 1958). Using the drop-size distribution data of Laws and Parsons, Wischmeier and Smith developed the following energy equation:

$$Y = 916 + 331 \log x \qquad (13\text{-}2)$$

where Y is the kinetic energy in foot-tons per acre-inch of rain and x is the rainfall intensity in inches per hour. The curves that relate kinetic energy to rainfall intensity level off at the higher intensities, probably due to changes in drop-size distribution. Rose (1960) proposed that the detachment of soil particles by raindrop impact was related to two equations:

$$M = kR \frac{mv}{m} \qquad (13\text{-}3)$$

and

$$M = kR \frac{mv^2}{m} \qquad (13\text{-}4)$$

where M is the rainfall momentum per unit area and time, k is the rainfall kinetic energy per unit area and time, R is the rainfall rate in inches per hour, v is the impact velocity in meters per second, and m is the drop mass in kilograms. Rose concluded that the rate of detachment depended more on m than k, since m represented the pressure exerted by the rainfall. It is the force per unit area and is a mechanical stress.

Raindrop Impact and Detachment

SPLASH EROSION. The investigations of Ellison (1944a, 1944b) led him to conclude that a considerable amount of soil was eroded by the simple process of splashing; it was the initial step in the erosion process. He developed a special type of splash board to measure these effects. He showed that there was a higher percentage of sand and gravel and of aggregates smaller than 0.105 mm in the splash than in the original soil. The runoff material, on the other hand, contained 95 percent silt and clay. It was interesting to note that the runoff also contained more aggregates smaller than 0.105 mm than the original soil. Apparently, the impact of the raindrops broke down the larger aggregates in the soil. The quantities of soil in the raindrop splash increased with drop size, drop velocity, and rainfall intensity. The maximum amount of material in the splash occurred about 2 to 3 min after the rainfall began, after the surface was covered with a film of water. It was significant to note that on a 10 percent slope, three-fourths of the soil in the splash from a given rain was downhill. Ellison suggested the following equation to express the amount of splash erosion:

$$E = KVdI \tag{13-5}$$

where E is the relative amount of soil splashed in 30 min, K is a soil constant, V is the drop velocity in feet per second, d is the drop diameter in millimeters, and I is the rainfall intensity in inches per hour.

The percentage of sand transported downhill by raindrop impact depends upon the degree of slope (Eckern and Muckenhirn, 1947). About 50 percent moves downhill on a 0 percent slope and 60 percent on a 10 percent slope. The percentage of downhill transport is equal to 50 plus the percentage of slope. It was estimated that it only took the energy of a 1-mm drop of water falling at its terminal velocity to move fine sand by impact (Eckern, 1950). Consequently, 100 percent of the fine sand in soil is eroded by raindrop splash. Smaller percentages of larger particles are moved because of their size. Finer particles are compacted to form surface seals. The erosivity of soils due to raindrop impact is a function of the amount of rain and raindrop parameters (Eckern, 1953).

$$E = f \text{ (precipitation intensity} \times \text{time)} \left(\frac{\text{drop mass} \times \text{velocity}^2}{\text{drop cross section}} \right) \tag{13-6}$$

Ellison observed that the maximum distance of splash for a 5.9-mm drop was 60 in.; the height was about 15 in. Mihara (1952) found that the distance and height depended upon the condition of the soil surface as well as the velocity of the drop. A 6-mm drop splashed to a distance of 37 in. and a height of 12 in. in a cultivated soil. The splash traveled 59 in. on a compacted soil. Ellison reported that maximum splash occurred soon after the soil became wet and then decreased with increasing time as a deeper water layer was formed. There is an increase in splash with water depth and then a decrease (Kuron and Steinmetz, 1957). However, the total soil movement increases because of the turbulence that is produced by the transfer of raindrop energy to the whirling action of the surface water layer. McIntyre (1958) proposed four phases of splash erosion at the surface. First, there is a rapid wetting of the surface which causes low soil cohesion and high splash rates. This is followed by a compacting effect of the raindrops to form a thin surface crust, which decreases splash and causes an accumulation of water. The turbulence then created removes part of the crust. This brings about an increase in water infiltration and an increase in the splash rate.

SURFACE FLOW. Once the soil is detached by the impact of raindrops, sheet and microchannel erosion become the second phase in the erosion

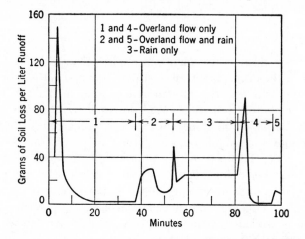

FIG. 13-1. Variation of soil loss with time, applying rainfall and overland flow separately and in combination on bare Keene silt loam. (Ellison, 1945.)

process. This might be termed the *transport phase*. Ellison (1945) studied the combined effect of the dispersive action of raindrops and of surface flow on soil movement during a storm. He found that surface flow over a plot of soil at the rate of 20 in./hr without any rainfall action removed a considerable amount of soil for a very few minutes (see Figure 13-1). This erosion probably resulted from the removal of loose particles on the surface as well as particles that were brought into suspension by slaking. After this first removal the rate of erosion was very small and constant. The water was not separating many particles from the soil as it flowed over the surface. (It should be emphasized that there were no rills or gullies present.) At the end of about 38 min rainfall was applied at the rate of 3.3 in./hr along with surface flow. It may be noted that erosion was increased considerably, especially for a few minutes following the rain. When overland flow was discontinued at about 55 min, erosion again increased momentarily and then was constant. In other words, as the depth of the water was decreased, apparently there was an increase in dispersion due to the impact of the raindrops as they established closer contact with the soil. The instantaneous stopping of rainfall after about 80 min and the beginning of surface flow caused a heavy removal of particles for a few minutes, after which erosion fell to almost zero. Evidently, the overland flow removed all particles that had been dispersed by the rain and then did not bring more particles into suspension. The addition of rain after about 96 min again increased erosion. It is important to note that the second application of rain did not disperse as much soil as the original. This was due to surface compaction arising from the continued beating of the raindrops. These results indicate most convincingly that there must be detachment before there can be transportation.

The loss of soil per cubic foot of runoff from surface flow from a laboratory plot with slopes of 8 percent or less was observed to be only 10 percent of that from simulated rain (Woodruff, 1947). When the slope was changed to 16 percent, erosion from surface flow was about 60 percent of that from simulated rain. The raindrop impact was the major soil detachment factor on the lesser slopes. The energy of higher velocity water caused more soil detachment on the steeper slopes.

FACTORS CONTRIBUTING TO EROSION LOSSES

Having established the fact that the erosion process consists of a detachment and a transportation phase, let us evaluate the four major factors that contribute to the extent of soil losses.

The Climatic Factor

The major climatic factors that influence runoff and erosion are rainfall, temperature, solar energy, and wind. Rainfall is the most important. Temperature in the temperate zone exerts its primary influence through changes in the absorptive properties of the soil for water. In the winter months, the soil freezes to varying depths, which prevents infiltration. Temperature determines whether the precipitation will be rain or snow. It also causes snow to melt to produce runoff, which can result in micro-channel erosion on an unprotected shallow, thawed surface layer. During the remainder of the year, temperature, as an index of solar energy, plays a significant role in evapotranspiration processes that regulate the amount of water in the soil at the time of precipitation. Soil losses from erosion plots correlate with rainfall intensity only when the effective soil moisture at the time of precipitation is considered as one of the parameters (Baver, 1937). Effective soil moisture varies inversely with the square of the temperature (F°). Data from Hawaii showed that evapotranspiration was more highly correlated with solar energy than with temperature (Baver, 1954). Solar energy determined the moisture content of the soil between rains.

Wind affects the erosion process primarily through the angle and velocity of impact of raindrops. It also influences evapotranspiration and consequently the moisture content of the soil. Confining our attention to precipitation as rain, we find that the amount, intensity, and distribution of the rainfall help to determine the dispersive action of the rain upon the soil, the amount and velocity of runoff, and the losses due to erosion. A large total rainfall may not cause excessive erosion if the intensity is low. Similarly, an intensive rain of extremely short duration may not cause much soil loss, because there is not enough rainfall to produce runoff. On the other hand, when both the capacity and the intensity factors are high in a given storm, both runoff and erosion will be serious. This is especially true if the precipitation is so distributed that the rains fall at a time during the cropping cycle when the soil has no plant protection.

There has been a great deal of experimental evidence to emphasize the importance of the rainfall-intensity factor on soil losses. Controlled experiments on a laboratory plot to evaluate the effect of intensity where slope, soil, and amount of rainfall could be regulated demonstrated that erosion increased according to a power function of the intensity (Neal, 1938). Perhaps the most thorough investigations of soil losses in relation to rainfall characteristics have been reported by Wischmeier and his associates (Wischmeier, Smith, and Uhland, 1958; Wischmeier and Smith, 1958; Wischmeier, 1959). Results from these investigations produced poor correlations between amounts of rainfall, or even maximum short-period intensities, and soil losses. Rainfall energy, however, gave a high correlation. The best correlation was obtained between soil losses and the product of the kinetic energy of the storm times the maximum 30-min intensity. The total energy of the storm was calculated according to equation 12-2. This product was called the erosion index, *EI*. This means that the erosion potential of a given storm is a function of the amount of rain, raindrop velocities, and maximum sustained intensity. It represents the integrated effects of the impact energy of raindrops and the rate and turbulence of runoff. There is a high degree of linear correlation between soil loss and the erosion index.

The Topographic Factor

Slope characteristics are also important factors in determining the amount of runoff and erosion. Erosion is usually not a problem on extremely flat lands. As soon as the topography becomes slightly rolling, erosion begins to be serious. The degree and length of the slope are the two essential features of topography that are concerned in runoff and erosion. The uniformity of slope is often important in determining the relative ease or difficulty of establishing suitable erosion-control practices.

Of the two characteristics of slope, degree is usually more important than length from the standpoint of the severity of erosion. Experiments have shown that on slopes below about 10 percent the amount of erosion approximately doubled as the degree of slope doubled. The curve that relates erosion as a function of slope for any given rain is slightly S-shaped (Neal, 1938). The losses from the steeper slopes do not increase in the same proportion as the losses from the more gentle slopes. Neal's data showed that erosion varied as the 0.7 power of the percentage slope. Analysis of the erosion data from five different Soil Conservation Experiment Stations (Zingg, 1940) showed that erosion varied according to the equation

$$X_c = 0.65S^{1.49} \qquad (13\text{-}7)$$

where X_c is the coded total soil loss and S is the land slope in percent. He found that doubling the degree of slope increased the total soil loss 2.8 times. Although the total runoff increased with the degree of slope, the effect of slope on runoff was not as great as on erosion. Smith and Wischmeier (1957) proposed a parabolic equation to express the relationship between slope and soil loss:

$$A = 0.43 + 0.30S + 0.043S^2 \qquad (13\text{-}8)$$

where A is the soil loss in tons per acre and S is the percent slope. The curve from this equation is similar to that obtained with Zingg's equation. However, data from other investigators show variations from these relationships.

The effect of length of slope on erosion varies considerably with the type of soil. Generally speaking, longer slopes have less runoff than shorter ones. This is especially true of permeable soils.

Musgrave (1935) offered some valuable suggestions as to the effect of length of slope on runoff and erosion. He found that the intensity of rainfall greatly influenced these losses. Runoff and erosion on the highly permeable Marshall silt loam increased with the length of slope with high intensity rains. With rains of an intensity only slightly higher than the infiltration rate, there was a greater total infiltration with the longer slopes. With intense rains, there is less time for infiltration, which results in more total runoff and a greater velocity of runoff. Moreover, the intense rains may cause a partial destruction of the structure in the surface of the soil, which leads to more rapid runoff and greater erosion.

Erosion data from five Soil Conservation Experiment Stations (Zingg, 1940) indicated that soil loss is a function of the length of slope according to the equation:

$$X_c = 0.0025 L^{1.53} \qquad (13\text{-}9)$$

where X_c is the coded total soil loss and L is the length of the land slope in feet. Doubling the slope increased the erosion about three times.

The Vegetation Factor

A good vegetative cover, such as a thick sod or a dense forest, offsets the effects of climate, topography, and soil on erosion. This fact is particularly emphasized by the experimental results of the Federal Soil and Water Conservation Experiment Stations. A good grass sod has permitted less than 1 ton of soil loss per year on soils ranging from the highly permeable Marshall silt loam to the fairly impermeable Shelby loam, and on slopes varying from 4 percent at Temple, Texas to 30 percent at LaCrosse, Wisconsin. Naturally, the agricultural production of various crops cannot be maintained by having all the land covered with trees and grass. But even under conditions of cropping, vegetation effects play a significant role in controlling erosion.

The major effects of vegetation may be classified into at least four distinct categories: (1) the interception of rainfall by the vegetative canopy; (2) the decreasing of the velocity of runoff and the cutting action of water; (3) the root effects in increasing granulation, porosity, and biological activities associated with vegetative growth and their influence on soil porosity; and (4) the transpiration of water leading to the subsequent drying out of the soil.

Interception of Rainfall

The interception of raindrops by the canopy of vegetation affects soil erosion in two ways. First, part of the intercepted water never reaches the soil but is evaporated directly from the leaves and stems. This water therefore cannot contribute to runoff and is not a factor in erosion. Second, the vegetative canopy absorbs the impact of the raindrops and thereby minimizes the destructive effects of the beating action of the rain on soil structure. In 1880, Wollny (Baver, 1938) showed that from 12 to 55 percent of the total rainfall was intercepted by plant canopies and prevented from reaching the land surface directly. The interception percentage depended on the type of crop and the number of plants per unit area. Later data (Haynes, 1948; Smith et al., 1945) reported interception percentages that ranged from 7 to 43 percent. These results point out that the canopy effect of vegetation against the impact of raindrops can be an important factor in decreasing soil erosion losses.

Decreasing the Amount of Velocity of Runoff

Vegetation increases the storage capacity of the soil for rainfall because transpiration by vegetation is a major causative factor for the removal of water from soils. Consequently, the amount of runoff from a given rain is decreased. A vegetative cover is most effective in decreasing the amount of runoff when it is growing. During the winter months in the temperate zone, when large areas of leaf surface are not present to intercept the raindrops and to transpire water, vegetation functions primarily by decreasing the rate of runoff. It may also protect the soil from freezing and thereby maintain a fairly good infiltration rate for the soil in question. Any vegetative cover is an impediment to runoff water. A well-distributed, close-growing surface growth will not only slow up the rate at which water travels down the slope but also will tend to prevent a rapid concentration of water. Reducing the velocity of runoff and preventing the concentration of this water lessen the cutting action of the water.

The fact that vegetation exerts a greater influence upon the velocity than upon the total amount of runoff is seldom appreciated. These effects are clearly shown in the data in Table 13-1. These results point out that the difference between grass and corn with regard to rates of runoff is about twice as great as the difference between these two crops with respect to total runoff. Moreover, even though the total runoff from oats and bluegrass is approximately the same, the rate of runoff from the former is over twice as great as that from the bluegrass. When the rate of runoff is decreased by vegetation, there is more time for infiltration which reduces total runoff.

A good grass sod resists the cutting action of water. This is evidenced by the fact that sod-stabilized terrace outlets are among the best and

TABLE 13-1
The Effect of Vegetation on the Amount and Velocity of Runoff
(Norton and Smith, 1937)

Year	Number of rains	Crop	Maximum rate of runoff (inches of rain per hour)	Total runoff (inches of rain)
1933–1936	50	Corn	0.67	0.57
		Clover and timothy	0.27	0.38
		Corn / Clover and timothy	2.50	1.50
1933–1934	36	Corn	0.79	0.34
		Bluegrass	0.20	0.13
		Corn / Bluegrass	3.95	2.60
1935	19	Oats	1.23	0.52
		Bluegrass	0.53	0.50
		Oats / Bluegrass	2.30	1.00

that many terraces are emptied into old pastures. The use of grass water-ways to conduct water down the slope testifies also to the resisting qualities of sod to runoff.

Root and Biologic Effects

Root effects and biologic influences need to be discussed concurrently since both contribute to stable granulation and greater porosity, as discussed in Chapter 4. Moreover, when there is good vegetative cover, such as sods and forests, there is abundant biotic life, especially earthworm activities. The binding influences of plant roots on the erodibility of soils are clearly evident in Table 13-2. Weaver's experimental technique for measuring the erodibility of soils consisted in subjecting small samples of soils to the eroding action of water that was applied at the rate of 12.7 gal/min under a pressure of 1 lb/1.5 in.^{2} from an open hose. The samples of soils in their natural structure were taken to a depth of about 10 cm (4 in.); the total surface area was $\frac{1}{2}$ m^2.

These data point out the differences that may be expected between various crops with respect to their effects on holding the soil together

TABLE 13-2
The Effect of Plant Roots on the Erodibility of Soils (Weaver, 1937)

Crop	Dry weight (g)	In percentage of native grasses	Time of erosion (min)
Big bluestem	462	—	60*
Bluegrass	282	—	60*
Hungarian brome grass, 4-yr old	220	48	120 plus 25 min with nozzle attached
Alfalfa, 4-yr old (15 percent of roots fibrous)	196	43	21
Sweet clover, 2-yr old	66	18	15
Sudan grass	77	17	20
Winter wheat	75	16	17
Hegari sorgo	74	20	10

* Bare soil eroded in 18 min; erosion time for bare soil in other experiments was 7 to 10 min.

against the eroding action of the rain and runoff water. The sorghum plants were 3.5 ft tall when the sample was taken and were thick in the row. After the tops were removed, however, the soil eroded almost as easily as if it were bare. Most intertilled crops, such as corn, tobacco, cotton, and soybeans, afford little protection for the soil after the tops are removed. The roots of winter wheat extended the erosion time to 17 min.

The Sudan grass was unpastured and was 5 ft high when the soil was sampled. The root system, however, was only about 17 percent as extensive as native grasses. The erosion time was increased to 20 min. Two-year-old sweet clover eroded in about the same time as Sudan grass. Four-year-old alfalfa, with a total root system equivalent to 43 percent of that under native grasses, extended the erosion time to only 21 min. This small resistance to erosion in the presence of such a large root system was probably due to the fact that only about 15 percent of the total roots were of a fibrous nature.

The interesting grass in this table is the Hungarian brome grass. Although the root system was only 48 percent as large as the native grasses, the erosion time was more than double that of the bluestem sods. This high resistance to erosion was due to the dense mat of interwoven rhizomes. Such root systems are effective soil binders.

The Soil Factor

The effects of soil properties on water erosion are manifested in two ways. First, there are those properties that determine the rate with which rainfall enters the soil; second, there are those properties that resist dispersion and erosion during rainfall and runoff. Although these two phases of the soil factor are definitely related, the former is by far the more important.

Rainfall Acceptance

The ability of a soil to accept rainfall depends on (1) the condition of the soil surface as represented by its aeration porosity, (2) the moisture content of the soil at the time of the rain, and (3) the permeability of the soil profile. The first two factors are important for determining the infiltration rates of soils. The permeability of the profile expresses the transmission rate of water through soils and is also related to the aeration porosity.

Horton (1933) divided storms into two classes, A and B, upon the basis of their occurrence in relation to other rains. This classification was designed to take into consideration the moisture content of the soil at the time of the rain. Class A storms were those that followed two or more rainless days; the class B storms were designated as those that occurred after one or less rainless days. The 8-year average of infiltration rates of a loessial soil in Iowa for the months of May through August was 1.25 and 0.46 in./hr for class A and B storms, respectively. The lower infiltration rates for class B storms were due to raindrop impact compaction and increased saturation of the pores in the surface horizon.

The soil-moisture content at the beginning of a rain has a greater effect upon the rate of infiltration during the first 20 min than any other factor (Neal, 1938; Diseker and Yoder, 1936). The rate of infiltration varies approximately inversely as the square root of the soil-moisture content at the beginning of the rain.

Wischmeier and Smith (1958) recognized the significance of soil moisture at the time of precipitation and included an "antecedent precipitation index" variable in their soil-loss equation.

Resistance to Dispersion

The major soil properties that affect the amount of erosion when runoff occurs are related to the ease of dispersion. There are two structural conditions of the soil surface that affect dispersion. When the soil is dry and somewhat compact, the first increment of rain causes a slaking action at the surface and a high density of runoff. As the rain continues after this layer of loose soil has been removed, a wet, compact surface is produced which decreases the density of runoff in spite of a greater total runoff. The resistance of the wet layer apparently increases with the clay content.

If rainfall intensity exceeds the infiltration capacity when the soil is loose and friable, as in freshly cultivated fields, soil losses are usually high. There is no binding together of the granules, and the soil erodes to the bottom of the loosened layer. Granulation in this case expedites erosion instead of hindering it.

Therefore it seems that at the beginning of large, intense storms or throughout short storms soil is eroded as a result of the slaking and beating action of raindrops on the soil. Fine material is carried away during this phase of the erosion process. The more granular and resistant the soil to slaking and dispersion, the lower will be the density of runoff. As the storm continues, however, and runoff increases, both in amount and velocity, the erosion of the soil depends upon the coherence of the particles in the immediate surface with those underneath. In this case a highly granular soil will probably erode more per given amount of runoff than one that has a smooth compacted surface.

The greater erodibility of loose aggregates as compared with a compacted surface is well illustrated in exposed clay subsoils. When such exposed areas are wet, there are rather extreme compaction and coherence between particles; the density of runoff is not high. However, after a prolonged dry spell with only occasional showers the surface of these areas will slake down into a coarsely aggregated condition. The next heavy rain will easily remove this loosened layer down to the compacted zone.

In addition to these types of surface conditions, there is the situation of extreme wetness that is often encountered with highly silty soils on rather moderate slopes. During periods of continued rainfall with low intensities, these soils become saturated with water; cohesion is very small. If a torrential rain should fall on such a soil, runoff would be high and erosion would be extremely severe. This condition exists often with the rolling claypan soils of the midwest.

Middleton (1930) sought to obtain an index of soil erodibility based upon the physical properties of soils. He measured a variety of physical properties of the soils from the various Federal Erosion Experiment Stations and searched for some correlations between these properties and the erosion that was determined in the field. He suggested the dispersion ratio as an index of the ease with which the particles would be brought into suspension by the action of the rain or by runoff water. This ratio was obtained by dividing the amount of silt plus clay that was easily suspended by shaking the soil in pure water by the total quantity of silt plus clay that was present. The greater this ratio, the more easily the soil could be dispersed.

The colloid-moisture equivalent ratio was used to express the relative permeability of the soil for water. Permeability was considered to increase with this ratio. On the theoretical assumption that erosion should increase directly with the dispersion ratio and inversely with the colloid-moisture equivalent ratio, the erosion ratio was obtained. The erosion ratio was simply equal to dispersion ratio/colloid-moisture equivalent ratio. Middleton and his associates found satisfactory qualitative correlations between the erosion ratio and the erodibility of most of the soils that were investigated.

Field observations in North Carolina definitely pointed out that the Iredell sandy clay loam is an erosive and the Davidson clay a nonerosive soil. The two soils occur in adjacent areas; the factors that generally affect erosion, except soil properties, are generally the same on the two sites. Laboratory studies of the physical properties of these soils showed conclusively that one of the principal differences between the Davidson and Iredell from the standpoint of soil erosion was the extent to which the soils were aggregated (Lutz, 1934). The Davidson contained a large amount of aggregates, almost all of which were larger than 0.25 mm in diameter; the Iredell contained a somewhat lower quantity, most relatively small compared with those of the Davidson. Microscopic examinations showed that the Davidson aggregates were composed of clusters of smaller granules and therefore were permeable, whereas the Iredell aggregates were dense and impermeable. The greater permeability of the Davidson profile was the direct result of the higher content of large aggregates, which were more friable, porous, and stable than those of the Iredell. The greater stability of the Davidson aggregates was indicated by the extent to which the soil was naturally granulated and by the resistance of the soil to dispersion as evidenced by a lower dispersion ratio.

Studies with the extracted colloids showed a high state of flocculation for the Davidson. Swelling data indicated that the Davidson colloidal material did not swell, irrespective of the nature of the exchangeable cations present. Colloids extracted from the Iredell were highly hydrated and exhibited considerable swelling. The erosivity of the Iredell was attributed to the ease with which it is dispersed and its impermeability to water. The nonerosive nature of the Davidson was explained by its high state of aggregation into large, porous, and stable granules. These granules resist dispersion and permit a rapid percolation of water through the soil profile.

The erodibility of Russian soils was observed to be directly proportional to the amount of dispersion (particles <0.05 mm) times the water-holding capacity and inversely proportional to the quantity of water-stable aggregates (>0.25 mm) (Voznesensky and Artsruul, 1940). Erodibility indexes that range from 45.6 in an illuvial serozem to 0.84 in a leached chernozem were obtained; these differences reflected the observed differences in erodibility of the two soils. The runoff and erosion from various cropping systems on Marshall silt loam have been related to the amount of aggregates >2.0 mm (Wilson and Browning, 1945). A linear inverse relationship of soil loss to the percentage of aggregates was found. Soil losses from continuous corn, rotation corn, rotation oats, and rotation meadow were 38.3, 18.4, 10.1, and 0.3 T/A annually, respectively. The relative percentages of aggregates >2.0 mm were 100, 340, 450, and 580, respectively. Rai, Raney, and Vanderford (1954) prepared stable aggregates of Sharkey clay with a soil conditioner and measured the amount of erosion from a 3-in. rain on a 0 percent slope. They observed an increase in erosion of about 1.4 times as the aggregate size decreased from a mean diameter of 3.0 to 0.75 mm. There was a 5.7-fold increase in eroded material as the size of the aggregates decreased from 0.75 to 0.355 mm. A further doubling of erosion losses occurred as the aggregate size diminished to 0.21 mm. These data emphasize the importance of large-sized aggregates in resisting soil detachment and transportation.

CONTROLLING SOIL EROSION

Adequate control of water erosion requires an integrated use of the best agronomic and engineering practices that will protect the soil and reduce runoff.

Agronomic Practices

Agronomic practices involve the planning and management of crops and crop sequences to provide maximum plant cover, tillage practices that will give optimum absorption of rainfall, and the use of residues to assure the best possible protection from raindrop impact.

Cropping System

The classical soil erosion experiments at the University of Missouri (Miller and Krusekopf, 1932) definitely established the tremendous significance of plant cover from commonly practiced cropping systems for controlling soil erosion. Continuous corn, for example, reduced erosion losses 50 percent from open fallow. A rotation of corn, wheat, and clover further diminished soil losses by 86 percent. The importance of the sod crop in reducing erosion from corn was vividly emphasized by the fact that runoff and soil losses from corn following sod was only about $\frac{1}{3}$ and $\frac{1}{5}$, respectively, of the amounts from continuous corn. The sod crop raised the infiltration rate and increased the resistance of the soil to detachment and transportation. The density of runoff from the continuous corn plot was 0.88 lb/ft^3; that for corn following sod was 0.48 lb/ft^3. Continuous bluegrass sod only lost 0.3 tons of soil per acre. The major vulnerability of this rotation occurred in the fall at wheat-seeding time when the soil was unprotected, torrential rains occurred frequently, and the soil had a high moisture content. Similar results from various soil conservation experiment stations have confirmed the reduction of erosion from row crops by incorporating them in a sod-based rotation.

Minimum Tillage

Minimum tillage prepares a seedbed-rootbed that has a minimum of compaction and a maximum infiltration rate and aggregate or clod size. This leaves the soil more receptive for rainfall absorption and more resistant to detachment and transportation. Harrold (1960) studied the hydrology of small watersheds under three different tillage practices in a 4-year rotation of corn, wheat, and 2 years of meadow. The prevailing practice consisted of straight-row tillage across the slope, a low level of fertilization, liming to pH 5.4 and an alsike-red clover-timothy mixture as the meadow. The improved practice involved contour tillage, a high level of fertilization, liming to pH 6.8, and a clover-alfalfa-timothy mixture for the meadow. Minimum tillage consisted of plow-planting in the corn year of the rotation, handled according to the improved practice. The corn was planted behind the plow. Under the prevailing and improved practices, the meadow was plowed, disked twice, harrowed, planted and given two or three cultivations for weed control. A 3-year average of total runoff and soil erosion indicated that relative runoff losses for the prevailing, improved, and minimum tillage practices were 100, 85, and 54, respectively; the corresponding relative soil losses were 100, 70, and 10, respectively. The tremendous impact of plow-plant on reducing runoff is clearly illustrated in Figure 13-2, which shows a hydrograph reading of a major storm of 1.34 in. that fell in two short periods within less than 1 hr. The maximum intensity of the first period was 5 in./hr; that of the second was about 3 in./hr. Note that the peak

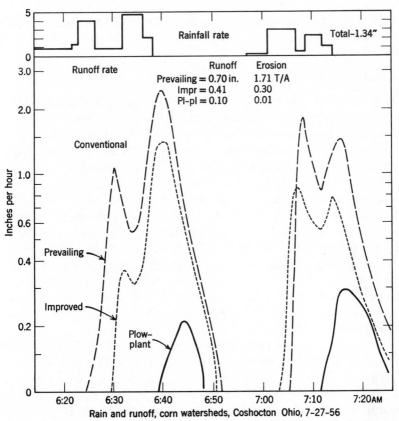

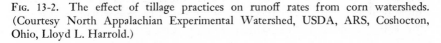

Fɪɢ. 13-2. The effect of tillage practices on runoff rates from corn watersheds. (Courtesy North Appalachian Experimental Watershed, USDA, ARS, Coshocton, Ohio, Lloyd L. Harrold.)

flow from the plow-plant watershed for the first part of the storm was about $\frac{1}{12}$ that of the prevailing practice; the peak flow for the second part of the storm when the soil was wet was about $\frac{1}{6}$ that of conventional. The total runoff for the entire storm was $\frac{1}{7}$ of the prevailing practice. The soil loss from the prevailing practice was 171 times that of the plow-plant operation. Even though the slopes on the plow-plant watershed averaged about 6 percent less than the other two, it is obvious that minimum tillage was more favorable for erosion control than conventional tillage methods for corn.

Plow-plant and wheeltrack planting of corn have been equally effective in increasing infiltration and decreasing soil losses over the conventional method (Mannering, Meyer, and Johnson, 1966). There are substantial benefits from cultivating minimum-tillage plots that have become crusted from raindrop impact.

Mulch Tillage

Stubble-mulch farming was introduced by Duley and Russell (1941) to serve as a protective covering for the soil which would control runoff and erosion. In other words, since the falling raindrop is the precursor of water erosion, as has been discussed thoroughly earlier in this chapter, then a mulch on the surface of the soil should provide the necessary protection. The major problem consisted of being able to obtain a mulch while preparing the land for the planting of crops. Duley and Russell solved this problem by using a duckfoot type of cultivator that stirred the soil under the trash blanket without turning under the organic matter serving as the mulch. This practice has been termed "subsurface tillage." It has received great emphasis throughout the country as a soil and moisture conservation measure. Although most stubble-mulch farming experiments exhibited considerable reductions in soil losses when compared with ordinary practices, it has not been as effective as contour listing on the permeable Marshall silt loam soils of western Iowa.

Hydrograph readings from a 1.54-in. rain that fell on dry soil showed the maximum peak runoff from plowed land to be in excess of 2 in./hr (Zingg and Whitfield, 1957); that from the mulched area was about 0.2 in./hr. The total runoff for the plowed and the mulched land was about 0.54 and 0.07 in., respectively. A subsequent storm falling on wet soil caused more total runoff from the mulched area but the rate of runoff was retarded by the residues. Soil erosion from concentrated water flow will take place with stubble mulching during rains of high intensity that fall on soils with low infiltration capacities. McCalla and Army (1961) have summarized the pertinent data on stubble-mulch farming and erosion control.

Engineering Practices

The engineering aspects of erosion control are primarily concerned with changing slope characteristics so that the amount and velocity of runoff are lessened. For example, the length of slope may be controlled by the spacing of terraces across the slope. The effect of degree of slope on the concentration of runoff water similarly may be controlled by terracing. The cutting action of concentrated water in terrace channels, gullies, or drainage ways may be controlled by a combination of suitable engineering structures and vegetative protection. In addition, contour farming and strip cropping change slope characteristics. Contour farming seeks to make each row of a particular crop serve as a minimum terrace to decrease the rate of water movement from a cultivated field. Strip cropping functions as a means to reduce the length of slope for tilled crops.

Terracing

The major effect of terraces on the control of erosion is to cut down the length of the watershed and to conduct the water off the field at

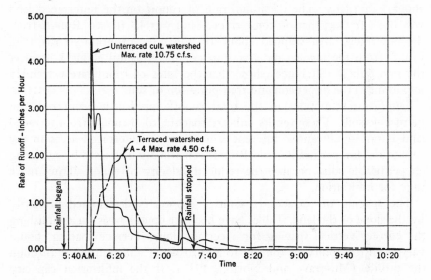

FIG. 13-3. Hydrograph curves showing runoff from one storm at LaCrosse, Wisconsin. (Hays and Palmer, 1935.)

a low velocity. Reducing the velocity of the water in the terrace channel causes a deposition of a large portion of the silt load in the channel. Consequently, only a small part of the soil that is eroded between terraces leaves the field in the water that caused the erosion. When terracing is coupled with good agronomic practices, erosion can be decreased to a minimum.

Terraces are not as effective in controlling the total runoff as they are in slowing up the rate of runoff. Hydrograph charts from an unterraced cultivated watershed and from a variable grade terrace, for the rain of July 5, 1935, at LaCrosse, Wisconsin, clearly illustrate this point (see Figure 13-3). The total runoff from the unterraced watershed was 68.5 percent of the rainfall; this runoff produced 51.16 tons of soil loss per acre. Runoff from the variable grade terrace was 66.3 percent of the rainfall; 7.8 tons of soil were lost per acre. What were the reasons for such large differences in soil loss when the total runoff was nearly the same? First, it is granted that the amount of soil deposited in the silt boxes at the end of the terrace did not represent the total amount of soil that was moved during the rain. Nevertheless, the soil that was deposited in the terrace channel was not lost from the field. The answer to this difference is found in comparing the rates of runoff from the two areas. Rapid runoff occurred on the unterraced area about 18 min after the rain began; rather slow runoff occurred on the terraced area about 7 min later. The maximum rate of runoff on the unterraced land was 10.75 ft^3/sec; that on the terraced land was 4.5 ft^3/sec. Runoff from the unterraced area stopped about 18 min after cessation of the rain; runoff from the terrace did not stop completely until about 3 hr after the rain. These differences show that the rates of runoff are extremely important in the dispersion and transportation of soil.

Significant reductions in the total runoff apparently occur only on permeable soils. These results indicate that the total runoff from terraced land is not significantly different from that on unterraced areas in the case of the less permeable soils. Runoff is markedly decreased by terracing on permeable soils, because the reduced velocity of runoff allows more time for infiltration.

The spacing of terraces to control erosion is determined primarily by the slope of the land. Tables have been set up as a basis for establishing the correct vertical spacing for a certain slope. Terrace spacing should vary with the ability of the soil to absorb water and transmit it through the profile (Musgrave and Norton, 1937). If the infiltration capacity of a soil is twice that of another soil under comparable slope conditions, a wider spacing of terraces on the former than on the latter should not cause increased runoff losses. Their data on the Marshall silt loam showed that the infiltration capacity of this soil was so high that runoff from terraces with a vertical spacing of 6 ft on a 9 percent slope did not even average 0.2 percent of the total rainfall during the first 5 yr of experimentation. The usual recommendation of this type of slope is a vertical spacing of 4 to 5 ft. It is possible that the normal recommendations for an extremely impervious soil may often prove too high because of the low infiltration capacity of such soils.

The key to the successful functioning of a system of terraces is the terrace outlet. Since terracing concentrates water in the channels, the outlets must handle this concentrated flow as it is discharged down the slope. This means that outlets must be completely stabilized to prevent gully erosion. The ease and economy of stabilization depend considerably upon the properties of the soil over which this concentrated water must flow. The major difficulties in obtaining suitable outlets are found in connection with the terracing of moderately sloping soils that have a high silt content and soils that possess heavy clay subsoils.

Many silty soils do not contain sufficient clay to cause much cohesion between particles. Such soils erode easily under the cutting action of concentrated water. They present a difficult problem in the construction and maintenance of good terrace outlets. Since the soil has little cohesion, binding power can be provided through the roots of suitable grasses.

A different problem exists with claypan soils. Unfavorable fertility and physical conditions make the establishment of vegetation difficult. Even when good vegetative protection has been provided, a prolonged drought may kill a considerable portion of the vegetation. The stability of the outlet is immediately reduced. The occurrence of cracks during dry weather also weakens the stability of the outlet.

Contour Farming

Contour farming is essentially a system of small terraces around the slope. Therefore the success of contour farming for causing increased infiltration of water should be greater on permeable soils, even as terraces are more successful on these types of soils. This fact is evidenced in Table 12-3. The Shelby loam is a rather impervious soil. The percentage decrease in runoff from contouring is (of the five locations shown in Table 13-3) the lowest at Bethany, Missouri. These data emphasize the importance of contouring for reducing soil losses on all soils. Erosion from storms only of normal or below normal intensity and of short duration can be controlled effectively by contouring alone (Hill and associates, 1944; Smith and co-workers, 1945). Hill observed as much erosion under contouring as with the rows running up and down the slope when the land was in cotton on slopes from 4 to 6 percent. He attributed the sizable erosion on the contoured land to the gullies that were formed when the contour lines broke and allowed the water to concentrate in a channel down the slope. He recommended strip cropping or terracing as necessary supplements to contouring. It should be emphasized that stabilized waterways are essential to remove water from contoured slopes if gullies are to be prevented.

Strip Cropping

Strip cropping is a means of reducing the length of slope planted to a given crop, thereby reducing the erosion potential on that field. It is especially effective when intertilled crops are interspersed with sod-

TABLE 13-3
The Relation of Runoff and Erosion to Contour Farming and Strip Cropping

Location	Cropping system	Rows up and down slope Runoff (percent)	Rows up and down slope Erosion (T/A)	Cultural practice Rows on contour Runoff (percent)	Cultural practice Rows on contour Erosion (T/A)	Land strip cropped Runoff (percent)	Land strip cropped Erosion (T/A)
Bethany, Mo.*	Corn-wheat-clover	—	—	7.7	1.95	7.6	0.85
Southern New York†	Potatoes-oats-clover	9.3	49.2	7.5	23.6	—	—
		15.7	15.5	5.2	1.25	1.9	0.15
Zanesville, Ohio‡	Cotton	7.2	58.0	4.4	16.2	—	—
Temple, Texas§	Corn-oats	13.6	15.7	4.6	5.9	—	—
Southern Illinois‖	Corn-soybeans	11.7	4.8	8.3	3.0	—	—
	Corn-oats (sw.cl.)	12.1	6.1	5.5	2.5	—	—
Western Iowa¶		11.5	25.2	7.4	10.1	—	—

* Smith, Whitt, Zingg, McCall, and Bell (1945).
† Free, Carleton, Lamb, and Gustafson (1946).
‡ Borst, McCall, and Bell (1945).
§ Hill, Peevy, McCall, and Bell (1944).
‖ Van Doren, Stauffer, and Kidder (1950).
¶ Moldenhauer and Wischmeier (1960).

type crops. Strip cropping is also more effective in controlling erosion in areas of moderate to low rainfall intensities. Where the slopes of a field are more or less uniform, contour stripping is generally practiced. Where the slopes are irregular, a combination of so-called field strip cropping and grassed waterways is usually employed. The strips are placed across the general slope of the field. As with contour farming, good stabilized waterways are essential to help remove water from the slopes without causing gullies.

The most effective system of strip cropping employs a meadow strip between the intertilled strips. For example, it is seen in Table 13-3 that the corn-wheat-clover rotation at Bethany, Missouri, and the potatoes-oats-clover rotation in southern New York use a meadow strip as the strong link in the system. The results in New York clearly point out that strip cropping is very effective in controlling erosion and runoff with such an intensively tilled crop as potatoes. It has been far more effective than contouring. There is a much greater effect of strip cropping on erosion than on runoff (Smith and associates, 1945). One must remember, however, that these results were obtained on a rather impervious soil. Meadow strips absorbed little of the runoff water from the cultivated strips but did filter out the soil from the silt-laden water.

SOIL-LOSS EQUATIONS

Previous discussions have shown the development of equations for storm kinetic energy, erosion index, and slope effects. Attention was called to the importance of soil moisture content and the condition of the soil surface in affecting infiltration rates, runoff, and erosion. Wischmeier, Smith, and Uhland (1958) recognized the role of the soil moisture and soil compaction variables in erosion losses and proposed that they be included in soil-loss prediction equations. They suggested the term "antecedent precipitation index" to characterize the soil moisture status at the time of precipitation. This index is based upon the amount of rain during the previous 30-day period. It actually characterizes the starting point on the time-infiltration rate curve. The accumulated rainfall energy since the last tillage operation was offered as a measure of soil compaction since much of the raindrop energy is expended in dispersing and sealing the soil surface and in direct compaction. They developed the following equation for predicting soil losses from climatic parameters:

$$Y_c = b_0 + b_1 X_e + b_2 X_1 + b_3 X_p + b_4 X_c \qquad (13\text{-}10)$$

where Y_c is the computed soil loss (T/A), X_e is the kinetic energy of the storm, X_1 is the erosion index, X_c is the accumulated rainfall

TABLE 13-4
R^2 Values for Correlations between Rainfall Parameters and Erosion (Wischmeier and Smith, 1958)

Parameter	Shelby loam R^2	Marshall subsoil R^2
$E(X_e)$	78.2	54.9
I (30 min max)	59.8	56.0
$E \times I$ (X_i)	89.2	70.7
$(E \times I) + X_p + X_c$	92.1	78.6

energy since the last tillage operation, X_p is the antecedent precipitation index, and b_0, b_1, . . . , are constants depending upon the soil and slope situations. Wischmeier and Smith (1958) analyzed the contributions of the rainfall parameters to erosion losses on Shelby loam and Marshall silt loam subsoil. A summary of their findings is shown in Table 13-4. These data point out that the erosion index (X_1) accounted for about 89 and 71 percent of the erosion on the Shelby and Marshall soils, respectively. Inclusion of antecedent precipitation (X_p) and soil compaction between tillage operations (X_c) increased these values to about 92 and 79 percent, respectively.

Analyses of cropping management and erosion-control practice factors led Wischmeier and Smith (1960) to propose the universal rainfall-erosion equation:

$$A = RKLSCP \qquad (13\text{-}11)$$

where A = the computed average soil loss in tons per acres from a given field under given rainfall parameters, crop-management plans, and erosion-control practices,

R = the rainfall factor, characterized primarily by the erosion index and is a measure of the erosion potential of the average annual rainfall of the area,

K = the soil-erodibility factor, which is the average soil loss in tons per acre per unit of erosion index from a cultivated fallow plot 72.6 ft long with a 7 percent slope (values range from 0.02 to 0.05 tons per acre per unit erosion index),

S and L = topographic factors of slope length and percent (a graph has been developed to give the combined effects of these two variables; Wischmeier et al., 1958),

C = the crop-management factor which is expressed as the ratio of soil loss under the conditions of the specific cropping system to that of the fallow plot [Ratios have been proposed for many cropping systems (Wischmeier, 1960). These ratios vary with the stage of development of the crop. For example, continuous corn with the residues returned showed ratios of 63, 50, 26, and 30 percent for the stages of seedbed preparation, establishment of the crop, growing the crop, and the residues, respectively. Note the erosion-preventing effects of the growing crop.], and

P = the erosion-control practice factor accounts for the positive values of contour farming, strip cropping, and terracing. [There are tables that give the factors to be used for these practices under varying slope conditions (Smith and Wischmeier, 1957). For example, the factors for contour farming, strip cropping and terracing on a 5 percent slope were 0.50, 0.25, and 0.10, respectively.]

This equation can be used in planning farm programs and conservation practices to minimize soil erosion.

References

Baver, L. D. (1937). Rainfall characteristics of Missouri in relation to runoff and erosion. *Soil Sci. Soc. Am. Proc.*, **2**:533–536.

Baver, L. D. (1938). Ewald Wollny—A pioneer in soil and water conservation research. *Soil Sci. Soc. Am. Proc.*, **3**:330–333.

Baver, L. D. (1954). The meteorological approach to irrigation control. *Hawaiian Planters' Rec.*, **54**:291–298.

Borst, Harold L., A. G. McCall, and F. G. Bell (1945). Investigations in erosion control and reclamation of eroded land at the Northwest Appalachian Conservation Experiment Station, Zanesville, Ohio, 1934–1942, *U.S. Dept. Agr. Tech. Bull.* 888.

Diseker, E. G., and R. E. Yoder (1936). Sheet erosion studies on Cecil clay. *Alabama Agr. Exp. Sta. Bull.* 245.

Duley, F. L., and J. C. Russell (1941). Crop residues for protecting row-crop land against runoff and erosion. *Soil Sci. Soc. Am. Proc.*, **6**:484–487.

Eckern, P. C., Jr., and R. J. Muckenhirn (1947). Water drop impact as a force in transporting sand. *Soil Sci. Soc. Am. Proc.*, **12**:441–444.

Eckern, Paul C. (1950). Raindrop impact as the force initiating soil erosion. *Soil Sci. Soc. Am. Proc..* **15**:7–10.

Eckern, Paul C. (1953). Problems of raindrop impact erosion. *Agr. Eng.*, **34**:23–25.

Ellison, W. D. (1944a). Two devices for measuring soil erosion. *Agr. Eng.*, **25**:53–55.

Ellison, W. D. (1944b). Studies of raindrop erosion. *Agr. Eng.*, **25**:131–136.

Ellison, W. D. (1945). Some effects of raindrops and surface-flow on soil erosion and infiltration. *Trans. Am. Geophys. Union*, **26**:415–429.

Free, G. R., E. A. Carleton, John Lamb, Jr., and A. G. Gustafson (1946). Experiments in the control of soil erosion in central New York. *Cornell Agr. Exp. Sta., Bull.* 831.

Harrold, Lloyd L. (1960). The watershed hydrology of plow-plant corn. *J. Soil and Water Conserv.*, Vol. 15., No. 4.

Haynes, J. L. (1938). Interception of rainfall by vegetative canopy. *U.S. Dept. Agr. Soil Conserv. Ser. Mimeo. Rept.* 2668.

Hays, O. E., and V. J. Palmer (1935). Soil and water conservation investigations. *Soil Conserv. Ser. Progress Rept. Upper Miss. Valley Soil Conserv. Exp. Sta.*, 1932–1935.

Hill, H. O., W. J. Peevy, A. G. McCall, and F. G. Bell (1944). Investigations in erosion control and reclamation of eroded land. *Blackland Conservation Experiment Station, Temple, Tex., 1931–41. U.S. Dept. Agr. Tech. Bull.* 859.

Horton, R. E. (1933). The role of infiltration in the hydrologic cycle. *Trans. Am. Geophys. Union*, 14:446–460.

Kuron, H., and H. J. Steinmetz (1957). Die Plantschwirkung von Regentropfen ais ein Factor der Bodenerosion. *Assemblie Générale de Toronto*, Tome I, pp. 115–121.

Laws, J. Otis, and Donald A. Parsons (1943). The relation of raindrop-size to intensity. *Trans. Am. Geophys. Union*, 24:452–459.

Lutz, J. F. (1934). The physicochemical properties of soils affecting erosion. *Missouri Agr. Exp. Sta. Research Bull.* 212.

McCalla, T. M., and T. J. Army (1961). Stubble mulch farming. *Advances in Agronomy*, 13:125–196. Academic Press, New York.

McIntyre, D. S. (1958). Soil splash and the formation of surface crusts by raindrop impact. *Soil Sci.*, 85:261–266.

Mannering, J. V., L. D. Meyer, and C. B. Johnson (1966). Infiltration and erosion as affected by minimum tillage for corn (Zea Mays L.) *Soil Sci. Soc. Am. Proc.*, 30:101–105.

Middleton, H. E. (1930). Properties of soils which influence erosion. *U.S. Dept. Agr. Tech. Bull.* 178.

Mihara, Y. (1952). Raindrops and Soil Erosion. *Nat. Inst. Agr. Sci.*, Tokyo, Japan.

Miller, M. F., and H. H. Krusekopf (1932). The influence of systems of cropping and methods of culture on surface runoff and soil erosion. *Missouri Agr. Exp. Sta. Research Bull.* 177.

Moldenhauer, W. C., and W. H. Wischmeier (1960). Soil and water losses and infiltration rates on Ida silt loam as influenced by cropping systems, tillage practices and rainfall characteristics. *Soil Sci. Soc. Am. Proc.*, 24:409–413.

Musgrave, G. W. (1935). Some relationships between slope-length, surface-runoff and the silt-load of surface-runoff. *Trans. Am. Geophys. Union*, 16:472–478.

Musgrave, G. W., and R. A. Norton (1937). Soil and water conservation investigations at the *Soil Conservation Experiment Station Missouri Valley Loess Region, U.S. Dept. Agr. Tech. Bull.* 558.

Neal, J. H. (1938). The effect of the degree of slope and rainfall characteristics on runoff and soil erosion. *Missouri Agr. Exp. Sta. Research Bull.* 280.

Neal, J. H., and L. D. Baver (1937). Measuring the impact of raindrops. *J. Am . Soc. Agron.*, **29**:708–709.

Norton, R. A., and D. D. Smith (1937). Effect of density of vegetation on the rate of runoff of surface water. Paper presented before *Ann. Meeting Am. Soc. Agron.*, December, 1937.

Rai, K. D., W. A. Raney, and H. B. Vanderford (1954). Some physical factors that influence soil erosion and the influence of aggregate size and stability on growth of tomatoes. *Soil Sci. Soc. Am. Proc.*, **18**:486–489.

Rose, C. W. (1960). Soil detachment caused by rainfall. *Soil Sci.*, **89**:28–35.

Smith, D. D., D. M. Whitt, Austin W. Zingg, A. G. McCall, and F. G. Bell (1945). Investigations in erosion control and reclamation of eroded Shelby and related soils at the Conservation Experiment Station, Bethany, Mo., 1930–1942. *U.S. D. A. Tech. Bull.* 883.

Smith, Dwight D., and Walter H. Wischmeier (1957). Factors affecting sheet and rill erosion. *Trans. Am. Geophys. Union*, **38**:889–896.

Smith, Dwight D., and Walter H. Wischmeier (1962). Rainfall erosion. *Advances in Agronomy*, **14**:109–148.

Van Doren, C. A., R. S. Stauffer, and E. H. Kidder (1950). Effect of contour farming on soil loss and runoff. *Soil Sci. Soc. Am. Proc.*, **15**:413–417.

Voznesensky, A. S., and A. B. Artsruui (1940). A laboratory method for determining the anti-erosion stability of soils. *Soils and Fert.*, **10**:289, 1947 (Abst.).

Weaver, J. E. (1937). Effects of roots of vegetation in erosion control. *U.S. Dept. Agr. Soil Conserv. Ser. Mimeo. Paper* 2666.

Wilson, H. A., and G. M. Browning (1945). Soil aggregation, yields, runoff and erosion as affected by dropping systems. *Soil Sci. Soc. Am. Proc.*, **10**:51–57.

Wischmeier, Walter H. (1959). A rainfall erosion index for a universal soil-loss equation. *Soil Sci. Soc. Am. Proc.*, **23**:246–249.

Wischmeier, W. H. (1960). Cropping management factor evaluations for a universal soil-loss equation, *Soil Sci. Soc. Am. Proc.*, **24**:322–326.

Wischmeier, Walter H., and Dwight D. Smith (1958). Rainfall energy and its relationship to soil loss. *Trans. Am. Geophys. Union*, **39**:285–291.

Wischmeier, Walter H., and Dwight D. Smith (1960). A universal soil-loss equation to guide conservation farm planning. *Trans. 7th Int. Cong. Soil Sci.*, Madison, VI:418–425.

Wischmeier, W. H., D. D. Smith, and R. E. Uhland (1958). Evaluation of factors in the soil-loss equation. *Agr. Eng.*, **39**:458–462.

Woodruff, C. M. (1947). Erosion in relation to rainfall, crop cover and slope on a greenhouse plot. *Soil Sci. Soc. Am. Proc.*, **12**:475–478.

Zingg, Austin W. (1940). Degree and length of land slope as it affects soil loss in runoff. *Agr. Eng.*, **21**:59–64.

Zingg, A. W., and C. J. Whitfield (1957). A summary of research experience with stubble-mulch farming in the Western States. *U.S.D.A. Tech. Bull.* 1166.

CHAPTER 14

Soil Erosion— Wind Erosion

Wind, as well as water, is a natural force that can transport soil. Erosion by wind is especially important in dry regions. The dust storms of the middle 1930s in the western plains of the United States were convincing evidence of the significance of wind erosion in the agriculture of an area. Man had disturbed the natural equilibrium between climate, vegetation, and soil by plowing the prairie grasslands and planting them to cultivated crops. Vegetative protection of the soil against the driving forces of the wind was destroyed. There was a deterioration of the soil structure formed by roots of the previous vegetation. The dry soil was vulnerable to transportation by wind.

Until about 1940, little research had been carried out on the basic physical factors affecting wind erosion. Free (1911) studied the problem of soil movement by wind as early as 1911. He introduced the terms "saltation" to denote the movement of soil by a series of short bounces along the surface of the ground, and "suspension movement" to designate the particles carried by the wind, more or less parallel to the soil surface. Bagnold (1941) made comprehensive studies on the movement of sand by wind and suggested the term "surface creep" to explain the rolling or sliding of particles along the surface through the impact of wind. Chepil and Milne (1939) were among the first to give special attention to the dynamics of wind erosion of soils. They used both field and wind tunnel experiments to establish basic principles. Chepil and Woodruff (1963) summarized existing research, analyzed the mechanics of the wind erosion process, and discussed methods of control.

THE MECHANICS OF WIND EROSION

The mechanics of wind erosion involves the three factors of wind, nature of the surface, and the soil.

The Wind Factor

Wind erosion is centered around the wind pattern near the soil surface and its effect upon the initiation of soil movement, the transportation of soil particles either along the surface or in the air, and the deposition of soil at a new location.

Wind Velocity

Soil movement caused by the turbulent flow of wind at the surface. This turbulence consists of eddies that are moving in all directions at varying velocities. The curves in Figure 14-1 point out that the average

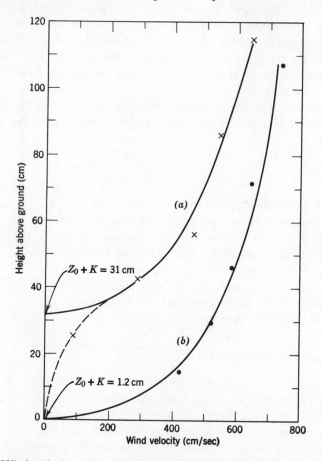

FIG. 14-1. Wind velocity above sorghum stubble 53 cm high (a) and growing wheat 5 cm high (b). (After Chepil and Woodruff, 1963.)

forward velocity of a turbulent wind increases exponentially with the height above the mean aerodynamic surface, Z_0. The height Z_0 above the soil surface is primarily dependent upon the height of the vegetation or other surface features that provide roughness. It is often called the zero displacement height. The air is usually calm or slow moving in this zone. The total surface roughness is equal to the zero displacement height plus the aerodynamic surface roughness. The latter depends upon the variations in height, density, and other characteristics of the surface features. Zero velocity of the wind is found at a height k above the mean aerodynamic surface. Its value increases with the roughness of the aerodynamic surface. Curve b in Figure 14-1 represents young wheat plants, 5 cm high, which bend in the wind and are quite porous to the wind. In this case, $Z_0 + k$ is nearly at ground level. Curve a was obtained with sorghum stubble, 53 cm high. There was some air movement through the stubble (see dotted line) and $Z_0 + k$ was zero at about 31 cm above the ground. Z_0 separates the fast-moving "free-flow" above the vegetative cover and the slow-moving "restricted flow" below the tops of the cover.

The rate of increase of velocity with the logarithm of the height can be expressed in terms of the drag velocity according to the following equation:

$$V = \frac{v_z}{5.75 \log (z/k)} \qquad (14\text{-}1)$$

where V is the drag velocity in centimeters per second, v_z is the velocity at any height z above the mean aerodynamic surface (Z_0), and k is the height above the mean aerodynamic surface where the velocity is zero. This equation shows that the drag velocity increases with the strength of the wind. Consequently, the drag velocity within 5 ft above Z_0 can serve as an index of the general atmospheric wind force. This relationship needs to be corrected for the reduced surface velocity of the wind caused by sand and soil movement.

Initiation of Soil Movement

Soil particles are moved in the field by direct pressure of the wind and by the impact of eroded particles moving in short bounces along the surface of the ground (saltation). A certain minimum velocity is needed to initiate the movement of the most easily erodible particles. This velocity has been called the "minimal fluid threshold velocity" (Chepil, 1945, II). As the strength of the wind increases, larger and larger particles are moved until a velocity is reached that causes move-

ment of all sizes of erodible particles present. This is the "maximum fluid threshold velocity." The minimal threshold requires a lower wind velocity on account of the impact of moving particles against those not yet disengaged. Chepil called this critical velocity the "minimal impact threshold." It is more clearly defined as "the minimal velocity required to initiate soil movement by the impact force of the descending grains, carried in saltation, rather than by the direct pressure of the fluid against the most erosive grains resting on the ground." The threshold velocity varies from day to day, depending upon the condition of the field. Erosion spreads like a fan to the leeward parts of the field under the combined effects of the wind and the bombarding action of particles in motion. The minimal threshold for the most erosive spots determines the threshold velocity for the field. Material in dunes and hummocks have the lowest threshold velocity. The erosive power of the wind to initiate soil movement is a function of the drag velocity.

Movement of Soil Particles

SALTATION. About 50 to 75 percent of the movement of soil particles takes place through saltation (see Table 14-1). Chepil (1945, I) vividly

TABLE 14-1
Relative Importance of Saltation, Surface Creep, and Movement in Suspension in the Wind-erosion Process
(Chepil, 1945, I)

Soil type	Soil removed in		
	Saltation (percent)	Surface creep (percent)	Suspension (percent)
Sceptre heavy clay	71.9	24.9	3.2
Haverhill loam	54.5	7.4	38.1
Hatton fine sandy loam	54.7	12.7	32.6
Fine drive sand	67.7	15.7	16.6

described this process as follows:

After being rolled by the wind, the particles suddenly leaped almost vertically to form the initial stage of the movement in saltation. Some grains rose only a short distance, others leaped 1 foot or more, depending directly on the initial velocity of rise from the ground. They also gained considerable forward momentum from the pressure of the wind acting upon them, and acceleration of horizontal velocity continued from the time grains began to rise to the time they struck the ground. In spite of this acceleration, the grains descended in almost a straight line invariable at an angle between 6 and 12 degrees from the horizontal. On striking the surface they either rebounded and continued their movement in saltation, or lost most of their energy by striking other grains, causing these to rise upward and themselves sinking into the surface or forming part of the movement in surface creep. Irrespective of whether the movement was initiated by impact of descending particles or by impact of rolling grains, the initial rise of a grain in saltation was generally in a verticle direction.

Chepil concluded that "the movement of soil by wind is dependent, not so much on the force of the wind acting on the surface of the ground, as on the velocity distribution to such height as the grains rise in saltation."

SURFACE CREEP. Movement of particles by "surface creep" is occasioned by the energy derived from smaller grains descending and hitting them in saltation. Very few are moved through the direct pressure of the wind. Consequently, there is a much smaller percentage of particles moved by creep than by saltation (see Table 14-1). Particles moved by the process of surface creep can have an abrasive action both on surface crusts, which are generally resistant to erosion by direct wind impact, and on normally nonerodible clods, which are broken down by the impacts of the moving grains.

SUSPENSION. Movement of fine dust in suspension is the most spectacular mode of transport. Dust storms may create considerable interest miles from the source of the fine particles. It is significant to point out, however, that fine dust itself is rather resistant to wind erosion. Dust is brought into wind currents through the saltation process. The original impact that causes movement is not due to the wind but to the energy imparted by impacts of larger particles. Once they are lifted off the ground, they are carried to great heights by the upward eddies of the erosive winds.

AMOUNT OF MOVEMENT. The rate and amount of wind erosion is determined largely by the drag velocity and turbulence of the wind. The rate of movement increases with the third power of the drag velocity (Chepil, 1945, 111). The amount of erosion varies according to the fifth power of the drag velocity.

DEPOSITION OF PARTICLES. Particles being moved by the wind will deposit either when the drag velocity gradient decreases, or when the wind velocity falls below the threshold limit, or when the particles move into depressions along the surface or behind ridges. As previously stated, there is a movement of particles to the leeward side of a field. Chepil observed that the progressive accumulation of erosive particles toward the leeward side of the field produces a condition more conducive to erosion. Moreover, as these erosive grains proceed leewardly by saltation, their impacts on clods and surface crusts cause abrasion and subsequent wind movement of heretofore nonerosive materials. Moreover, there is a gradual decrease in surface roughness as the erosive particles accumulate between soil clods, in small depressions, and behind small surface obstructions. The coarser particles tend to settle near the windward edge of the field and the finer near the leeward side.

The Soil Factor

Soil Moisture

It is well known that the severity of wind erosion increases with periods of drought and decreases with favorable moisture conditions. This is associated with changes in the protective influences of vegetative cover as well as the direct effect of soil moisture on decreasing the erodibility of the particles. Moisture films between individual particles provide the cohesive forces to hold them together. This has been discussed in Chapter 3. Wind velocities must create a force in excess of these film forces in order to cause soil movement. There are very few winds that have sufficient velocities to overcome the cohesive forces of moisture films at a tension of 15 atm (approximately the permanent wilting percentage) (Chepil, 1956).

Soil Structure

Soil structure effects on wind erosion are manifested primarily through the size and stability of the aggregates and clods. Although textural separates, sand, silt, and clay, play a significant role in the wind erosion process because of their impact in the formation of nonerodible structural units, it is the latter that resist soil movement. For example, high sand percentages are not conducive to clod formation and generally undergo high erodibility. Silt and clay, on the other hand, seldom are found as primary particles since they serve as binding agents in the formation of nonerodible clods.

SIZE OF STRUCTURAL UNITS. Wind tunnel experiments have shown that few units >0.84 mm in actual diameter are moved by most erosive winds. Although the size of water-stable aggregates may be as small as 0.02 mm or less, those soils with a high percentage of aggregates >1 mm offer considerable resistance to wind erosion. The smaller aggregates usually form larger structural units called clods. The amount of clods that are produced is highly correlated with the percentages of water-stable aggregates <0.02 mm and >0.84 mm in diameter (Chepil, 1953). These larger particles shield the erodible particles from the erosive wind forces.

STABILITY OF STRUCTURAL UNITS. The structural units of the soil may be broken down by abrasion from wind-driven material, tillage operations, impact of raindrops, alternate freezing and thawing or wetting and drying. The first wind erosion on a field usually takes place after a surface crust has been formed by the impact of raindrops (refer to Chapter 5). This thin crust originally offers considerable resistance to the force of the wind and a higher drag velocity is required to initiate soil movement than in subsequent windstorms. The abrasive action of the first eroded particles cuts through the crust and exposes more erodible particles. Even nonerodible clods on the surface are disintegrated by these abrasive impacts. The amount of abrasion varies directly as the square of the wind velocity and inversely as the modulus of rupture (Chepil and Woodruff, 1963). The latter is an index of the cohesive forces in the soil.

Tillage of dry soils tends to break down these cohesive forces between structural units and to increase soil erodibility. If tillage operations bring subsurface clods to the surface, there will be a decrease in wind erosion.

Alternate wetting and drying and freezing and thawing decrease soil cloddiness and reduce the mechanical stability of the surface clods by producing smaller granules that can be moved by the wind. It has been observed that the freezing and thawing of moist soils during the winter increased the erodibility of the surface in the spring.

The Surface Factor

The surface conditions that affect wind erosion are the surface roughness, the degree of protection by surface cover, and the sheltering of surfaces from direct wind impact. The rate of erosion decreases with increasing surface roughness because of the diminishing wind velocity that hits the ground. Surface roughness can be produced through tillage operations that form ridges and furrows or that bring clods to the surface.

These roughness features are effective only if they are constituted of nonerodible structural units. For example, a ridge of sand would soon be moved and flattened by the wind.

Vegetation not only adds to surface roughness but also provides cover for the soil surface. As previously stated, the height of the vegetative cover determines the zero displacement height, which is the main contributing factor in surface roughness. Tall vegetation increases the roughness factor more than short plants. Plants that are flattened by the wind have a lower total surface roughness than erect ones. Vegetation that is thick, such as grasses and small grain stubble, provide greater surface cover than coarse stubble, as corn or cotton.

Smooth or bare soil surfaces can be protected from wind erosion by sheltering them from direct wind impact with artificial barriers.

CONTROLLING WIND EROSION

Erosion control measures are based upon protecting the erodible soil fractions from the major erosive impacts of the wind and trapping the eroded particles either among the surface roughness barriers or on the leeward of these barriers. Live vegetation or residues from previous plant cover constitute the major control effort because of their effectiveness and permanence.

Vegetative Protection

Vegetative cover has proved to be the most economical and effective wind erosion control measure. It was only after the natural vegetation on the land was destroyed that wind erosion became a problem. The stabilization of soils under agricultural operations practically demands that vegetation serve as the key protective factor against wind erosion. The value of vegetation depends upon the density of the cover and the resistance to decomposition of the plant residues left on the surface. As previously mentioned, established grasses are the most effective for controlling wind erosion and row crops the least. Similarly, wheat residues give greater control than sorghum stover. Chepil and Woodruff (1963) reported that wind erosion of 500 lb/A from standing wheat stubble was only 17.5 percent of that from a bare soil; the corresponding losses under the same weight of flat straw were 53 percent of the bare soil. This difference was due primarily to the greater surface roughness of the erect stubble. The corresponding values for sorghum residues were about 81 and 90 percent, respectively. The differences between the sorghum and wheat residues were related to the higher density of the wheat. Whereas it took 1 ton of erect wheat stubble to reduce soil movement to a trace, 3 tons of standing sorghum residues were required. Stubble mulching and minimum tillage can be used to produce crops and keep the residues on the surface. Such practices can reduce the direct impact of the wind on the surface as much as nearly 100 percent in certain cases.

Vegetative Barriers

Vegetation can be employed to serve as (1) wind barriers that absorb or deflect the direct impact of erosive winds to the extent that drag velocities on their leeward are decreased below the minimal fluid threshold velocity necessary to initiate soil movement, and (2) traps for moving particles. Their effectiveness as wind barriers depends primarily upon their density, width, and porosity as well as orientation with respect to the direction of the prevailing winds. Barriers provide greatest protection when aligned at right angles to the wind. A porous barrier will cause a smaller reduction of the wind velocity than a dense one. However, the former will give greater protection over longer leeward distances. Although wind-velocity reduction may extend to leeward distances equivalent to 40 or 50 times the height of the barrier, effective erosion control requires relatively close spacing of the barriers. Protection at distances from 9 to 12 times the height are more realistic (Chepil and Woodruff, 1963).

Tree windbreaks were used extensively during the dust storm era of the 1930s. They were not as effective as planned because of the lack of growth under arid conditions and the competition they created with field crops for available soil moisture.

Tall annual crops such as sudan grass and sorghums make effective crop barriers if they are planted densely enough.

Strip cropping, using alternate strips of erosion-susceptible crops (mostly intertilled crops or fallow) and erosion-resistant crops (cereals or other close-seeded plants), is an effective farming method to trap eroded soil particles. The particles eroded from the tilled strips are caught in the closely seeded crops. Consequently, avalanching, or the acceleration of soil flow with distance downwind, is reduced. Strip cropping must be accompanied by stubble-mulch farming to be completely effective as a wind erosion control measure. The fact that the rows of the tilled crops in the strips are at right angles to the prevailing wind adds to reduced soil movement.

Tillage Operations

Soil tillage may have a harmful or beneficial effect on wind erosion. Tillage of dry fallow fields in semiarid regions to control weeds breaks down soil structure and favors increased soil movement. Tillage that produces a rough, cloddy surface increases the surface roughness and the amount of nonerodible fractions on the surface. These effects tend to decrease wind erosion. Moldboard plows, listers, and chisel cultivators are effective tools for generating rough, cloddy surfaces. Deep plowing that brings clay subsoil to the surface will produce clods that are highly resistant to wind impact. All these tillage operations, when possible, should be perpendicular to the prevailing wind direction. Any tillage operation should be considered a temporary, emergency measure and must be supplemented with other preventive practices if wind erosion is to be controlled.

WIND EROSION EQUATION

A wind erosion equation which expresses mathematically the many factors involved in soil losses from erosive wind effects has been suggested (Chepil and Woodruff, 1963):

$$E = f(ICKLV) \tag{14-2}$$

where E = annual soil loss in tons per acre,

I = soil erodibility, determined from the percentage of nonerodible soil fractions > 0.84 mm in diameter as measured by dry sieving [T/acres per yr],

C = local wind erosion climatic factor (percent), which varies directly with the cube of the wind velocity and inversely with the cube of the soil moisture content,

K = soil surface roughness (in.), which is equal to the average height of the clods or ridges constituting the surface,

L = equivalent unsheltered field width (ft) across the field along the prevailing wind direction,

V = equivalent quantity of vegetative cover, which includes the quantity of vegetation above ground (lb/A), the kind of cover as expressed in its total cross-sectional area (obtainable from tables), and the orientation of the cover which includes the surface roughness factor (obtainable from charts).

This equation can be used to estimate the potential amount of wind erosion for a given field under local climatic conditions. It also can serve as a guide to reduce potential wind erosion to a minimum.

References

Bagnold, R. A. (1941). *The Physics of Blown Sand and Desert Dunes.* Methuen and Co., London.

Chepil, W. D. (1945). Dynamics of wind erosion: I, Nature of movement of soil by wind. *Soil Sci.,* 60:305–320. II, Initiation of soil movement. *Soil Sci.,* 60:397–411. III, The transport capacity of the wind. *Soil Sci.,* 60:475–480.

Chepil, W. D., and R. A. Milne (1939). Comparative study of soil drifting in the field and in a wind tunnel. *Sci. Agr.,* 19:249–257.

Chepil, W. S. (1953). Factors that influence clod structure and erodibility of soil by wind: II. Water-stable structure. *Soil Sci.,* 76:389–399.

Chepil, W. S. (1956). Influence of moisture on erodibility of soil by wind. *Soil Sci. Soc. Am. Proc.,* 20:288–292.

Chepil, W. S., and N. P. Woodruff (1963). The physics of wind erosion and its control. *Advances in Agronomy,* 15:211–302. Academic Press, New York.

Free, E. E. (1911). The movement of soil material by the wind. *U.S. D. A. Bur. Soils Bull.* 68.

Author Index

Subject Index